NATURAL Tips & Techniques

James T. Wisdom

WH&O International Publications

NATURAL Study Guide

NATURAL Developers Handbook

Advanced NATURAL Study Guide

NATURAL Tips & Techniques

Developing Systems in NATURAL

Inside ADABAS

Introduction to PREDICT

PREDICT and Applications Development

NATURAL for DB2 Study Guide

NATURAL for DB2 Developers Handbook

NATURAL CONSTRUCT Study Guide

NATURAL CONSTRUCT Application Development Users Guide

NATURAL CONSTRUCT Tips & Techniques

ISBN 1-868960-02-4

NATURAL Tips & Techniques

Table of Contents

Chapter 3
NATURAL Data Structures

Chapter 4
NATURAL 2 Enhancements

Chapter 5

Maps and Windows

Chapter 6

NATURAL ADABAS Communication

Chapter 7
NATURAL in Batch

Chapter 8
Structured Programming in NATURAL

Preface

Who is this Book For?

This document is not meant to replace any documents produced by Software AG, but to supplement them. It is not a beginners teaching guide, although I had staff student programmers in our department not skilled in NATURAL read the first edition it and critique it for clarity. I realized that much of what is needed by programmers is in too many manuals (if at all) on too many other staff members shelves. Boston University's staff needed to clarify some issues for its own productivity and I hope other shops may find these chapters of interest.

Disclaimer

The version of NATURAL for the basis of these documents is 2.2.8. I cannot assure that everything mentioned here will be reflected in later version of NATURAL. It is my intent to revise this document to reflect future releases of NATURAL.

Acknowledgements

This document would not have been possible if not for the requests of Boston University's Information Systems management to provide our staff with documentation on several issues relating to NATURAL, i.e., NATURAL tips and techniques, NATURAL standards and guidelines. Also, several of my colleagues at Boston University and at other ADABAS shops have made many suggestions on content, style, and structure. To these people I am deeply grateful for their help.

I want to thank the management at Boston University's Information Systems and Technologies department for their indulgence in letting this project fly. I want to thank my colleagues and the many staff individuals who were more than happy to lend an ear and red ink to the development of tips and standards. Without their support, their keen eye for detail, their expertise, and their tireless efforts in proofing, this publication would not have been possible.

There are many individuals to numerous to name who have made invaluable contributions over the years and to them a gracious thank you. Many of the contributions have come by way of the Internet and specifically, SAG-L. Many thanks to those contributions and to the University of Arkansas for generously providing this wonderful venue for information sharing.

Comments

If you have any comments, please take the time to fill out the comments forms in the back of the book and mail it to the publisher. All comments are welcome.

You can reach me directly by email

Or visit my website: www.wizinc.com

Or contact my publisher, WH&O, at

Or visit their website: www.whobooks.com

Chapter 1

Information for the NATURAL Administrator

Constructing a Shared Nucleus

There's a new capability with NATURAL 2.2 to set up the NATURAL as two main portions - one nucleus which is invoked by both online and batch NATURAL dependent portions. To do this, the shared nucleus must reside in the LPA or ELPA (in MVS), or SVA in VSE. If you look at your organization's NATURAL link edit, you will notice modules that are taken from the NATURAL load library. These can be linked into what is to be named the shared nucleus. There are pros and cons for engaging in this approach, especially considering install and zap issues. This also reduces the size of the object required to sit within the TP region.

NATURAL Exits - NATUEX1, NATSREX2/3

NATUEX1 is designed to provide user control upon activating a NATURAL session. It is called using standard calling conventions. The parameter list, the address of which is in register 1, contains two address - a pointer to an 8 byte user id (becomes *INIT-USER) and a pointer to an 8 byte field containing an ADABAS transaction id. If register 15 does not contain zero, the NATURAL session is terminated.

Two other exits - NATSREX2 and NATSREX3 - are for sort processing. NATSREX2 gains control when NATURAL passes a record to SORT, and NATSREX3 gains control when a record is passed back after the sort operation.

There are several other aspects of the NATURAL environment which are important to know in order to use it properly. The four topics presented in this chapter are:

(1) NATURAL NATPARM Parameters
(2) NATURAL Buffers
(3) NATURAL Buffer Pool
(4) NATURAL Utilities and User Exits

I. NATURAL NATPARM Parameters

Parameters are linked to the NATURAL nucleus via a NATPARM module to help define a default startup environment. Figure 1.1 lists the NATPARM definition for the NATURAL 2.2 online environment at Boston University. The default set is defined by the NTPRM macro with possible override combinations for multiple NATURAL sessions via the NTSYS macro.

At Boston University, a front-end COBOL program establishes the appropriate session parameters by assigning the proper NTSYS macro set to a particular application LOGON. Another alternative would be to use the SYSPARM facility to establish particular dynamic overrides for a user application, e.g., printer assignments, WH parameter, etc. This table would be under the control of a security officer, or a NATURAL administrator.

I suggest that the *entire* parameter set be defined in a NATPARM module. This is important for documentation of sessions and preventing unintended defaults as well as making it easier to determine the parameters controlling each session.

If there is any question as to what can be affected by parameters, talk to the staff DBA. It is possible that a parameter adjustment might solve a problem that the applications or maintenance staff may be experiencing.

Some parameters can be overridden at LOGON time either through NATURAL's dynamic parameter facility, the Sysparm facility, or by using NATURAL Security. In order to use DBLOG to look at system files access, I LOGON dynamically with LOG=ON. Other examples are MENU=OFF, which allows a session without the startup menu, and BPID=0 which attaches the local buffer pool to the NATURAL session. All the specified NATURAL system files are defined as separate physical files for all of our NATURAL environments. Although this is not mandatory, it is certainly recommended.

The order is important in both the NATPARM module and the dynamic parameter facility. You will not see any errors but you may not get the proper assignments you expected. To avoid any problems, specify the parameters in this order:

> DBID, FNR, FUSER, FSEC, FDIC, FSPOOL.
> FUSER, FSEC, FDIC, FSPOOL may appear in any order immediately after FNR with no other intervening parameters.

There are some other interesting parameters which I think deserve some consideration.

UDB

If you have DDMs in your environment with the DBID value of 0, you default to the DBID for the session. In this case, UDB will act as the default if an ADABAS call is generated pointing to DBID=0. You can point to other DBIDs by generating the DDM out of PREDICT and establishing a non-zero DBID. Then, the UDB does not override an explicit DBID. Also, you can start up the session with a non-zero UDB value which then controls data base access. This capability can reduce the need to install many different system files.

An important issue relative to UDB/OPEN commands was discussed on the SAG-L listserv.

This parameter lets you work on DBID 1 and NATURAL will access DBID 2 for the user data files (not the NATURAL system files). Up until NATURAL 2.2.4, NATURAL security issued the 'OP' command against the 'UDB' data base. If the OPRB in NATURAL Security says 'acc=1,upd=2.' the user can only read file number 1 in DBID 2. In NATURAL 2.2.7, we found out that NATURAL issued the open against DBID 1 and not DBID 2. That means that the user now can UPDATE file 1 in DBID 2.

**NOTE This was not the original intended use by the user. I also have not verified if this is still true under 2.2.8.

RECAT

This parameter affects the decision to modify the object module header and to rebuild the symbol table for a loaded program which has a date/time stamp earlier than an external data area it uses. At first glance, I thought this parameter might be resource intensive. Theoretically, at least, I believe that this is a good parameter to set in a production environment.

```
NTPRM ADASVC=223,              ADABAS SVC                         -
      ADAPRM=ON,                                                  -
      AUTO=ON,                                                    -
      AVERIO=5,                                                   -
      BPID=229,                COM-PLETE LOCAL BUFFER POOL        -
      BPTEXT=2,                BUFFER POOL TEXT BLOCK SIZE        -
      CLEAR=R,                                                    -
      CSTATIC=(NATIN3,DBRRAY,AOSASM),                             -
      DATSIZE=64,              BUFFER SIZE FOR LOCAL DATA         -
      DELETE=OFF,                                                 -
      DSIZE=20,                SIZE OF DEBUG BUFFER               -
      DBID=240,                DEFAULT DBID                       -
      DBUPD=ON,                UPDATE TO ADABAS ALLOWED           -
      DTFORM=I,                USE DATE DEFAULT MM/DD/YY          -
      DU=OFF,                  TURN DUMP ON TEMPORARILY           -
      DYNPARM=ON,              ALLOW DYNAMIC PARMS ONLINE         -
      EJ=ON,                   DO PAGE BREAK FOR LGCAL PAGE       -
      ESIZE=76,                NAT886 WHEN NOT ENUF               -
      ET=OFF,                  ET/BT ONLY FOR CURRENT TRANS.      -
      EXTBUF=16,                                                  -
      FNR=043,                                                    -
      FNAT=(240,143),          SECURED FNAT FILE                  -
      FS=ON,                                                      -
      FSEC=(240,145),                                             -
      FSIZE=64,               DDM/SYMBOL TABLE BUFFER SIZE        -
      FSPOOL=(240,246),  NEW ADABAS SPOOL FILE FOR ADV. FAC.      -
      FUSER=(240,147),                                            -
      FDIC=(240,148),                                             -
      IKEY=ON,                DEFAULT ENTER IF NOT DEFINED        -
      INTENS=3,               NUM. OF OVERPRINTS TO HARDCOPY      -
      LC=OFF,                 TRANS. LOWER TO UPP FROM TERM.      -
      LE=OFF,                 ERROR MESS. WHEN OVER LIMIT         -
      LS=132,                 USE OUTPUT DEVICE LINE SIZE
      MADIO=9999,             MAX. ADA. CALLS BETW.SCREEN I/O     -
      MAXCL=9999,                                                 -
      MENU=ON,                MENU MODE                           -
      ML=T,                   MESSAGE LINE TOP/BOTTOM             -
      NC=OFF,                                                     -
      NISN=24,                                                    -
      OPRB=NOOPEN,                                                -
      PC=ON,                                                      -
      PD=100,                 MAX. # PAGES ON FNAT W/NATPAGE      -
      PRINTER=(,,,,,,PC3,,,,,,,,360,66,,,,,,,,,,,,,               -
      PC3,PC3,60),                                                -
      PROFILE=PROGRAM,                                            -
      PS=60,                  # LINES PER PAGE                    -
      RCA=OFF,                                                    -
      RJESIZE=8,              NATRJE BUFFER                       -
      SA=OFF,                                                     -
      SORTMAX=100,                                                -
      SORTSZE=32,             MAX SORT BUFFER                     -
      STACK=OFF,              NO LOGON SCREEN                     -
      STEPLIB=BUSYSTEM,       NEW DEFAULT STEPLIB                 -
      SYNERR=ON,                                                  -
      UDB=227,                                                    -
      USIZE=64,               USER BUFFER AREA                    -
      WORK=(PRDSHR99,WK2SHR22,WK3SHR200,WK4SHR22,WK5SHR200,       -
      WK6SHR22,PC3,WK8SHR22,WK9SHR22,WK0SHR22,,,,,,,,,,           -
      ,,,,,,,,,PC3,PC3,WK1SHR100),                                -
      XREF=ON,                                                    -
      ZP=ON
 NTSYS DEBUG,'DSIZE=64,DU=ON,LOG=ON'
 NTSYS DEVL,'DBID=242,LFILE=(227,241,121)'
 NTFILE ID=229,DBID=240,FNR=120     ELITE COURSEWARE
 NTFILE ID=227,DBID=240,FNR=121     NATURAL CONSTRUCT
 NTFILE ID=197,DBID=240,FNR=126     REVIEW DC/FILE
 NTFILE ID=241,DBID=240,FNR=125     REVIEW DB/FILE
 END
```

Figure 1.1. NATPARM Sample

RI
If this parameter is set to ON, any records in hold status which are rejected by NATURAL because of ACCEPT/REJECT logic are released. NATURAL issues an RI command; you trade off extra commands to reduce the number of hold queue elements for any user. The default is OFF.

ETA
This parameter provides the name of a program which gets control if an error is detected by NATURAL. This provides the mechanism to define a common error transaction handler.

ETID
This value fills the Additions-1 field in the ADABAS open call. This value also makes restart data possible. For batch jobs, it should be the jobname. The default is *INIT-USER. This value is important for restart information. If you want online users to run batch jobs with INIT-USER id on the job card, you want a blank value defined to NATURAL Security.

SM
Set structured mode for the session. The default is report.

HI
Defines the character which can invoke help. The default is '?'. If it is blank, then help can only be requested with PF keys.

ID
Sets the character as the input delimiter character.

PARM
This value is specified synamically only. It allows one to point to an alternate parameter module. Under CICS, you need a PPT entry.

SENDER
This value specifies the destination for output for an asynchronous process.

OUTDEST
This value also relates to asynchronous processing. It specifies the destination to receive any error messages.

MADIO
This value specifies the maximum value of ADABAS calls between two screen I/Os. Over the years, people have adopted various workarounds, such as the 'dancing dots' screen, or trapping the error (NAT1009) and FETCHing the module from some start point.

MAXCL
This value specifies the maximum number of program calls between two screen I/Os.

PD
This value controls the number of screen pages which are saved by the Natpage facility.

PROGRAM
This value specifies a non-NATURAL program to receive control at the successful completion of a NATURAL session. By calling the subroutine CMPGMSET (located in SYSEXTP) you can dynamically set this within the session.

SORTOPT
This buffer allows a user to specify additional parms in batch to the system SORT program (Syncsort in many shops). This is very useful in Year 2000 conversion considerations because it could be used to supply some of the new parms,eg. CENTWIN.

The WORK parameter under COM-PLETE provides COM-PLETE SD files to an online session with two options: 'TID' and 'DYN'. An SD file is described by an 8-character string.

```
WORK=(xxxyyyzz,...), where    xxx   names the file
                              yyy   defines the access
                              zz    defines blocksize
```

The access can be SHR, a 3-digit terminal id, 'TID', which assigns the current COM-PLETE terminal id, allowing the respective user exclusive access to the work file, and 'DYN' which is similar to 'TID'. This last option overwrites the third byte of the name with COM-PLETE's stack level so that different NATURAL sessions by the same user ensure different work files.

In batch, you can pass parameters in as part of the data string (up to 80 characters) which is passed to NATURAL. As another thought you might try:

```
//STEP010  EXEC NATURAL,ENVIRO=TEST,
// PARM=('SYS=DEBUG',
//          'STACK=(LOGON LIB1;EXEC PROG1;PARM1;PARM2;FIN)')
//
```

Under TSO however, this is an issue. "PARM=" information is sent to IKJEFT01, which will execute it as the first TSO command. One user suggests in this scenario to write a program that takes a given PARM and creates a control card. If it is so, write the program in assembler, C or PL/1 (but avoid COBOL as it doesn't handle variable character strings).

II. NATURAL Buffers

There are a number of buffers whose sizes are established in either the NATPARM module or dynamically. Some of the various NATURAL buffers are USIZE, ESIZE, FSIZE, DSIZE, DATSIZE, RUNSIZE, SSIZE, WSIZE, SRTSIZE, RJESIZE, USERBUF and EXTBUF. The sizing of the buffers depends on the way the invoked version of NATURAL is to be used (query only, exec only, creating/compiling programs, etc.). Each buffer has a specific purpose. Note that not all of the buffer assignments are done at initialization. RJESIZE, SORTSZE, and WSIZE are dynamically allocated. NATURAL fills RJESIZE in chunks of the original allocation. Therefore, you must remember to size accordingly so that the buffers are available when needed.

USIZE(User buffer)

At compile time, the generated object is NATURAL CONSTRUCTed in this buffer. In the event of the condition NAT0888, the resulting object is too big to fit in the USIZE buffer. I believe you are still limited to a 32K compilable object. The maximum allocation for USIZE at this time is 64K; the default is 32K.

At execution time, it contains the following types of information:

- BB (Benutzer Bereich - NATURAL's work area)
- User-specific statistics header
- Program stack per program level
- (ptrs to LDA, Break tables, etc.)
- IF stack control info
- System Variables
- NSC profile
- Standard System Function Table (SFT)
- Break Tables
- Constant table
- Loop Table
- ADABAS Control Table (SBT)

Any variables prefixed with '+' like the NATURAL 1.2 style global variables are maintained in this buffer if they are *not* defined in a GDA. This is very important to understand if you decide to continue to use these global definitions.

ESIZE (Extended user buffer)

The ESIZE buffer has a compile time use and a run time use. At compile time, this buffer holds the NATURAL source for the editor. At run time, this buffer contains the global data requirements for a NATURAL object. It also contains PF key information, the NATURAL stack and other work areas.

At execution time, the first object referencing a GDA establishes the addresses of the variable definitions. Any FETCHed object, however, will share the GDA if it references one of the same name.

DATSIZE (Data buffer)

This buffer replaced some of the functionality of the USIZE buffer in 2.1. At execution time, it contains:

Control information
GDA initial values
LDA of main program and all subordinate objects
PDA
One can get a NAT0888 during execution. You'll need to examine how the application is structured, since DATSIZE overflow is based on an aggregate of the data requirements for all of the objects called from level 1 and below. NATURAL CONSTRUCT often generates code which requires a higher DATSIZE than one normally use. I usually find an application can go about 7 or 8 levels deep, possibly more. Theoretically, you can go 32 levels deep. Returning up a level frees up the DATSIZE requirements for the current object. Recursive calls can get one into trouble as well, since NATURAL really isn't suited for recursion.

FSIZE (File buffer)

The FSIZE buffer is more important at compile time. The DDMs required for compilation are loaded into this buffer. At compile time, you may encounter a problem loading a DDM, reported as NAT0346. If a program is accessing a number of DDM's and some represent very large files (number of fields exceed 250), you might reconsider regenerating the DDM's without the comment lines. This buffer contains both the symbol table and all DDM's and format buffers at compile time.

The maximum allocation for FSIZE 64K; the default is 10K. If your environment can afford the space, you should increase the FSIZE buffer. The symbol table requirements in NATURAL 2 are different than in 1.2. For example, we needed to up our FSIZE to 36K to accommodate the compile of a particular NATURAL 1.2 program in NATURAL 2. The same program compiled with an FSIZE of 30K in the 1.2 environment. This buffer can be minimized in an execution only environment.

SRTSIZE (Sort buffer)

This parameter indicates the amount of storage to be reserved for use by the SORT program. The maximum setting is 64K; the default is 10K. If NATURAL cannot find the storage requested, it will default to zero.

DSIZE (Debug buffer)

The DSIZE buffer indicates the requirements for TEST DBLOG. The greater the amount reserved, the more command logging can be reported by TEST DBLOG. The maximum amount for this buffer is 64K; its default is 2K. If NATURAL cannot find the storage requested, it will default to 2K. This may be far to small to do adequate debugging.

SSIZE (SAG Editor buffer)

Under NATURAL 2.2.8 and PREDICT 3.3, this buffer takes on added meaning. Software AG recommends 45K to start, but users of it or ISPF have noticed the need for a comfort zone of 55 - 60K.

RUNSIZE (Size of runtime buffer)

The NATURAL runtime buffer sized accordingly aids in the NATURAL buffer pool's quick locate algorithm. It contains info such as:

- STEPLIB info
- The NTTF macro
- Information on the most recent command
- Table of invoked subroutines
- Buffer pool addresses of invoked objects

USERBUF (Size of incore trace buffer)

This parameter controls the size of a buffer area used to hold trace records record by the SYSRDC facility.

EXTBUF (Extra Buffer)

This parameter controls the size of a buffer area used to hold the format and search buffers needed to communicate to ADABAS. However, with the capability of running the buffer pool above the 16 Mb line, the ADABAS buffers had to be placed below the line. NATURAL used to use a portion of FSIZE, but this had its own complications. Therefore, this buffer was added in SM4. The maximum setting is 64 K; the default is 0.

This buffer is not required with ADABAS 5 and above. A NAT3010 is issued if this buffer is sized too small to hold the control blocks.

File Translation

Under version 2.2, file translation is better handled than in prior releases. There are two techniques which allow for file translation - the TF parameter and the NTTF macro. The TF parameter is allowed only dynamically. One must remember that the total length of a dynamic parameter is 256 bytes. Also, the SYSPARM facility allows only three files to be overriden. The best method for the greatest number of file translations is to define a NTTF macro.

The NTTF macro is at the end of a NATPARM (NTPRM) module, just like NTSYS or NTFILE. The format is:

- NTTF old-DBID,old-fnr,new-DBID,new-fnr
- NTTF old-DBID,old-fnr,new-DBID,new-fnr
- NTTF old-DBID,old-fnr,new-DBID,new-fnr

The number of entries is controlled by the ISIZE setting. ISIZE is stored in the **P**(ermanent) **C**(ontrol) **B**(lock). It defaults to 4K and can have a maximum of 32K. At Boston University, approximately 3K is used by NATURAL. Sixteen bytes are required for each file translation entry. One can define 64 entries with the remaining 1K of space.

You can enable this feature by defining a NTPRM set with a NTTF macro definition and allowing a logon (under COM-PLETE's Compass) with *STNAT2 PARM=label. This method allows a conversion team to access files located in another database to test and compare results where the files are originally located. This is useful to compare before you decide to generate the DDMs and programs are re-compiled to pick up the new database id.

III. NATURAL Buffer Pool

The *local* NATURAL Buffer Pool is an area of storage taken from the TP monitor's or the batch region (GETVIS) storage. A global buffer pool is storage assigned out of CSA (below the 16MB line for non XA/ESA), or it can be initialized above the line in MVS/XA or ESA. The NATURAL Buffer Pool is to NATURAL what the ADABAS Buffer Pool is to ADABAS.

The purpose of the buffer pool is to provide an area of storage into which NATURAL objects (programs, maps, subprograms and data areas) can be loaded. The programs are executed with re-entrancy. This is possible as long as multiple users invoke the same object from the same application library. Objects located in the STEPLIB library are also shared. This is a relief for organizations that use a central library.

A Global Buffer Pool is available. The advantage is that the Global Buffer Pool can be shared by multiple TP environments, therefore reducing the storage requirements and CPU resources by running several local buffer pools. You can load NATURAL objects stored in various system files in several databases into the same Global buffer pool. Running a Global Buffer Pool does not, however, preclude running a local buffer pool. The Global Buffer Pool is activated by running a load module named BPMNUC in the NATURAL 2 load library. To get allocation out of CSA, you must link the module to an authorized load library with the linkage editor parameter AC=1. The MVS buffer pool responds to a system modify command - REFRESH. There are five parameters to this load module:

1. BPID=n - > 0 but different from any ADABAS data base ID
2. IDLE=n - elapsed time for BPMNUC to check for operator commands
3. SVC=nnn - installed ADABAS SVC
4. RESIDENT - value 'Y' or 'N', indicating whether or not BPMNUC is to remain active or not.
5. NATBUFFER=(n1,n2,XA) - where n1 = amt. of storage in Kbytes n2 = number of directory entries XA (opt), allocated above 16Mb line

At startup of your NATURAL session, either through the NATURAL parameter module or the dynamic parameter facility, NATURAL attempts to locate the global buffer pool using parameters BPID and ADASVC. BPID=0 will provide NATURAL with a local buffer pool.

The inclusion of steplibs in the buffer pool forces a look up in the following order. First, it searches its directory for a program with the tag of library name, program name, DBID and FNR. If not found, then the buffer pool manager issues calls to ADABAS to the appropriate system file with the same key. If not found, the buffer pool manager searches its directory with a key of steplib name, program name, DBID and FNR. If not found, then it issues calls to ADABAS to the appropriate system file with a key of steplib name, program name, DBID and FNR. In my opinion, the search technique that the buffer pool manager now uses for steplib definitions is superior to how it previously handled objects in a steplib. This is one instance for which I would trade off ADABAS calls for the benefits and management of steplib objects versus duplicating shared modules in various application libraries.

You can analyze and manually manage the buffer pool through the SYSBPM facility. You can prevent sets of objects from loading into the buffer pool, which provides an interesting technique to stop an application from running.

This is manageable in batch by invoking BPMBLBAT and 'locking out' a library with:

```
FUNC=LOCK,BPID=xxx,LIB=name,DBID=xxx,FNR=xxx
```

Followed with:

```
OBJ1
OBJ2… the list of objects which you wish to add to the blacklist.
Executing 'FUNC=RLS,BPID=xxx' releases the blacklist.
```

The SYSBPM main menu offers several options:

```
12:34:56          ***  NATURAL BUFFER POOL MAINTENANCE  ***          97-04-01
BPID 229                            -  Main Menu  -                Type Global

            Code Function
            ---- -------------------------
             G   General Statistics
             L   Load/Locate Statistics
             F   Function Usage Statistics
             R   Fragmentation Overview
             S   Individual Statistics
             I   List Directory Information
             O   Display Buffer Pool Object
             D   Delete Buffer Pool Objects
             B   Blacklist Maintenance
             ?   Help
             .   Exit
            ---- -------------------------
    Code ... _   Library ... *_______
                 Object .... *_________________________________________
                 DBID ...... 0__      FNR ...... 0__

Command ===>
```

Figure 1.2. SYSBPM Main Menu

Here is the window that appears by selecting General statistics, useful for monitoring the performance of the NATURAL buffer pool:

```
12:34:56          ***  NATURAL BUFFER POOL MAINTENANCE  ***          97-04-01
BPID 229                      - General Statistics -               Type Global

Buffer Pool Address ..... 047A9000    Buffer Pool Size (KB) ... 5000
Directory Section ....... 047A9100    Directory Size .......... 104
Text Record Section ..... 047E5778    Text Record Size ........ 2048
                                      Number of Text Records .. 2379

Initialisation .......... 01:49:40    1997-04-12
Last Refresh ............ 01:55:38    1997-04-12      by User .. NATGBPM2


Total Use Count ......... 52          Active Text Records (%) ...  99.78
Active Objects .......... 970         Active Text Records ....... 2374
Max Active Objects ...... 1107        Max Active Text Records ... 2379
Total Object Size ....... 3939683     Total Text Record Size .... 4861952
Buffer Pool Usage (%) ...  76.94      Average Record Usage (%) ..  81.03
Average Lifetime (min) .. 226



Command ===>
```

Figure 1.3. SYSBPM General Statistics.

Here are a few screens from various choices within SYSBPM. First, load and locate statistics display snapshots upon pressing the ENTER key. The most crucial indicator is the ratio of Total executes to Load calls. It indicates how much the Buffer pool manager did not have to load the NATURAL object from the system file (FUSER/FNAT).

```
12:34:56          ***  NATURAL BUFFER POOL MAINTENANCE  ***           97-04-01
BPID 229                 -  Load/Locate Statistics  -              Type Global

Load Calls ........... 173384          Locate Calls .......... 4673872
Program Loads .... ... 28032
Finished Loads ....... 27995           Normal Locates ........ 675094
                                         successful .......... 321553
Algorithm 1 .......... 27199             failed .............. 348255
Algorithm 2 .......... 833
Allocation Failure ... 0               Quick Locates ......... 3998778
                                         successful .......... 3988380
Largest Allocation ... 36                succ. normal Locates  5286
Last Failing Sizes ... 0                 failed .............. 10398
                   ... 0
                   ... 0               Total succ. Locates ... 4315219

Buffer Pool Performance Indicator      Total Executes ........ 4673872
                                       Execs/Program Loads ...  166.73
```

Figure 1.4. SYSBPM Load Statistics.

Next, there is a display of buffer pool fragmentation. The '.' and the '_' indicate an object's text record in memory and that object has a use count of 0. The other symbols specify a use count > 0. 'X' denotes end of buffer pool, and a blank specifies an unused text record. If you have an indication that a request for an application shows continual READs to the FUSER file, this screen can reveal some interesting observations.

```
12:34:56          ***  NATURAL BUFFER POOL MAINTENANCE  ***           97-04-01
BPID 229            -  Buffer Pool Fragmentation Overview  -       Type Global
                                                                   Top of Data
Buffer Pool Address ..... 047A9000    Buffer Pool Size ........ 5120000
Number of Text Records .. 2379        Text Record Size ........ 2048
Text Record Section ..... 047E5778

          1---+----10---+----20---+----30---+----40---+----50
047E5778  .............._.___.................._...__.______
047FE778  ______________...._._________________________...__
04817778  ._____......_..._._..._._..+++++_..__._.______.._.
04830778  _._.._*...............................__..__.__._.
04849778  ._._..____...______.__._._._+._______.*_________._
04862778  ._.__....._.._......_.__.__._..________.._.._.____
0487B778  _..__._.__...__++.._____.__________._____...______
04894778  _..._._______._._.___.._..__._....._______**.......
048AD778  .._._..._....__.__......_......._.__.._.________..
048C6778  _.....___._.____............_._.____________.___..


Command ===>
```

Figure 1.5. SYSBPM Fragmentation Report.

IV. NATURAL Utilities

SYSMAIN and Its User Exits

The newer version of SYSMAIN has significantly improved capabilities over past versions. It provides maintenance online or batch for all NATURAL 2 objects. These objects include programming objects, NATURAL DDMs, editor / map / device profiles, and error messages. You can perform functions copy, move, find, delete, list and rename on these objects. You can invoke SYSMAIN in batch by providing a direct command after 'MENU'.

For example:

```
LOGON SYSMAIN
GLOBALS IM=D
MENU
COPY,SAVED,*,WITH,XREF,N,REP,%
FM,GALAXY,WHERE,DBID,240,FNR,147,%
DIC,(240/148),SEC,(240/145),TO,GALAXY,WHERE,DBID,242,%
FNR,142,DIC,(242/148),SEC,(242/145)
```

A callable subprogram named 'MAINUSER' is also provided. It can be called from a NATURAL program, subroutine or subprogram. Since it invokes SYSMAIN processing, all of your SYSMAIN exits are good as well your NATURAL Security definitions. The subprogram must be located in the library from which the program invoking it is executed. The parameters for the CALLNAT are the following:

```
CALLNAT 'MAINUSER' #COMMAND #ERR-NUM #MESSAGE #LIB
```

where the parameters are defined as:

```
DEFINE DATA LOCAL
1 #COMMAND  (A250)  /* SYSMAIN DIRECT COMMAND
1 #ERR-NUM  (N4)    /* SYSMAIN ERROR NUMBER
1 #MESSAGE  (A72)   /* SYSMAIN ERROR MESSAGE
1 #LIB      (A8)    CONST <'SYSMAIN2'> /* SYSMAIN LIBRARY
END-DEFINE
```

For example:

```
IF #COMMAND = 'MOVE ALL * WITH XREF N FM ABC
  WHERE DBID 123 FNR 10 TO XYZ WHERE DBID 210 FNR 15 REPLACE'
```

The result is to copy all objects from library ABC with XREF off from one system file in one data base to another system file in another data data base. If XREF is turned on, be sure PREDICT 2.3.2 is installed; otherwise, a not so obvious error may occur.

The complete direct command syntax is

```
FUNCTION OBJECT-NAME AS NEW-NAME
  FROM LIBRARY/APPL-ID WHERE-CLAUSE
  TO LIBRARY/APPL-ID WHERE-CLAUSE WITH-CLAUSE
```

Both the WHERE and WITH clauses are optional. The WHERE clause is defined as:

```
      ⎧ DBID dbval                            ⎫
      ⎪ FNR file                              ⎪
      ⎪ NAME name                             ⎪
WHERE ⎨ CIPHER cipher                         ⎬
      ⎪ PASSWORD value                        ⎪
      ⎪ DIC (dbid,fnr,passw,cipher)           ⎪
      ⎪ SEC (dbid,fnr,passw,cipher)           ⎪
      ⎩ LANGUAGE language/language code       ⎭
```

The WITH clause is defined as:

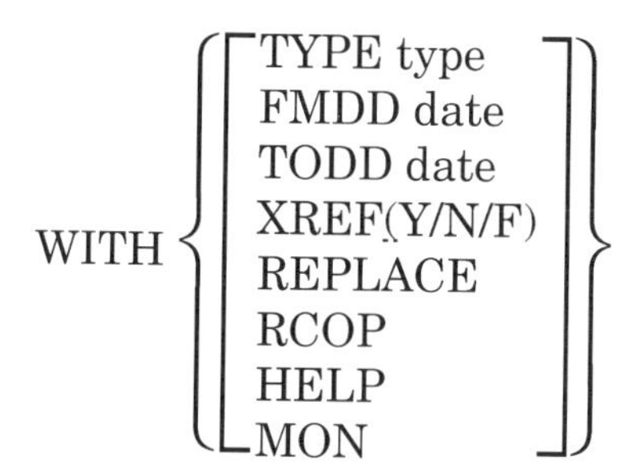

Currently, seven user exits are available - MAINEX01 through MAINEX07. To use these user exits, they must be stowed in the SYSMAIN library. MAINEX01 is invoked before object processing and MAINEX02 is invoked after object processing. MAINEX03 allows you to view the call to SYSMAIN before the command is processed and to potentially reject it. MAINEX04 allows you to interface before the ADABAS call and allow you to modify certain information in the control blocks. It is invoked before any request is processed. MAINEX05 takes control of a direct command before it is processed. MAINEX06 takes control at the start of a SYSMAIN session. MAINEX07 takes control at the termination of a SYSMAIN session. Each user exit which needs a local data area utilizes a local data area named MAINEXLx. The layout for the area MAINEXTL is shown in Figure 1.6.

In order to use the SYSMAIN with NATURAL Security, you need to be aware what programs affect what objects with what function. You can then allow the person in charge of defining your environment to decide what, if any, programs are disallowed, if you only want specific functions to be more widely distributed. In Figure 1.7 is a table acquired from Software AG which outlines these relationships.

```
   1 EXIT-PARMS                           A  250 /* EXIT PARAMETERS
 R 1 EXIT-PARMS                                  /* REDEF. BEGIN : EXIT-PARMS
   2 FIXED-AREA                                  /* PARAMETER SETTINGS
   3 IMAGE                                A    1 /* BEFORE OR AFTER IMAGE TYP
   3 OBJECT-REQUEST                       A    1 /* PROGRAM, VIEW, ERROR, PRO
   3 FUNC                                 A    1 /* MOVE, COPY, DELETE, RENAM
   3 FUNC-TYPE                            A    1 /* ALL, CAT, SAVED, STOWED
   3 SOURCE-LIB                           A    8 /* SOURCE LIB/APPLICATION
   3 SOURCE-DB                            N    3 /* SOURCE DBID
   3 SOURCE-FILE                          N    3 /* SOURCE FNR
   3 SOURCE-NAME                          A    8 /* SOURCE FCT
   3 TRGT-LIB                             A    8 /* TARGET LIB/APPLICATION
   3 TRGT-DB                              N    3 /* TARGET DBID
   3 TRGT-FILE                            N    3 /* TARGET FNR
   3 TRGT-NAME                            A    8 /* TARGET FCT
   3 OBJECT-NAME                          A   32 /* OBJECT NAME BEING PROCESS
   3 OBJECT-TYPE                          A    1 /* TYPE OF OBJECT
   3 RENAME-TYPE                          A    1 /* RENAME TYPE (MOVE OR COPY
   3 DATE-FROM                            A    8 /* SAVE/STOW FROM DATE
   3 DATE-TO                              A    8 /* SAVE/STOW TO DATE
   3 ERRNO-TYPE                           A    1 /* TYPE OF ERROR(USER OR SYS
   3 ERRNO-FROM                           N    4 /* ERROR TEXT NO. FROM
   3 ERRNO-TO                             N    4 /* ERROR TEXT NO. TO
   3 ERR-LANG-FM                          A    9 /* ERROR TEXT LANG CODES (FR
   3 ERR-LANG-TO                          A    9 /* ERROR TEXT LANG CODES (TO
   3 ERR-REN-FM                           N    4 /* REN. ERROR TEXT NUMBER FR
   3 ERR-REN-TO                           N    4 /* REN. ERROR TEXT NUMBER TO
   3 REPLACE-OPT                          A    1 /* REPLACE OPTION
   3 ERROR-CODE                           N    4 /* ERROR CODE
   3 XREF-IND                             A    1 /* XREF INDICATOR
   3 SRC-FSEC-DB                          N    3 /* SOURCE FSEC DBID
   3 SRC-FSEC-FILE                        N    3 /* SOURCE FSEC FNR
   3 TGT-FSEC-DB                          N    3 /* TARGET FSEC DBID
   3 TGT-FSEC-FILE                        N    3 /* TARGET FSEC FNR
   3 SRC-FDIC-DB                          N    3 /* SOURCE FDIC DBID
   3 SRC-FDIC-FILE                        N    3 /* SOURCE FDIC FNR
   3 TGT-FDIC-DB                          N    3 /* TARGET FDIC DBID
   3 TGT-FDIC-FILE                        N    3 /* TARGET FDIC FNR
   2 VARIABLE-INFO                               /* DETAILED INFORMATION
   3 OBJ-ID                               A   32 /* DEPENDENT ON OBJECT-REQUE
 R 3 OBJ-ID                                      /* REDEF. BEGIN : OBJ-ID
   4 OBJ-ERRNO                            N    4 /* ERROR TEXT NUMBER
   4 OBJ-ERR-IND                          A    3 /* S (SHORT) E (EXTENDED) IN
   4 OBJ-ERR-TYPE                         A    1 /* ERROR TYPE (USER OR SYSTE
   4 OBJ-NEW-ERR                          N    4 /* NEW ERROR NUMBER FOR RENA
   4 OBJ-FILL1                            A   20
 R 3 OBJ-ID                                      /* REDEF. BEGIN : OBJ-ID
   4 OBJ-PROFILE                          A    8 /* PROFILE NAME
   4 OBJ-PROF-TYPE                        A    3 /* E (EDT) M(MAP) D(DEV)
   4 OBJ-FILL2                            A   21
 R 3 OBJ-ID                                      /* REDEF. BEGIN : OBJ-ID
   4 OBJ-PGM                              A    8 /* CAT OR SAVED OBJECT NAME
   4 OBJ-PGM-IND                          A    3 /* C (CAT) S (SAVED) IND
   4 OBJ-PGM-TYPE                         A    1 /* OBJECT TYPE
   4 OBJ-FILL3                            A   20
 R 3 OBJ-ID                                      /* REDEF. BEGIN : OBJ-ID
   4 OBJ-LIB                              A    8 /* NAME OF LIBRARY TO BE LIS
   3 DATE-TO                              A    8 /* SAVE/STOW TO DATE
   3 ERRNO-TYPE                           A    1 /* TYPE OF ERROR(USER OR SYS
   3 ERRNO-FROM                           N    4 /* ERROR TEXT NO. FROM
```

Figure 1.6.a. MAINEXTL, User Exit LDA

```
      3 ERRNO-TO                    N    4 /* ERROR TEXT NO. TO
      3 ERR-LANG-FM                 A    9 /* ERROR TEXT LANG CODES (FR
      3 ERR-LANG-TO                 A    9 /* ERROR TEXT LANG CODES (TO
      3 ERR-REN-FM                  N    4 /* REN. ERROR TEXT NUMBER FR
      3 ERR-REN-TO                  N    4 /* REN. ERROR TEXT NUMBER TO
      3 REPLACE-OPT                 A    1 /* REPLACE OPTION
      3 ERROR-CODE                  N    4 /* ERROR CODE
      3 XREF-IND                    A    1 /* XREF INDICATOR
      3 SRC-FSEC-DB                 N    3 /* SOURCE FSEC DBID
      3 SRC-FSEC-FILE               N    3 /* SOURCE FSEC FNR
      3 TGT-FSEC-DB                 N    3 /* TARGET FSEC DBID
      3 TGT-FSEC-FILE               N    3 /* TARGET FSEC FNR
      3 SRC-FDIC-DB                 N    3 /* SOURCE FDIC DBID
      3 SRC-FDIC-FILE               N    3 /* SOURCE FDIC FNR
      3 TGT-FDIC-DB                 N    3 /* TARGET FDIC DBID
      3 TGT-FDIC-FILE               N    3 /* TARGET FDIC FNR
      4 OBJ-LIB-IND                 A    3 /* S (SAVED) C (CATALOGED) I
      4 OBJ-FILL4                   A   21
    R 3 OBJ-ID                             /* REDEF. BEGIN : OBJ-ID
      4 OBJ-VIEW                    A   32 /* VIEW-NAME
      3 NEW-OBJ-NAME                A   32 /* NEW OBJECT NAME (RENAME)
      3 MESSAGE                     A   20 /* MESSAGE
```

Figure 1.6.b. MAINEXTL, User Exit LDA

FUNCTION	PGMS. in FUNCTION	FUNCTIONALITY	OBJECTS AFFECTED
MAINFUNC	MAINMCOP	COPY	PROGRAMMING OBJECTS
	MAINMDEL	DELETE	
	MAINMDIC		
	MAINMLIS	LIST	
	MAINMFIN	FIND	
	MAINMMOV	MOVE	
MAINEMO	MAINMECT	COPY	ERRORS
	MAINMEDT	DELETE	
	MAINMEFT	FIND	
	MAINMELT	LIST	
	MAINMEMT	MOVE	
	MAINMERT	RENAME	
MAINVIEW	MAINMVDW	DELETE	VIEWS
	MAINMVCW	COPY	
	MAINMVLW	LIST	
	MAINMVMW	MOVE	
MAINPROF	MAINMPCF	COPY	PROFILES
	MAINMPDF	DELETE	
	MAINMPLF	LIST	
	MAINMPMF	MOVE	
	MAINMPRF	RENAME	
	MAINPCMD		DIRECT COMMANDS

Figure 1.7. SYSMAIN Function/Program Relationships

SYSERR Utility

Maintaining NATURAL Text

This facility is useful for creating and maintaining your own error messages for an application. Short text up to 64 characters can be created and invoked via a NATURAL program by issuing a REINPUT *nnnn where nnnn is the error number.

There are four individual message types supported by the facility:

- NS - NATURAL short error text
- NL - NATURAL long error text
- US - User short error text
- UL - User long error text

There is a user exit in library SYSEXT named USR0020N. You can use this exit in your application programs to invoke the display of the long error text.

For example, you wished to create the error text to appear on the defined message line stating 'You are not authorized to use the function', followed by the function name. Assume the application name is NATOOLS. Log onto the application SYSERR and fill in the appropriate information. See Figures 1.8 and 1.9.

```
12:34:56                 ***** NATURAL SYSERR Utility *****              97-04-01
                                     - Menu -

                 Code  Function
                 ----  ----------------------------------------
                  AD   Add new messages
                  DE   Delete messages
                  DI   Display messages
                  MO   Modify messages
                  PR   Print messages
                  SC   Scan in messages
                  SE   Select messages from a list
                  TR   Translate messages into another language
                  ?    Help
                  .    Exit
                 ----  ----------------------------------------
          Code .. SE   Message type .... US
                       Library ......... DRMTOOLS
                       Message number .. 1___ - 9999
                       Language codes .. 1________

Command ===>
```
Figure 1.8. NATURAL Error Text Maintenance.

The error texts for the library DRMTOOLS appear on the next screen as follows:

```
12:34:56              ***** NATURAL SYSERR Utility *****                  97-04-01
                - List User Error Messages    1 - 9999 -

                                                            Languages
Se  Number        Short Message      (truncated)            short      long
--  ------------ -------------------------------------- --------- --------
__  DRMTOOLS0001  Error. The command:1:is not legal.     1          1
__  DRMTOOLS0002  The help document for option:1:is not r 1
__  DRMTOOLS0003  The application:1:is currently availabl 1
__  DRMTOOLS0004  The application:1:is disabled until con 1
__  DRMTOOLS0005  Request to:1:table is invalid.          1
__  DRMTOOLS0006  You are not authorized to move this tab 1
__  DRMTOOLS0100 YOU ARE NOT DESIGNATED TO ENTER THE CAST 1          1
__  DRMTOOLS0101 THIS IS A TEST.                          1
__
__
__
__
__
__
```

Figure 1.9. NATURAL Error Text Maintenance.

A selection allows you to issues several different commands, such as 'MO', which presents what you see in Fig. 1.10.

```
12:34:56                 ***** NATURAL SYSERR Utility *****               97-04-01
                             - Modify Short Message -


Number        Short Message (English)
------------  -------------------------------------------------------------
DRMTOOLS0006  YOU ARE NOT AUTHORIZED TO MOVE THIS TABLE.
              ....+....1....+....2....+....3....+....4....+....5....+..


 1 Tx.
 2
 3
 4 Ex.
 5
 6
 7
 8
18 Ac.
19
20

Enter-PF1---PF2---PF3---PF4---PF5---PF6---PF7---PF8---PF9---PF10--PF11--PF12--
Mod               Exit                     -     +                      Canc
```

Figure 1.10. NATURAL SYSERR example of 'MO' option for Short Text Messages.

In the NATURAL program, executing the following statement produces the desired result.

```
REINPUT *0020, #FUNCTION
```

The utility provides two unload/load programs named ERRULDUS and ERRLODUS in the library SYSERR. These can be used to unload error text of the NATURAL system or those created by users. When messages are unloaded they are written to work file 2. When messages previously unloaded are loaded, the input file file is also work file 2.

NATLOAD/NATUNLD

This function is used to unload NATURAL source, object or DDMs. The output is written to WORK FILE 1 in a format useful for reloading with INPL function. The facility is invoked from online or batch.

In batch, it is recommended to set MT=0 and MADIO=0 for parameters. The input mode should be set to delimiter mode (IM=D). The records generated can vary in length up to 254 characters per record. There is a bug for this in 226 and 227. To allow for records with length greater than 92, you will need to contact SAG support for a zap.

Upon error, NATLOAD/NATUNLD terminates with a condition code of 40.

Since Software AG provides most of their files for INPL with the same DCB information, I used that definition for my NATUNLD work file. It may not be optimum blocking, but it works.

```
//CMWKF01 DD DISP=(,CATLG,DELETE),DSN=NATURAL.UNLOAD.DATA,
//   UNIT=TEST,SPACE=(TRK,(400,80),RLSE),
//   DCB=(RECFM=VB,BLKSIZE=10000,LRECL=9996)
```

The input data is provided through SYSIN as follows:

```
GLOBALS IM=D ID=,
NATUNLD
SAVED,*,FM,LIB1,TO,LIB2,WITH,SETUSER,APDDBS,SETNO,2
SAVED,*,FM,LIB3,TO,LIB2
.
FIN
```

This run unloads a source objects from a library named LIB1 with the names contained in set 2. To the same dataset, the source from library 3. Both unloads are designated to load into LIB2. NATUNLD with sets did not work correctly on the mainframe until SM8.

```
NATUNLD ALL,*,FM,*
This unloads the entire FUSER file.
```

INPL

This is NATURAL's INitial Program Load facility. The INPL function loads DDMs to FDIC (the dictionary file), libraries starting with 'SYS' (except SYSTEM) to FNAT. Other libraries are loaded into the system file FUSER, such as NATURAL subsystems NATURAL CONSTRUCT, ELITE, etc. The newer NATLOAD facility functions equivalently to INPL.

Although not part of the INPL process, loading the error messages file is also part of initializing your NATURAL environment. Whereas the input file to INPL is WORK FILE 1, the input file for loading messages is work file 2, and overwrites existing error messages. The programs ERRLODUS and ERRULDUS in the library SYSERR loads and unloads the various message types supported by NATURAL.

SYSEXT - NATURAL User Exit Library

Ever since 2.1.5, NATURAL has supplied a library that contains user exits not connected to any facility. There are subprograms that are worth investigating. As of NATURAL 2.2.8 on the mainframe, here is the list of members in the SYSEXT library.

Source	User Exit	Description
USR0010P	USR0010N	Get 'SYSPROF' Information
USR0011P	USR0011N	Information about logical file
USR0020P	USR0020N	Read any error text from FNAT / FUSER
USR0040P	USR0040N	Get type of last error
USR0050P	USR0050N	Get 'SYSPROD' Information
USR0060P	USR0060N	Copy LFILE definition from 'FNAT' to 'FUSER'
USR0070P	USR0070N	Default Editor Profile 'SYSTEM'
USR0080P	USR0080N	Handle Type/Name of Editor Contents
USR0100P	USR0100N	Control LRECL
USR0120P	USR0120N	Read NATURAL Short Error Message
USR0210P	USR0210N	Save, cat or stow NATURAL object
USR0220P	USR0220N	Read NATURAL Long Error Message
USR0320P	USR0320N	Read User Short Error Message from FNAT or FUSER
USR0330P	USR0330N	Read NATURAL Object Directory
USR0340P	USR0340N	NATURAL Buffer Pool Interface 0340
USR0341P	USR0341N	NATURAL Buffer Pool Interface 0341
USR0350P	USR0350N	Read current Recording Flags
USR0360P	USR0360N	Modify User Short Error Message
USR0400P	USR0400N	Number of rows affected by searched UPDATE
USR0420P	USR0420N	Read User Long Error Message from FUSER
USR0421P	USR0421N	Update User Long Error Message on FUSER

Source	User Exit	Description
USR0500P	USR0500N	Display a string in the title bar of a window
USR0600P	USR0600N	Display Program Level Information
USR0610P	USR0610N	Display DB error information
USR0620P	USR0620N	Translate strings to ASCII char. set
USR0621P	USR0621N	Logical File Translation for RMS
USR0622P	USR0622N	Reset error counter
USR1001P	USR1001N	Close TPU mailbox temporarily
USR1002P	USR1002N	Save and restore NATURAL environment parameter
USR1003P	USR1003N	Rdb Transaction Mode Switch
USR1004P	USR1004N	Free RFA (Record File address) od DBKEY table
USR1005P	USR1005N	Information about some NATURAL System Parameters
USR1006P	USR1006N	Support skip-sequential processing
USR1007P	USR1007N	Display Work File and Printer File Assignments
USR1008P	USR1008N	Physical linesize may be 80 or 132.
USR1009P	USR1009N	Convert *TIMESTMP to numeric variable
USR1010P	USR1010N	Close Rdb Read-Only Transactions
USR1011P	USR1011N	Wildcard / Asterisk check (short)
USR1012P	USR1012N	Read dynamic error part :1:
USR1013P	USR1013N	Display EBCDIC character set
USR1016P	USR1016N	Display Error Level for Copycode
USR1017P	USR1017N	Add CATALL call to CATALL control list
USR1018P	USR1018N	Dynamic OPEN
USR1019P	USR1019N	Get 'SYSBUS' Information
USR1020P	USR1020N	Add User Short Error Message to FUSER
USR1021P	USR1021N	Wildcard / Asterisk check (long)
USR1022P	USR1022N	Type of Data Base
USR1024P	USR1024N	Read results of CATALL
USR1025P	USR1025N	Handle multiple steplibs
USR1026P	USR1026N	Display RETURN information
USR1027P	USR1027N	Search user short error message
USR1028P	USR1028N	Bit/byte conversion
USR1029P	USR1029N	Get type of NATURAL object
USR1030P	USR1030N	Language code conversion
USR1031P	USR1031N	Check object name
USR1032P	USR1032N	List cataloged NATURAL objects with type
USR1033P	USR1033N	Find DBID/FNR of a cataloged DDM
USR1034P	USR1034N	Display NTTF file table
USR1035P	USR1035N	Maintain objects via the SAG editor engine
USR1036P	USR1036N	Maintain the user profile of the SAG editor
USR1037P	USR1037N	Information about NATURAL ABEND data
USR1038P	USR1038N	Retrieve characteristica of the current platform
USR1040P	USR1040N	Get or set the UDB Parameter
USR1041P	USR1041N	Install Error Transaction (*ERROR-TA)
USR1042P	USR1042N	Get or set the value of the UPDATE command

Source	User Exit	Description
USR1043P	USR1043N	Perform ADABAS direct calls
USR1045P	USR1045N	Dynamic OPEN for VSAM/ISAM datasets
USR1046P	USR1046N	Dynamic OPEN for LEASY files
USR1047P	USR1047N	Dynamic switch of file name
USR1054P	USR1054N	List libraries
USR1055P	USR1055N	List objects in a library
USR1056P	USR1056N	List DDMs on the FDIC file or in a library
USR1057P	USR1057N	Read a NATURAL source into an array
USR1058P	USR1058N	Read a DDM source into an array

For example, USR0010P invokes subprogram USR0010N which returns 'SYSPROF' information through array structures. This library is plentiful and is increased with more and more interface modules every SM. This library ought to be public and available to the staff. Once you find an object which is useful, you can copy it to SYSTEM or somewhere within your steplib architecture.

SYSPROF

The SYSPROF command displays the system file assignments for your current NATURAL session. This information is also programmatically available through the USR0010N in the SYSEXT library.

```
NATURAL System File Assignments

 File Name          DBID  FNR  DB Type
 ----------------   ----  ---  -------
 FUSER               240  147  ADABAS5
 REVIEW-DC           240  126  ADABAS5
 NATURAL CONSTRUCT   240  121  ADABAS5
 CAI                 240  120  ADABAS5
 REVIEW-DB           240  125  ADABAS5
 FSPOOL              240  246  ADABAS5
 FDIC                240  148  ADABAS5
 FSEC                240  145  ADABAS5
 FNAT                240  143  ADABAS5
```

Figure 1.11. SYSPROF Command Screen.

SYSPROD

This command invokes a pop-up window which displays the current NATURAL subsystems installed in the environment.

```
Product Information

  Cmd Product Name                  V/R  SM  Date
  __  NATURAL                       2.2   8  97-03-15
  __  ADABAS ONLINE SERVICES        2.3   3  95-08-09
  __  NATURAL ELITE                 2.2   1  94-08-03
  __  NATURAL NATURAL CONSTRUCT     3.3   2  95-02-27
  __  (ELI)                         2.1   2  94-02-26
  __  NATURAL ADVANCED FACILITIES   2.2   8  97-03-27
  __  NATURAL SECURITY              2.2   8  97-03-15
  __  NATURAL CONNECTION            2.2   8  97-03-15
  __  PREDICT                       3.2   2  94-02-27
  __  REVIEW                        3.4   3  96-12-03
  __
  __
  __
  __
  __
```

Figure 1.12. NATURAL Error Text Maintenance.

SYSFILE

The SYSFILE facility displays the print and work files available for this session.
See SYSFILE's display as illustrated in Figure 1.13.

```
12:34:56                    *****  NATURAL DBA UTILITY  *****              97-04-01
User: APDDBS                  - Work/Print File Information -        Library SYSMAIN

Device   No    Type      Name    Recf  Recl   Blks              Status
-------- --- -------- -------- ---- ------ ------ --------------------------

Workfile   1 COMPLETE PRDSHR          126        Available for Input/Output
           2 COMPLETE WK2SHR          126        Available for Input/Output
           3 COMPLETE WK3SHR          126        Available for Input/Output
           4 COMPLETE WK4SHR          126        Available for Input/Output
           5 COMPLETE WK5SHR          126        Available for Input/Output
           6 COMPLETE WK6SHR          126        Available for Input/Output
           7 PCNEW                               Available for Input/Output
           8 COMPLETE WK8SHR          126        Available for Input/Output
           9 COMPLETE WK9SHR          126        Available for Input/Output
          10 COMPLETE WK0SHR                     Available for Input/Output
          11 Unknown  UNASSGND                   Not Available
          12 Unknown  UNASSGND                   Not Available
          13 Unknown  UNASSGND                   Not Available
          14 Unknown  UNASSGND                   Not Available
          15 Unknown  UNASSGND                   Not Available
          16 Unknown  UNASSGND                   Not Available
          17 Unknown  UNASSGND                   Not Available
          18 Unknown  UNASSGND                   Not Available
          19 Unknown  UNASSGND                   Not Available
          20 Unknown  UNASSGND                   Not Available
          21 Unknown  UNASSGND                   Not Available
          22 Unknown  UNASSGND                   Not Available
          23 Unknown  UNASSGND                   Not Available
          24 Unknown  UNASSGND                   Not Available
          25 Unknown  UNASSGND                   Not Available
          26 Unknown  UNASSGND                   Not Available
          27 Unknown  UNASSGND                   Not Available
          28 Unknown  UNASSGND                   Not Available
          29 PCNEW                               Available for Input/Output
          30 PCNEW                               Available for Input/Output
          31 COMPLETE WK1SHR                     Available for Input/Output
Printer    1 Unknown  UNASSGND                   Not Available
           2 Unknown  UNASSGND                   Not Available
           3 Unknown  UNASSGND                   Not Available
           4 Unknown  UNASSGND                   Not Available
           5 Unknown  UNASSGND                   Not Available
           6 Unknown  UNASSGND                   Not Available
           7 PCNEW    PCPRNT07                   Available for Output
           8 Unknown  UNASSGND                   Not Available
           9 Unknown  UNASSGND                   Not Availabl
          10 Unknown  UNASSGND                   Not Available
          11 Unknown  UNASSGND                   Not Available
          12 Unknown  UNASSGND                   Not Available
          13 Unknown  UNASSGND                   Not Available
          14 Unknown  UNASSGND                   Not Available
          15 COMPLETE 360                        Available for Output
          16 COMPLETE 66                         Available for Output
          17 Unknown  UNASSGND                   Not Available
          18 Unknown  UNASSGND                   Not Available
          19 Unknown  UNASSGND                   Not Available
          20 Unknown  UNASSGND                   Not Available
          21 Unknown  UNASSGND                   Not Available
          22 Unknown  UNASSGND                   Not Available
          23 Unknown  UNASSGND                   Not Available
Printer   24 Unknown  UNASSGND                   Not Available
          25 Unknown  UNASSGND                   Not Available
          26 Unknown  UNASSGND                   Not Available
          27 Unknown  UNASSGND                   Not Available
          28 Unknown  UNASSGND                   Not Available
          29 PCNEW    PCPRNT29                   Available for Output
          30 PCNEW    PCPRNT30                   Available for Output
          31 COMPLETE 60                         Available for Output
```

Figure 1.13. SYSFILE display.

SYSMENU Library

This library contains the sources to maps used by Software AG in the NATURAL environment. For example, MAINMAP1 (English) and MAINMAP2 (German) are the sources for MAINMENU. They can be modified and moved to SYSLIB (or the appropriate library).

```
12:34:56              ***** NATURAL LIST COMMAND *****              97-04-01
User APDDBS          LIST * *                                Library SYSMENU

Cmd  Name      Type         S/C  SM  Vers  Level  User-ID   Date      Time
---  --------  -----------  ---  --  ----  -----  --------  --------  --------
___  CATALL1H  Map          S    R   2.2   0000   RKE       91-03-28  10:18:44
___  CATALL10  Map          S    R   2.2   0000   RKE       91-03-28  10:19:58
___  CATALL11  Map          S    R   2.2   0000   RKE       91-03-28  10:21:47
___  CATALL12  Map          S    R   2.2   0000   RKE       91-03-28  10:27:17
___  CATALL13  Map          S    S   2.2   0003   BKA       92-04-28  18:29:24
___  CATALL14  Map          S    R   2.2   0000   RKE       91-03-28  10:35:39
___  CATALL2H  Map          S    R   2.2   0000   RKE       91-03-28  10:37:12
___  CATALL20  Map          S    R   2.2   0000   RKE       91-03-28  10:38:49
___  CATALL21  Map          S    R   2.2   0000   RKE       91-03-28  10:40:18
___  CATALL22  Map          S    R   2.2   0000   RKE       91-03-28  10:41:31
___  CATALL23  Map          S    R   2.2   0003   BKA       92-04-28  19:08:53
___  CATALL24  Map          S    R   2.2   0000   RKE       91-03-28  10:47:01
___  DBAMAP1   Map          S    R   2.2   0000   RKE       91-07-05  17:11:41
___  DBAMAP2   Map          S    R   2.2   0000   RKE       91-07-05  17:12:08
From ________ (New start value)                                  0
Command ===>
```

Figure 1.14.a. NATURAL Listing objects in Library SYSMENU.

```
12:34:56              ***** NATURAL LIST COMMAND *****              97-04-01
User APDDBS          LIST * *                                Library SYSMENU

Cmd  Name      Type         S/C  SM  Vers  Level  User-ID   Date      Time
---  --------  -----------  ---  --  ----  -----  --------  --------  --------
___  DELMAP1   Map          S    R   2.2   0000   RKE       91-02-25  12:07:00
___  DELMAP2   Map          S    R   2.2   0000   RKE       91-02-25  12:07:25
___  DEVMAP1   Map          S    R   2.2   0004   RKE       93-02-24  15:37:09
___  DEVMAP2   Map          S    R   2.2   0004   RKE       93-02-24  15:38:43
___  DICMAP1   Map          S    R   2.2   0000   RKE       91-07-05  17:13:12
___  DICMAP2   Map          S    R   2.2   0000   RKE       91-07-05  17:13:28
___  LISTMAP1  Map          S    R   2.2   0003   RKE       92-07-13  11:11:19
___  LISTMAP2  Map          S    R   2.2   0003   RKE       92-07-13  11:11:43
___  MAINMAP1  Map          S    R   2.2   0000   RKE       91-07-05  17:13:58
___  MAINMAP2  Map          S    R   2.2   0000   RKE       91-07-05  17:14:17
___  OBJMAP1   Map          S    R   2.2   0002   RKE       91-10-29  09:48:24
___  OBJMAP2   Map          S    R   2.2   0002   RKE       91-10-29  09:48:43
___  RENMAP1   Map          S    R   2.2   0000   RKE       90-11-15  14:07:39
___  RENMAP2   Map          S    R   2.2   0000   RKE       90-11-06  11:47:40
```

Figure 1.14.b. NATURAL Listing objects in Library SYSMENU.

SYSTP Facility

```
 12:34:56              ***  NATURAL/CICS INTERFACE  ***               97-04-01
TID: E238                         -  Menu  -                      User: APDDBS

                         Code   Function
                         ----   -------------------------------
                           U    NATURAL User Sessions
                           R    NATURAL Roll Facilities
                           G    NATURAL Thread Groups
                           T    NATURAL Storage Threads
                           C    NCI Global System Information
                           P    NCI Generation Options
                           D    NATURAL Thread Group Definitions
                           S    NATURAL Swap Information
                           L    NATURAL Local Server Information
                           O    Own NATURAL User Session
                           A    System Administration Facilities
                           Z    Applied NCI Zaps
                           .    Exit NATURAL Application
                         ----   -------------------------------
                 Code ..

Command ===>
Enter-PF1---PF2---PF3---PF4---PF5---PF6---PF7---PF8---PF9---PF10--PF11--PF12--
Cont  Help  Admin Exit  Grps  Defs  Roll  Users Thrds Syst  Parms Swap  Canc
```

Figure 1.15. SYSTP Main Menu. SYSTP Aids in Monitoring the NATURAL/CICS Interface

SYSPARM

The SYSPARM facility allows for the dynamic allocation of parameters by user, terminal or program id.

```
12:34:56       *** NATURAL PROFILE PARAMETER MAINTENANCE ***           97-04-01
Profile                        -  Main Menu  -                DBID,FNR  240 143

               Code  Function
               ----  -------------------------------------------
                1    Edit Dynamic Parameters I   (ADASVC - ETA)
                2    Edit Dynamic Parameters II  (ETDB - OPRB)
                3    Edit Dynamic Parameters III (OUTDEST - SORTMAX)
                4    Edit Dynamic Parameters IV  (SORTMIN - ZSIZE)
                5    Edit NATURAL System File Assignments
                L    List Profile
                S    Save Profile
                D    Delete Profile
                ?    Help
                .    Exit
               ----  -------------------------------------------
        Code .. _    Profile .... ________ DBID .. 240 FNR .. 143
                     Password ... ________ Cipher Key .. ________
                     Parameter .. ________

Command ===>
Enter-PF1---PF2---PF3---PF4---PF5---PF6---PF7---PF8---PF9---PF10--PF11--PF12---
Help        Exit  Last        Flip                                  Canc
```

Figure 1.16. SYSPARM Main Menu.

```
12:34:56          *** NATURAL PROFILE PARAMETER MAINTENANCE ***          97-04-01
Profile STNAT2              -  NATURAL System Files  -           DBID,FNR  240 143

             Log.ID    DBID      FNR       Password        Cipher Key
System .....           ___       ___       ________        ________
FDIC .......           ___       ___       ________        ________
FNAT .......           ___       ___       ________        ________
FSEC .......           ___       ___       ________        ________
FSPOOL .....           ___       ___       ________        ________
FUSER ......           ___       ___       ________        ________
LFILE (1) .. 197       240       126       ________        ________
LFILE (2) .. ___       ___       ___       ________        ________
LFILE (3) .. ___       ___       ___       ________        ________
LFILE (4) .. ___       ___       ___       ________        ________
LFILE (5) .. ___       ___       ___       ________        ________
LFILE (6) .. ___       ___       ___       ________        ________
LFILE (7) .. ___       ___       ___       ________        ________
LFILE (8) .. ___       ___       ___       ________        ________


Command ===>
Enter-PF1---PF2---PF3---PF4---PF5---PF6---PF7---PF8---PF9---PF10--PF11--PF12--
     Help         Exit  Last  Save  Flip                    Def+  Def-  Canc
```

Figure 1.17. SYSPARM screen after selection of code 5.

One weirdness occurred working in this facility. After modifying the LOG parameter on this screen, before saving this appeared - FC .... ' '. After saving this profile and logging on, I received a NATPARM error for FC character, which I didn't change. I had to re-enter the facility and code what you see below to eliminate the error.

```
12:34:56          *** NATURAL PROFILE PARAMETER MAINTENANCE ***          97-04-01
Profile STNAT2              -  Dynamic Parameters II  -          DBID,FNR  240 143

ETDB ...... 240          ETID ...... ' '_____     EXCSIZE ... 0__
EXRSIZE ... 0__          EXTBUF .... 16           FC ........ X'40'
FS ........ ON_          FSIZE ..... 64_          GRAPHIC ... OFF
HI ........ ?            IA ........ =            ID ........ ,
IDSIZE .... 0_           IKEY ...... ON_          IM ........ F
IMSG ...... ___          INTENS .... 3_           KC ........ OFF
LC ........ OFF          LE ........ OFF          LOG ....... ON_
LS ........ 80_          LSIZE ..... ___          LT ........ 99999999__
MADIO ..... 9999_        MAINPR .... 0_           MAXCL ..... 9999_
MENU ...... ON_          ML ........ T            MT ........ 60______
NC ........ OFF          NISN ...... 24__         OBJIN ..... R

KEY ....... ____________________________________________________________
            ____________________________________________________________
MCG ....... ____________________________________________________________
OPRB ...... ____________________________________________________________


Command ===>
Enter-PF1---PF2---PF3---PF4---PF5---PF6---PF7---PF8---PF9---PF10--PF11--PF12--
     Help         Exit  Last  Save  Flip                    Def+  Def-  Canc
```

Figure 1.18. SYSPARM screen after selection of code 2.

SYSTRANS

This facility is a little different than NATUNLD. The main use for this is to prepare for unloads to other platforms. It works well with NATURAL CONNECTION. For example, if you wanted to transfer a library to another platform, SYSTRANS unload, then FTP to the client. From what I understand about this facility, you should not have to worry about blocksize issues once you FTP to the client. This is a nice feature, very much how WINZIP wraps files and allows for binary file transfer without 'damaging' the type of file.

```
12:34:56              ***** NATURAL TRANSFER UTILITY *****             97-04-01
User APDDBS                         - Main Menu -                Library SYSTRANS




            Functions

            U  Unload TRANSFER Objects
            L  Load TRANSFER Objects
            T  Direct TRANSFER Functions
            S  Scan TRANSFER Work File
            R  Restart TRANSFER Load






Command ===>
Enter-PF1---PF2---PF3---PF4---PF5---PF6---PF7---PF8---PF9---PF10--PF11--PF12--
      Help        Exit              Flip                                 Canc
```

Figure 1.19. SYSTRANS Main Screen.

Sample JCL for MVS to run SYSTRANS in batch to unload NATURAL objects and maps prefixed with "TEST" in library DEVPUB for uploading to library 'TOOLS':

```
//APDDBS JOB CLASS=T
//DELETE   EXEC PGM=IEFBR14
//DD1  DD UNIT=DASD,DSN=APDDBS.SYSTRANS.UNLOAD,
//     DISP=(MOD,DELETE),SPACE=(TRK,1)
// EXEC NATURAL,ENVIRO=DEVL
//CMWKF01 DD DISP=(,CATLG,DELETE),DSN=APDDBS.SYSTRANS.UNLOAD,
//  SPACE=(TRK,(100,10),RLSE)
//SYSIN DD *
LOGON SYSTRANS
GLOBALS IM=D
MENU
U
N,N,N,Y,N,N,N
N
DEVPUB,TEST*,*,TOOLS
M
DEVPUB,TEST*,Y,Y,TOOLS   ← Y,Y see text box below
FIN
/*
//
```

```
Incorporate all PREDICT Rules .. N  (Y/N)
Unload Associated Free Rules ... N  (Y/N)
```

HELP Facility Under PREDICT

Many users have discussed how to build centralized help facilities for some time now. A help request may be of different varieties: (1) what values are legal for a field; (2) what values are currently stored in the data base; (3) description of the field. More generally, it may be a request on how a function works, or any sub-function, or a description of any specific map.

As an option for developing help, PREDICT now allows you to view descriptive text stored in its repository. There are a set of routines which can make this text available to a programmer. A library named SYSDICH contains a subprogram SYSHELP and several help routines named SYSHLP1, SYSHLP2, SYSHLP3, SYSHLP4, SYSHLP5, SYSHLP6, SYSHLP7, SYSHLP8, and SYSHLP9.

The routine SYSHELP is delivered in object form with PREDICT to assure compatibility, whereas the help routines are provided in source to modify as you require. If you execute the program EXAPGM in the library SYSDICH you can see the routines work. Then you can modify them as you see fit. Here is what the routines do.

SYSHLP1 displays comments for a field and for associated verifications. SYSHLP2 display permissible values for associated verifications and allows one value to be selected by marking it with the cursor. SYSHLP3 displays extended descriptions for associated verifications. SYSHLP4 displays the extended description for the field. SYSHLP5 displays the short program descriptions. SYSHLP6 displays the extended program descriptions. SYSHLP7 displays as help comments of the file description (INFTYPE='F'), comments of the database description (INFTYPE='D'), and comments of the system description (INFTYPE='S'). SYSHLP8 displays the extended (long) description for a file (INFTYPE='G'), for a database (INFTYPE='B'), and for a system (INFTYPE='Y'). SYSHLP9 displays the comments for the relationship description.

The routines are activated after the help routine is assigned to a field via the HE parameter. During the map edit session, it is assigned by editing the field for the HE parameter, e.g., HE='SYSHLP1',=. A parameter can be passed to the helproutines. It may be passed as a constant or a user-defined variable with the value you wish to pass.

If an '=' is specified, as above, the name of the field as defined in the map is passed to the help routine. According to the rules of the HE parameter, if '=' is specified then the parameter in the HELPROUTINE must be defined with format A65. If the help routine is defined to the map, the name of the map is passed. Upon entering '?' into the field of question, as long as PREDICT is available, the required help information will be taken from the dictionary and displayed. Any change in PREDICT is then immediately available.

There are nine parameters which are passed in the call to SYSHELP. Figure 1.20 is a listing of the parameter area named SYSHELPA.

```
  1 FIELDNAME               A   32 /*NAME OF THE FIELD (IN)
  1 FILENAME                A   32 /*NAME OF THE FILE (IN)
*                                  /*OR NAME OF THE MAP/PROGRAM
  1 VENAME                  A   32 /*NAME OF THE VERIFICATION (OUT)
  1 VETYPE                  A    1 /*TYPE OF THE VERIFIC. (OUT)
  1 INFTYPE                 A    1 /*TYPE OF THE INFORMATION (IN)
*                                  /* C FIELD COMMENTS
*                                  /* E FIELD EXTENDED DESCRIPTION
*                                  /* V VERIFICATION COMMENTS
*                                  /* T VERIFICATION VALUES
*                                  /* W VERIFIC. EXTENDED DESCRIPT.
*                                  /* P PROGRAM COMMENTS
*                                  /* Q PROGRAM EXTENDED DESCRIPT.
  1 STARTLN                 N    4 /*STARTLINE (IN-OUT)
  1 RSP                     B    1 /*ERRORCODES (OUT)
*                                  /* H'00' OK
*                                  /* H'01' FUNCTION NOT AVAILABLE
*                                  /* H'03' END OF INFORMATION
*                                  /*       FOR THIS OBJECT
*                                  /* H'04' END OF INFORMATION
*                                  /*       FOR THIS TYPE/CLASS OF
*                                  /*       INFORMATION
*                                  /* H'FF' NO INFORMATION
*                                  /*       FOR THIS FIELD
*                                  /*       FOUND IN DICTIONARY
  1 HELPINF                 A  240 /*HELPINFORMATION (OUT)
R 1 HELPINF                        /* REDEF. BEGIN : HELPINF
  2 HELP-TEXT21             A   76 (1:3)
R 1 HELPINF                        /* REDEF. BEGIN : HELPINF
  2 HELP-TEXT1              A   30 (1:8) /*FOR SHORT COMMENTS
*                                        /*AND VERIFIC. VALUES
R 1 HELPINF                        /* REDEF. BEGIN : HELPINF
  2 HELP-TEXT2                     (1:3) /*FOR DESCRIPTION
  3 LNR                     A    4
  3 HELP-TEXT3              A   72
  1 VE-INDEX                N    2 /*INDEX FOR VERIFICATION
*                                  /*FOR INTERNAL USE ONLY
```

Figure 1.20. SYSHELPA, Parameter Data Area

You might ask what to do about fields on maps which do not belong to views. You can set up a special file with the same name as the application. This way, if SYSHELP cannot find a field using the DDM, it always tries to find the field using a file with the same name as the application.

Helpful Tip on Using NATURAL Security (SYSSEC) in Batch.

Often, the question is asked how to delete NATURAL Security records in batch. Here is sample input to NATURAL to enable this.

```
SYSSEC,USERID,PASSWORD
MENU
A
L
D,U,*,00-00-00,97-04-02 (This will purge records for a certain date range)
FIN
```

To list logon records in batch:

```
SYSSEC,USERID,PASSWORD
COMMAND LOGREC
L,U,,
FIN
```

You can process directly against the NATURAL Security file, but this requires knowing in what fields the relevant information is maintained on the NATURAL FSEC file. As of 2.2.7, NATURAL stores the logon records in the fields LO, LP and LV.

One can write a NATURAL program to read the logon records by:

```
READ FSEC by LO-FIELD starting from 'QLO' ending at 'R'
```

To read the logon error records:

```
READ FSEC by LO-FIELD starting from 'QER' ending at 'QES'
```

This way you could process these records and put them to another storage medium.

Performing SECULD/SECLOAD

The instructions in the manual are not totally correct on how to use NATURAL Security's migration facility. In the example JCL below, SECULD is instructed to unload a library definition for SYSPRINT. This definition is loaded into another Security file on another database. The user is linked as required to SYSSEC.

```
//DRMJTK JOB CLASS=T
//* BATCH JOB TO UNLOAD SECURITY OBJECTS
//DELETE EXEC PGM=IEFBR14
//DD1 DD DSN=DB.TEST.SECULD,UNIT=DASD,
//       DISP=(MOD,DELETE),SPACE=(TRK,1)
//UNLOAD EXEC NATURAL,ENVIRO=TEST
//CMWKF05 DD DISP=(,CATLG,DELETE),DSN=DB.TEST.SECULD,
// DCB=(RECFM=VB,LRECL=4620,BLKSIZE=4624),
// SPACE=(TRK,(100,20),RLSE)
//SYSIN DD *
LOGON SYSSEC
GLOBALS IM=D ID=,
SECULD
O,N,LI,SYSPRINT,1,,
.
FIN
//LOAD EXEC NATURAL,ENVIRO=DEVL
//CMWKF01 DD DISP=SHR,DSN=DB.TEST.SECULD
//SYSIN DD *
LOGON SYSSEC
GLOBALS IM=D ID=,
SECLOAD
N,N
FIN
//
```

A Very Useful NATURAL Security Tip

On SAG-L, a request was posted to answer:

How to obtain a list of all NATURAL Users together with their Groups and Libraries as defined in NATURAL Security. The following code in which one can see how to use two of NATURAL Security's subprograms was offered by Ken McAskill. <SAKCM@MDCC.EDU>

"I have written two NATURAL programs that user the NSCXR and NSCUS subprograms to produce the output you desire. Run both of them in batch mode to make looking at the output easier. Hope this is what you are looking for and that it might help anyone who is interested. The first program I call XREFUSER follows:"

```
0290 *
0300 DEFINE DATA LOCAL
0310 01 POBJ-TYPE                  (A2)        /* Object type
0320 01 REDEFINE POBJ-TYPE
0330   02 PFUNCTION                (A2)        /* Function code
0340 01 POBJ-ID                    (A54)       /* Object Id
0350 01 REDEFINE POBJ-ID
0360   02 PUSERID                  (A8)        /* UserId
0370 01 PRC                        (N4)        /* ReturnCode
0380 *
0390 1 NCUS
0400    2 PUSERID2                 (A8)        /* UserId 2 (for copy func)
0410    2 PPFKEY                   (N2/1:24)/* PF-key assignments
0420    2 PPARM                    (A50)
0430    2 REDEFINE PPARM
0440      3 PNOMAP                 (L)      /* TRUE= without map,
0450 *                                         FALSE=with map
0460      3 PNOET                  (L)      /* FALSE=ET performed,
0470 *                                         TRUE=no ET performed
0480      3 PSELTYPE               (A2)
0490      3 PPARM-DU1              (A6)
0500      3 PPFPRESSED             (A4)     /* Last pressed PF-key
0510      3 PCLASSID               (A2)     /* PFUNCTION='XR' --> Class ID
0520      3 PPARM-REST             (A34)
0530 *
0540    2 PPARM1                   (A250)
0550    2 REDEFINE PPARM1
0560      3 PUSERTYPE              (A1)     /* User type
0570      3 PDEFAP                 (A8)     /* Default library
0580      3 PETID                  (A8)     /* ETID
0590      3 PLSTAP                 (A8)     /* Last library used
0600      3 PLSTETID               (A8)     /* Last ETID
0610      3 PBATCH                 (A8)     /* Batch user ID
0620      3 PPASSW                 (A8)     /* Password (only input)
0630      3 PPARM1-DUMMY           (A8/1:4) /* (reserved for future use)
0640      3 PDAYS                  (N3)     /* Password change interval
0650      3 PTDLANG                (N2)     /* Language code or time diff
0660      3 PGROUP                 (A8/1:5) /* Priv groups or first 5 mems
0670      3 PMAIL                  (A8/1:5) /* Mailboxes
0680      3 PCGROUP                (N6)     /* Number of groups/members
0690      3 PPRVLIB                (A1)     /* User has a private library
0700      3 PPARM1-REST            (A77)
```

```
0710 *
0720   2 PPARM2              (A250)
0730   2 REDEFINE PPARM2
0740     3 PCREMOD           (A96)
0750     3 REDEFINE PCREMOD
0760       4 PCREID          (A8)      /* Created by
0770       4 PCREDAT         (A8)      /* Creation date
0780       4 PCRETIME        (A8)      /* Creation time
0790       4 PCRESIGNER      (A8/1:3) /* Co-owners
0800       4 PMODID          (A8)      /* Last modified by
0810       4 PMODDAT         (A8)      /* Modification date
0820       4 PMODTIME        (A8)      /* Modification time
0830       4 PMODSIGNER      (A8/1:3) /* Co-owners
0840     3 REDEFINE PCREMOD
0850       4 PCREDATA        (A48)
0860       4 PMODDATA        (A48)
0870     3 POWNERA           (A80)
0880     3 REDEFINE POWNERA
0890       4 POWNER          (1:8)
0900         5 POWNER-ID     (A8)      /* Owner ID
0910         5 POWNER-NR     (N1)      /* Number of co-owners
0920         5 POWNER-TYPE   (A1)      /* Owner type
0930 *                                    ('A'dministrator, 'G'roup)
0940     3 PNAME             (A32)     /* User name
0950     3 PPARM2-REST       (A42)
0960   2 PPARM3              (A250)
0970   2 REDEFINE PPARM3
0980     3 PLOCKTYPE         (A1)      /* 'L'=Logon error,
0990 *                                    'C'=COUNTERSIGN ERROR
1000     3 PLOCKDATE         (A1)      /* 'Y'=User has lock dates
1010     3 PLOCKDATES        (D/1:10,1:2) /* Lock dates;
1020 *                                       only if PLOCKDATE='Y'
1030     3 PLDATE            (D)       /* Date of locking
1040     3 PLTIME            (T)       /* Time of locking
1050     3 PLAPPLID          (A8)      /* Library ID
1060     3 PLTERMINALID      (A8)      /* Terminal ID
1070     3 PLTPUSERID        (A8)      /* TP user ID
1080     3 LOCKTYPE-CL       (A100)
1090     3 REDEFINE LOCKTYPE-CL        /* Only if PLOCKTYPE='C'
1100       4 PLOBJECTID      (A32)     /* Object ID
1110       4 PLOWNER         (A8)      /* Owner ID
1120     3 REDEFINE LOCKTYPE-CL        /* Only if PLOCKTYPE='L'
1130       4 PLERRNR         (N4/1:5) /* Error numbers
1140       4 PLERRUSER       (A8/1:5) /* User IDs
1150       4 PLERRAPPLID     (A8/1:5) /* Library IDs
1160     3 PPARM3-REST       (A33)
1170   2 PPARM4              (A250)    /* (reserved for future use)
1180   2 PTEXT               (A60/1:8)/* Security notes
1190 *
1200 01 NSCXR
1210   02 PLINKID            (A8)         /* GroupID
1220   02 SUBFUNC            (A2)         /* See text
1230   02 PPARM1             (A20)
1240   02 REDEFINE PPARM1
1250     03 PNOMAP           (L)
1260     03 PSLINES          (B1)         /* Number of output lines
1270     03 PRCOUNT          (B3)         /* Number of returned lines
1280     03 PPFKEY           (N2/1:4)     /* PF-Keys
```

```
1290 *
1300   02 PPARM2                  (A80/1:21)  /* Return Data
1310   02 REDEFINE PPARM2
1320     03 PGROUP-ID             (1:21)
1330       04 GROUP-ID            (A8)
1340       04 PRIV-GR             (A10)
1350       04 REDEFINE PRIV-GR
1360         05 PRIV              (A4)
1370       04 DUMMY               (A62)
1380 *
1390   02 REDEFINE PPARM2
1400     03 PPARM2-LIB            (1:21)
1410       04 LIBRARY-ID          (A8)
1420       04 LIBRARY-NAME        (A32)
1430       04 LINK-ID             (A8)
1440       04 LINK                (A1)
1450       04 TERMINAL-PROTECTION  (A1)
1460       04 STARTUP-PGM         (A8)
1470       04 TIMEWINDOWS         (A1)
1480       04 DEVICE              (A8)
1490       04 FILLER              (A13)
1500 *
1510 1 I                        (N2)
1520 1 #USER-NAME               (A32)
1530 1 #PRCOUNT                 (N3)
1540 1 #LIST-COUNT              (N3)
1550 1 #END-OF-LIST             (L)
1560 1 #LINK-TYPE               (A7)
1570 1 #MSG                     (A60)
1580 END-DEFINE
1590 *
1600 *
1610 DEFINE PRINTER (1) OUTPUT 'CMPRT01'
1620 FORMAT (1) LS=132 PS=55
1630 READ WORK FILE 1 PUSERID
1640 *
1650   ASSIGN PFUNCTION = 'DI'
1660   ASSIGN NCUS.PNOMAP = TRUE
1670   ASSIGN NCUS.PNOET = TRUE
1680   CALLNAT 'NSCUS' PFUNCTION PUSERID NCUS.PUSERID2 PRC NCUS.PPFKEY(*)
1690     NCUS.PPARM NCUS.PPARM1 NCUS.PPARM2 NCUS.PTEXT(*)
1700     NCUS.PPARM3 NCUS.PPARM4
1710   IF PRC = 0
1720     ASSIGN #USER-NAME = NCUS.PNAME
1730   ELSE
1740     WRITE (1) PUSERID 'User-ID not found in NATURAL SECURITY'
1750     ESCAPE BOTTOM IMMEDIATE
1760   END-IF
1770   AT TOP OF PAGE (1)
1780     WRITE (1) 'User GROUP cross reference for' PUSERID #USER-NAME //
1790     WRITE (1) 'GROUP-ID'
1800     WRITE (1) '--------'
1810   END-TOPPAGE
```

```
1820 *
1830   ASSIGN #END-OF-LIST = FALSE
1840   RESET #LIST-COUNT NSCXR.PPARM2 (*)
1850   REPEAT
1860     ASSIGN NSCXR.SUBFUNC = 'GR'
1870     PERFORM CALL_XREF
1880     FOR I 1 20
1890       IF NSCXR.GROUP-ID (I) = '        '
1900         ASSIGN #END-OF-LIST = TRUE
1910       ELSE
1920         WRITE (1)
1930           1T NSCXR.GROUP-ID (I)
1940         ASSIGN #LIST-COUNT = #LIST-COUNT + 1
1950       END-IF
1960     END-FOR
1970     UNTIL #END-OF-LIST
1980   END-REPEAT
1990 *
2000   COMPRESS 'User' PUSERID 'belongs to' #LIST-COUNT 'groups'
2010     INTO #MSG
2020   WRITE (1) / '-'(70)
2030   WRITE (1) #MSG
2040   SKIP (1) 2
2050   WRITE (1) 'User LIBRARY cross reference for' PUSERID #USER-NAME //
2060   WRITE (1) 'Library-ID Library Name                               '
2070     ' Link Type Linked Via Startup Program'
2080   WRITE (1) '---------- --------------------------------            '
2090     ' --------- ---------- ---------------'
2100 *
2110   ASSIGN #END-OF-LIST = FALSE
2120   RESET #LIST-COUNT NSCXR.PPARM2 (*)
2130   ASSIGN NSCXR.LIBRARY-ID (21) = 'A'
2140   REPEAT
2150     ASSIGN SUBFUNC = 'LA'
2160     PERFORM CALL_XREF
2170     ASSIGN #PRCOUNT = NSCXR.PRCOUNT
2180     FOR I 1 #PRCOUNT
2190       IF NSCXR.LINK (I) = 'L'
2200         ASSIGN #LINK-TYPE = 'Link'
2210       ELSE
2220         IF NSCXR.LINK (I) = 'S'
2230           ASSIGN #LINK-TYPE = 'Special'
2240         ELSE
2250           ASSIGN #LINK-TYPE = 'Public'
2260         END-IF
2270       END-IF
2280       WRITE (1)
2290         NSCXR.LIBRARY-ID (I)
2300         ' ' NSCXR.LIBRARY-NAME (I)
2310         ' ' #LINK-TYPE
2320         ' ' NSCXR.LINK-ID (I)
2330         ' ' NSCXR.STARTUP-PGM (I)
2340       ASSIGN #LIST-COUNT = #LIST-COUNT + 1
2350     END-FOR
2360     IF LIBRARY-ID (21) = '        '
2370       ASSIGN #END-OF-LIST = TRUE
2380     END-IF
2390 *
2400     UNTIL #END-OF-LIST
2410   END-REPEAT
```

```
2420 *
2430   COMPRESS 'User' PUSERID 'has access to' #LIST-COUNT 'libraries'
2440     INTO #MSG
2450   WRITE (1) / '-'(70)
2460   WRITE (1) #MSG
2470   NEWPAGE (1)
2480 *
2490 END-WORK
2500 *
2510 DEFINE SUBROUTINE CALL_XREF
2520 *
2530 ASSIGN NSCXR.PSLINES = 21
2540 ASSIGN POBJ-TYPE = 'US'                      /* Xref for user
2550 ASSIGN NSCXR.PNOMAP = TRUE
2560 CALLNAT 'NSCXR' POBJ-TYPE POBJ-ID NSCXR.PLINKID PRC
2570   NSCXR.SUBFUNC NSCXR.PPARM1 NSCXR.PPARM2(*)
2580 *
2590 END-SUBROUTINE
2600 *
2610 END
**************************************************************
*
*
*  JCL TO RUN PROGRAM LISTED ABOVE                              *
*
*
**************************************************************
//SAKCMNAT JOB (M119,4HA6),'KEN',CLASS=P,MSGLEVEL=(1,1),
//       USER=SAKCM
//NAT01L01 EXEC PGM=NATBATCH,
//      PARM=('ECHO=OFF,CC=OFF,INTENS=1,MT=0,MAINPR=1')
//CMPRINT   DD  SYSOUT=A
//CMPRT01   DD  SYSOUT=A
//CMWKF01   DD  *
SULCF
/*
//SYSUDUMP  DD  DUMMY
//DDCARD    DD  *
ADARUN PROGRAM=USER
/*
//CMSYNIN   DD *
LOGON DBATOOLS
XREFUSER
FIN
/*
//
```

The second program called XREFGRPS to list a cross reference for groups follows. Again, I run this in batch mode:

```
0190 *
0200 DEFINE DATA
0210 LOCAL
0220 1 POBJ-TYPE                  (A2)        /* Object type
0230 1 POBJ-ID                    (A54)       /* Object Id
0240 1 REDEFINE POBJ-ID
0250   2 PUSERID                  (A8)        /* UserId
```

```
0260 1 PLINKID                      (A8)        /* GroupID
0270 1 SUBFUNC                      (A2)        /* See text
0280 1 PFUNCTION                    (A2)        /* Function code
0290 1 PGROUPID                     (A8)        /* User ID
0300 1 PUSERID2                     (A8)        /* Dummy
0310 1 PRC                          (N4)        /* Return code
0320 1 PPFKEY                       (N2/1:24)   /* PF-key assignments
0330 1 PPARM                        (A50)
0340 1 REDEFINE PPARM
0350   2 PNOMAP                     (L)         /* TRUE= no map, FALSE=with map
0360   2 PNOET                      (L)         /* FALSE=ET,TRUE=no ETperformed
0370   2 PSELTYPE                   (A2)        /* ad / de / li
0380   2 PPARM-DU1                  (A6)
0390   2 PPFPRESSED                 (A4)        /* Last pressed PF-key
0400   2 PCLASSID                   (A2)
0410   2 PPARM-REST                 (A34)
0420 1 PPARM1                       (A250)
0430 1 REDEFINE PPARM1
0440   2 PMEMBER  (A8/1:30)                     /* GRUPPEN MITGLIEDER
0450   2 PPARM1-REST (A10)
0460 1 REDEFINE PPARM1
0470   2 PUSERTYPE                  (A1)
0480   2 PDEFAP                     (A8)
0490   2 PETID                      (A8)
0500   2 PLSTAP                     (A8)
0510   2 PLSTETID                   (A8)
0520   2 PBATCH                     (A8)
0530   2 PPASSW                     (A8)
0540   2 PPARM1-FILLER1             (A8/1:4)
0550   2 PDAYS                      (N3)
0560   2 PTDLANG                    (N2)
0570   2 PGROUP                     (A8/1:5)
0580   2 PMAIL                      (A8/1:5)
0590   2 PCGROUP                    (N6)
0600   2 PPRVLIB                    (A1)
0610   2 PTDHALF                    (N2)
0620   2 PTDSIGN                    (A1)
0630   2 PPARM1-FILLER2             (A74)
0640 *
0650 1 PPARM2                       (A250)
0660 1 REDEFINE PPARM2
0670   2 PCREMOD                    (A96)
0680   2 REDEFINE PCREMOD
0690     3 PCREID                   (A8)      /* Created by
0700     3 PCREDAT                  (A8)      /* Creation date
0710     3 PCRETIME                 (A8)      /* Creation time
0720     3 PCRESIGNER               (A8/1:3)  /* Co-owners
0730     3 PMODID                   (A8)      /* Last modified by
0740     3 PMODDAT                  (A8)      /* Modification date
0750     3 PMODTIME                 (A8)      /* Modification time
0760     3 PMODSIGNER               (A8/1:3)  /* Co-owners
0770   2 REDEFINE PCREMOD
0780     3 PCREDATA                 (A48)
0790     3 PMODDATA                 (A48)
0800   2 PPARM2-FILLER1             (A80)
0810   2 PNAME                      (A32)
0820   2 PPARM2-FILLER2             (A42)
```

```
0830 1 PPARM3                    (A250)       /* (reserved for future use)
0840 1 PPARM4                    (A250)       /* (reserved for future use)
0850 01 PPARM5                   (A20)
0860 01 REDEFINE PPARM5
0870    02 PPARM5-PNOMAP         (L)          /* false = NO map, true = map
0880    02 PSLINES               (B1)         /* Number of output lines
0890    02 PRCOUNT               (B3)         /* Number of returned lines
0900    02 PPARM5-PPFKEY         (N2/1:4)     /* PF-Keys
0910 1 PPARM6                    (A80/1:21)
0920 01 REDEFINE PPARM6
0930    02 PPARM6-LIB            (1:21)
0940       03 LIBRARY-ID         (A8)
0950       03 LIBRARY-NAME       (A32)
0960       03 LINK-ID            (A8)
0970       03 LINK               (A1)
0980       03 TERMINAL-PROTECTION (A1)
0990       03 STARTUP-PGM        (A8)
1000       03 TIMEWINDOWS        (A1)
1010       03 DEVICE             (A8)
1020       03 PPARM6-FILLER1     (A13)
1030 1 PTEXT                     (A60/1:8)/* Security notes
1040 1 GROUP-EDITOR              (A1)
1050 1 #END-OF-LIST              (L)
1060 1 #FIRST-CALL               (L)
1070 1 #LIST-COUNT               (N5)
1080 1 #I                        (N2)
1090 1 #PGROUPID                 (A8)
1100 1 #PMEMBER                  (A8/1:30)
1110 1 #LINE-COUNT               (P5)
1120 1 #LINK-TYPE                (A7)
1130 1 #MSG                      (A70)
1140 *
1150 END-DEFINE
1160 *
1170 IF *DEVICE = 'BATCH'
1180   DEFINE PRINTER (1) OUTPUT 'CMPRT01'
1190 ELSE
1200   DEFINE PRINTER (1) OUTPUT 'SOURCE'
1210 END-IF
1220 FORMAT (1) PS=60 LS=133
1230 *
1240 INPUT #PGROUPID
1250 *
1260 RESET PPARM
1270   PMEMBER (*)
1280   PTEXT (*)
1290   #LIST-COUNT
1300 MOVE ALL H'00' TO PPARM2
1310 MOVE ALL H'00' TO PPARM3
1320 ASSIGN #END-OF-LIST = FALSE
1330 ASSIGN #FIRST-CALL  = TRUE
1340 ASSIGN #LINE-COUNT = 60
1350 *
1360 *
```

```
1370 PERFORM CALL_NSCXR
1380 IF PRCOUNT > 0
1390   REPEAT
1400 *
1410     PERFORM WRITE_GROUP_LINKS
1420     ASSIGN LIBRARY-ID (1) = LIBRARY-ID (21)
1430     PERFORM CALL_NSCXR
1440 *
1450     UNTIL #END-OF-LIST
1460   END-REPEAT
1470   COMPRESS 'The Group' #PGROUPID 'has access to' #LIST-COUNT
1480     'libraries.' INTO #MSG
1490   WRITE (1) '-'(70)
1500   WRITE (1) #MSG
1510   RESET #LIST-COUNT
1520   ASSIGN #END-OF-LIST = FALSE
1530 ELSE
1540   WRITE (1) #PGROUPID 'is not linked to any libraries.'
1550 END-IF
1560 *
1570 PGROUPID  := #PGROUPID
1580 PSELTYPE  := 'LI'
1590 PFUNCTION := 'EG'
1600 PERFORM CALL_NSCUS
1610 ASSIGN #PMEMBER (*) = PMEMBER (*)
1620 ASSIGN PUSERID2 = PMEMBER (30)
1630 IF #PMEMBER (1) = '        '
1640   WRITE (1) #PGROUPID 'has no memebers.'
1650 ELSE
1660   REPEAT
1670 *
1680     PERFORM WRITE_GROUP_MEMBERS
1690     IF NOT #END-OF-LIST
1700       RESET PMEMBER (*)
1710       PGROUPID  := #PGROUPID
1720       PSELTYPE  := 'LI'
1730       PFUNCTION := 'EG'
1740       PERFORM CALL_NSCUS
1750       ASSIGN #PMEMBER (*) = PMEMBER (*)
1760       ASSIGN PUSERID2 = PMEMBER (30)
1770     END-IF
1780 *
1790     UNTIL #END-OF-LIST
1800   END-REPEAT
1810   COMPRESS 'The Group' #PGROUPID 'has' #LIST-COUNT 'members.'
1820     INTO #MSG
1830   WRITE (1) '-'(70)
1840   WRITE (1) #MSG
1850 END-IF
1860 *
1870 DEFINE SUBROUTINE WRITE_GROUP_LINKS
1880 *
```

```
1890 FOR #I 1 20
1900   IF LIBRARY-ID (#I) = '        '
1910     ASSIGN #END-OF-LIST = TRUE
1920     ESCAPE BOTTOM IMMEDIATE
1930   ELSE
1940     IF #LINE-COUNT > 55
1950       NEWPAGE (1)
1960       WRITE (1) 25X 'User Group Cross Reference for' #PGROUPID //
1970       WRITE (1) 'The group' #PGROUPID
1980         'has access to the following libraries:' /
1990       WRITE (1) 'Library-ID Library Name                     Link Type'
2000         'Start Program'
2010       WRITE (1) '---------- -------------------------------- ---------'
2020         '-------------'
2030       ASSIGN #LINE-COUNT = *LINE-COUNT
2040     END-IF
2050   END-IF
2060   IF LINK (#I) = 'L'
2070     ASSIGN #LINK-TYPE = 'Link'
2080   ELSE
2090     IF LINK (#I) = 'S'
2100       ASSIGN #LINK-TYPE = 'Special'
2110     ELSE
2120       ASSIGN #LINK-TYPE = 'Public'
2130     END-IF
2140   END-IF
2150   WRITE (1) LIBRARY-ID (#I) ' ' LIBRARY-NAME (#I) #LINK-TYPE
2160     ' ' STARTUP-PGM (#I)
2170   ASSIGN #LIST-COUNT = #LIST-COUNT + 1
2180   ASSIGN #LINE-COUNT = *LINE-COUNT (1)
2190 END-FOR
2200 IF LIBRARY-ID (21) = '        '
2210   ASSIGN #END-OF-LIST = TRUE
2220 END-IF
2230 *
2240 END-SUBROUTINE
2250 *
2260 DEFINE SUBROUTINE WRITE_GROUP_MEMBERS
2270 *
2280 IF #FIRST-CALL
2290   ASSIGN #FIRST-CALL = FALSE
2300   IF #LINE-COUNT > 55
2310     NEWPAGE (1)
2320     WRITE (1) 25X 'User Group Cross Reference for' #PGROUPID
2330     ASSIGN #LINE-COUNT = *LINE-COUNT (1)
2340   END-IF
2350   PGROUPID := #PMEMBER (1)
2360   PFUNCTION := 'DI'
2370   PERFORM CALL_NSCUS
2380   WRITE (1) // 'The Group' #PGROUPID 'has the following members:' /
2390   WRITE (1) 'UserID   User Name                          ETID    '
2400     '  User Group Num of Groups'
2410   WRITE (1) '------   --------------------------------   ------- '
2420     ' ---------- -------------'
2430   WRITE (1) #PMEMBER (1) PNAME ' ' PETID ' ' #PGROUPID
2440     ' ' PCGROUP
2450   ASSIGN #LIST-COUNT = #LIST-COUNT + 1
2460   ASSIGN #LINE-COUNT = *LINE-COUNT (1)
2470 END-IF
```

```
2480 FOR #I 2 30
2490   IF #PMEMBER (#I) = ' '
2500     ASSIGN #END-OF-LIST = TRUE
2510     ESCAPE BOTTOM IMMEDIATE
2520   ELSE
2530     PGROUPID  := #PMEMBER (#I)
2540     PFUNCTION := 'DI'
2550     PERFORM CALL_NSCUS
2560     IF #LINE-COUNT > 55
2570       NEWPAGE (1)
2580       WRITE (1) 25X 'User Group Cross Reference for' #PGROUPID //
2590       WRITE (1) 'UserID   User Name                          ETID  '
2600         '  User Group Num of Groups'
2610       WRITE (1) '------   --------------------------------   ------ '
2620         '  ---------- -------------'
2630     END-IF
2640     WRITE (1) #PMEMBER (#I) PNAME ' ' PETID ' ' #PGROUPID
2650       ' ' PCGROUP
2660     ASSIGN #LIST-COUNT = #LIST-COUNT + 1
2670     ASSIGN #LINE-COUNT = *LINE-COUNT (1)
2680   END-IF
2690 END-FOR
2700 *
2710 END-SUBROUTINE
2720 *
2730 DEFINE SUBROUTINE CALL_NSCXR
2740 *
2750 POBJ-TYPE := 'US'
2760 POBJ-ID   := #PGROUPID
2770 SUBFUNC   := 'LA'
2780 PPARM5-PNOMAP := TRUE
2790 CALLNAT 'NSCXR' POBJ-TYPE POBJ-ID PLINKID PRC SUBFUNC PPARM5 PPARM6(*)
2800 *
2810 END-SUBROUTINE
2820 *
2830 DEFINE SUBROUTINE CALL_NSCUS
2840 *
2850 PNOMAP    := TRUE
2860 PNOET     := FALSE              /* ET PERFORMED
2870 CALLNAT 'NSCUS' PFUNCTION PGROUPID PUSERID2 PRC PPFKEY(*)
2880   PPARM PPARM1 PPARM2 PTEXT(*) PPARM3 PPARM4
2890 *
2900 END-SUBROUTINE
2910 *
2920 END
```

```
************************************************************************
*                                                                      *
*  JCL TO RUN THE PROGRAM LISTED ABOVE                                 *
*                                                                      *
************************************************************************
//SAKCMNAT JOB(M119,4HA6),'KEN',CLASS=P,MSGLEVEL=(1,1),USER=SAKCM
//NAT01L01 EXEC PGM=NATBATCH,
// PARM=('FSEC=(3,252),ECHO=OFF,CC=OFF,INTENS=1,MT=0,MAINPR=1')
//CMPRINT   DD  SYSOUT=A
//CMPRT01   DD  SYSOUT=A
//SYSUDUMP  DD  DUMMY
//DDCARD    DD  *
ADARUN PROGRAM=USER
/*
//CMSYNIN   DD *
LOGON DBATOOLS
XREFGRPS USERAST1
FIN
/*
//
```

Notes

Chapter 2

Working In A NATURAL Session

I. NATURAL Session Commands

ADACALL Command

This command allows one to issue direct calls to ADABAS. It's useful for debugging scenarios, especially if you want to test a direct call. It's useful to view ADABAS control blocks. A user exit is supplied to provide additional processing and/or security, called ADAEXIT in SYSDBA.

```
Enter information and press 'PF10' to execute
18:53:19            ***** NATURAL ADACALL Utility *****          97-05-08
User APDDBS               - ADABAS Direct Calls -
Mode Char                                                  Call No. 0
 *** Control Block ***
  Cmd            Cmd ID               File      0          Database 240
 Resp     0        ISN          0      ISQ        0            ISL         0
  FBL 210          RBL 980             SBL 140     VBL 140     IBL 0
 COP1             COP2          User Area                  Cmd Time
 Addition1        Addition2  Addition3         Addition4        Addition5

 *** Buffer Areas ***
 Format

 Record

 Search

  Value

    ISN
Command ===>
```
Figure 2.1. ADACALL Main Menu Executed from SYSDBA.

```
DEFINE DATA LOCAL
1 CAR VIEW OF VEHICLES
  2 PERSONNEL-ID /* CNNNNNNN
  2 MAKE
  2 MODEL
  2 YEAR /* YEAR OF THE CAR'S MANUFACTURE
    /* END OF VIEW VEHICLES-VIEW
END-DEFINE
READ (3) CAR BY MAKE
  DISPLAY PERSONNEL-ID MAKE MODEL YEAR
END-READ
END
```

Execution yields:

```
MORE _
Page      1                                              97-04-01  12:34:56
PERSONNEL-ID          MAKE                  MODEL            YEAR
------------ -------------------- -------------------- ----
11100328     ALFA ROMEO           GIULIETTA 2.0         85
11400308     ALFA ROMEO           QUADRIFOGLIO          84
20004100     AMERICAN MOTOR       HORNET                77
```

By entering values for Cmd, Cmd Id, File, Database, Addition1, Format and Search buffers you issue a call to ADABAS. Other fields will display, such as the Record buufer, upon return from ADABAS (see Fig.2.1a)

```
Enter information and press 'PF10' to execute
12:34:56            ***** NATURAL ADACALL Utility *****           97-04-01
User APDDBS                - ADABAS Direct Calls -
Mode Char                                                     Call No. 0
 *** Control Block ***
  Cmd: L3        Cmd ID: WIZ2          File:    5         Database 240
 Resp      0          ISN: 205      0      ISQ         0          ISL          0
  FBL 210             RBL 980              SBL 140      VBL 140        IBL 0
 COP1                COP2         User Area                       Cmd Time
 Addition1           Addition2  Addition3          Addition4          Addition5
  AD
 *** Buffer Areas ***
 Format: AC,008,A,AD,020,A,AE,020,A,AG,002,U.

 Record: 11100328ALFA ROMEO          GIULIETTA 2.0        85

 Search: AD,020,A.

  Value

    ISN
Command ===>
```

Figure 2.1a. ADACALL Data to Duplicate Above Program.

Hitting PF10 repeatedly is equivalent to executing the program with a READ loop. Below is a program example

You can also write programs to act similarly to the ADACALL operation. Below is a listing of a program and the ADACB local data area description to CALL 'CMADA', NATURAL's entry point to issuing calls to ADABAS.

Program listing

```
DEFINE DATA
LOCAL USING ADACB
LOCAL
1 #FB (A200)
1 #RB (A77/100)
1 #SB (A200)
1 #VB (A200)
1 #IB (A200)
1 #RC-NUM (N4)
1 #FILE-NAME (A32) INIT<'VEHICLES'>
END-DEFINE
* FORMAT LS=250
ASSIGN #CID = 'WIZ3'
ASSIGN #CC    = 'L3'
ASSIGN #DBID  = 240
ASSIGN #FNR   = 005
/*Sag Vehicles file
ASSIGN #FBL   = 180
ASSIGN #RBL   = 980
ASSIGN #SBL   = 140
ASSIGN #VBL   = 140
ASSIGN #IBL   = 200
ASSIGN #CMD-OPT-2 = 'V'
ASSIGN #ADD-1 = 'AD'
/* same as search buffer, w/o get
RC 21
ASSIGN #SB = 'AD.'
ASSIGN #FB = 'AC,AD,AE,AG'
CALL 'CMADA' #CONTROL-BLOCK
     #FB #RB(*) #SB #VB #IB
ASSIGN #RC-NUM = #RC
DISPLAY
  / '=' #RB(1:3)
/* value matches request in format
/* buffer
  / #RC-NUM
END
```

ADACB listing

```
   1 #CONTROL-BLOCK                A   80
 R 1 #CONTROL-BLOCK                    /*
REDEF. BEGIN : CONTROL-BL
   2 #FILLER-1                     A    2
   2 #CC                           A    2
   2 #CID                          A    4
   2 #DBID                         B    1
   2 #FNR                          B    1
   2 #RC                           B    2
   2 #ISN                          B    4
   2 #ISN-LOW                      B    4
   2 #ISN-QTY                      B    4
   2 #FBL                          B    2
   2 #RBL                          B    2
   2 #SBL                          B    2
   2 #VBL                          B    2
   2 #IBL                          B    2
   2 #CMD-OPT-1                    A    1
   2 #CMD-OPT-2                    A    1
   2 #ADD-1                        A    8
   2 #ADD-2                        A    4
   2 #ADD-3                        A    8
   2 #ADD-4                        A    8
   2 #FILLER-2                     A    8
   2 #TIME                         B    4
   2 #USER-AREA                    A    4
```

MAINMENU Command

This command invokes the NATURAL startup menu with an optional parameter 'ON' or 'OFF'. This is useful for anyone who wishes to work at the NEXT prompt and not with a menu. The maps - MAINMAP1 (English), MAINMAP2 (German) - are retrievable from SYSMENU library and can be modified, then stored in SYSLIB.

PROFILE Command

If you are logged onto a session under NATURAL Security, you can execute the command PROFILE on the command line. This displays information about parameters affected by NATURAL Security, as seen in Fig. 2.2. The display includes system files (also available with the SYSPROF command); time windows; start/restart/user programs; session parameters; security limits; available commands and statement restrictions and allowable editors. The GLOBALS command and user-written programs can alter some of your session parameters.

```
12:34:56                *** NATURAL SECURITY ***                 97-04-01
                          - Security Profile -

User
  ID ....... APDDBS                                   Link ID ..
  Name ..... DATABASE SPECIALIST                      ETID .....
  Type ..... Person

Library                                               --- Steplibs ----
  ID ....... APDDBS                                   Library  DBID FNR
  Name .....

Transactions
  Startup .. MAINMENU
  Restart ..
  Error ....

                                                      BUSYSTEM ( *STEPLIB )
Additional Options ... N
```

Figure 2.2. PROFILE main screen.

Additional options display the following pop-ups:

```
Additional options

    _  + Security options
    _  + Security limits
    _  + Session paramcters
    _    Command restrictions
    _    Editing restrictions
    _  + Statement restrictions
    _    Time windows
    _  + System files
    _  + NATURAL version
```

Figure 2.3. PROFILE Additional Options

```
Security options


  Definition of command mode
    NEXT, MORE line allowed ...... Y
    System commands allowed ...... Y
  Execution of update programs ... Y
  Modules allowed/disallowed ..... D
  Device ........................ PC
  PC download ................... Y
  PC upload ..................... Y
```

Figure 2.4. PROFILE Display of Security Options

```
Security limits

  Non-activity logoff limit (seconds) ....... 0
  Maximum transaction duration (seconds) .... 0
  Maximum number of source lines ........... 99999
  Maximum amount of CPU time (MT;seconds) ... 60
  Maximum number of ADABAS calls (MADIO) .... 9999
  Maximum number of program calls (MAXCL) ... 9999
  Processing loop limit (LT) ............... 99999999
```

Figure 2.5. PROFILE Display of Security Limits.

GLOBALS Command

This command is the NATURAL system command equivalent of the statement SET GLOBALS (see Fig. 2.6). If the command is typed without parameters, a screen is displayed with all the current values set during NATURAL's initialization. The screen also serves as a menu to change certain parameters.

```
12:34:56                    *** NATURAL GLOBALS COMMAND ***                 97-04-01
Database ........ 242     File Number ..... 147
Version ......... 22      SM Level ........ 08      INPL Update Level ... 0
User Area (USIZE) 64  K   Ext. Area (ESIZE) 78  K   System Buffer (FSIZE) 64  K
Available Printers .... 07 15 16 29 30 31

Available Work Files .. 01 02 03 04 05 06 07 08 09 10 29 30 31

Dec. Character .(DC)    . Input Delimiter ..(ID)    , Input Assign ...(IA)   =
Line Size ......(LS)   80 Page Size ........(PS)   23 Input Mode .....(IM)   F
Sound Alarm.....(SA)  OFF Dump Generation ..(DU)  OFF Zero Printing ..(ZP)   ON
Control ........(CF)    % Page Eject .......(EJ)   ON Wait on Hold ...(WH)  OFF
Zero Division ..(ZD)   ON Source Line Length(SL)   72 Spacing Factor .(SF)    1
Default Format .(FS)   ON Page Dataset .....(PD)  100 Limit Error ....(LE)  OFF
Machine Code Generator ..(MCG) OFF                    Limit .....(LT) 99999999
   Function-Key Settings in Command Mode .. ON   Library .. APDDBS
PA1                 PA2                 PA3                 PF1
PF2                 PF3                 PF4                 PF5
PF6                 PF7                 PF8                 PF9
PF10                PF11                PF12                PF13
PF14                PF15                PF16                PF17
PF18                PF19                PF20                PF21
PF22                PF23                PF24
```

Figure 2.6. GLOBALS Command screen.

The GLOBALS Command to turn the Sound Alarm parameter to ('OFF') is issued as:

```
GLOBALS SA=OFF
```

The GLOBALS Command to turn of CPU time for a NATURAL session (esp. batch) is issued as:

```
GLOBALS MT=0
```

The GLOBALS Command to establish input as delimited and to specify a specific character (comma) is issued as:

```
GLOBALS IM=D ID=,
```

The GLOBALS Command to establish Structured Mode is issued as:

```
GLOBALS SM=ON
```

Setting Session Parameters Via a User-Written Program

The NATURAL System Variable *STARTUP contains the value of the program which gets control at the start of a session. The default is MAINMENU. It is modifiable by the user. This is how a program other than MAINMENU can serve as the startup transaction.

Here is a sample program which shows how you can alter your session global parameters programmatically. It also sets PF keys which are toggled at command level.

Warning: Be sure that the startup program you are defining exists, and provides the means to communicate to NATURAL. Remember you are replacing the NEXT COMMAND prompt, unless you stack to MAINMENU. Without this communication, you will find yourself in an endless loop with NATURAL Security.

```
0010 * STARTUP = 'MYGLOBS'
0020 SET GLOBALS WH=OFF  ZD=OFF LE=OFF LT=50
0030 STACK TOP COMMAND 'KEY PA1 = "LOGOFF" '
0040 STACK TOP COMMAND 'KEY PF1 = "CLOCK" '
0050 STACK TOP COMMAND 'KEY PF2 = "CATALL" '
0060 END
```

These settings will remain until either another GLOBALS or KEY command is issued at COMMAND, or another SET GLOBALS (note: SET GLOBALS not available in Structured Mode) or STACK COMMAND 'KEY...' is issued by another program. If you are not re-establishing the startup program, your key settings will not take effect because your session is under the control of MAINMENU, which establishes its own PF keys.

You can see the results of executing the program or by issuing the command by typing 'GLOBALS'. The display is shown in Figure 2.7.

Multiple GLOBALS settings can be established with the same SET GLOBALS statement, as is seen in the above example. There is a side effect about which you ought to be familiar. During an edit session, if you enter a SET GLOBALS statement into a program for a compile time parameter, such as LS, upon pressing the ENTER key your session global parameters will also change. This could have ill effects on later compiles.

Note that in batch these parameters take affect *after* the program runs. If GLOBAL settings are needed in batch, execute a GLOBALS command in-stream before the execution of the program! With this background knowledge one can write more substantial programs to establish any session parameters, including setting PF keys.

```
12:34:56                *** NATURAL GLOBALS COMMAND ***                     97-04-01
Database ........ 242    File Number ..... 147
Version ......... 22     SM Level ........ 08      INPL Update Level ... 0
User Area (USIZE) 64  K  Ext. Area (ESIZE) 78  K   System Buffer (FSIZE) 64 K
Available Printers .... 07 15 16 29 30 31

Available Work Files .. 01 02 03 04 05 06 07 08 09 10 29 30 31

Dec. Character .(DC)   . Input Delimiter ..(ID)   , Input Assign ...(IA)  =
Line Size ......(LS)  80 Page Size ........(PS)  23 Input Mode .....(IM)  F
Sound Alarm.....(SA) OFF Dump Generation ..(DU)  ON Zero Printing ..(ZP)  ON
Control ........(CF)   % Page Eject .......(EJ)  ON Wait on Hold ...(WH) OFF
Zero Division ..(ZD)  ON Source Line Length(SL)  72 Spacing Factor .(SF)   1
Default Format .(FS)  ON Page Dataset .....(PD) 100 Limit Error ....(LE) OFF
Machine Code Generator ..(MCG) OFF                  Limit .....(LT) 99999999
   Function-Key Settings in Command Mode .. ON   Library .. APDDBS
PA1  LOGOFF          PA2                 PA3                 PF1  CLOCK
PF2  CATALL          PF3                 PF4                 PF5
PF6                  PF7                 PF8                 PF9
PF10                 PF11                PF12                PF13
PF14                 PF15                PF16                PF17
PF18                 PF19                PF20                PF21
PF22                 PF23                PF24
```

Figure 2.7. Menu Presented When GLOBALS Command Executed

PF keys are set either programmatically ('SET KEY' statement) or with the system command 'KEY'. The system command affects the parameters for the session whereas the NATURAL statement 'SET KEY' affects a different set of keys which are program sensitive. If you wish to use a program to affect the session PF key settings you can do so by stacking the system command 'KEY' onto the NATURAL stack. The examples below illustrate this.

There are three sets of PF keys in the NATURAL environment - Global PF keys, program sensitive PF keys and Editor PF keys. Program sensitive PF keys are activated by the SET KEY statement:

```
SET KEY PFn = PGM
```

They can be assigned a transaction by:

```
SET KEY PFn = value
```

The value can be a literal constant or the contents of an alphanumeric variable. They can be flipped on or off by:

```
                ON
SET KEY PFn = { OFF}
```

Turning the keys on or off reactivates or deactivates program sensitivity. When assigning keys and navigating an application via CALLNATs, a value will stick unless it is set with another SET KEY statement. This problem has more to do with application design than with Pfkeys specifically. Deactivation returns control to the TP monitor. Reactivation reassigns the value; if none existed, it is set to [the] ENTER [key]. There are some existing PF key zaps under NATURAL.2.8. Remember: *A key setting is maintained across programs during a session.* There is also no way currently to clear all PF keys (what a pity). However, the closest one can come to making key setting 'go away' at the end of a routine is to start the routine with SET KEY ALL and to conclude with SET CONTROL 'ENTR'. However, upon returning to higher level subprograms one must re-establish keys (possibly by using INCLUDE code after a CALLNAT, for example).

II. The NATURAL Editors

The Program Editor

The Software AG's *NATURAL Reference Manual* and *NATURAL Programmers Guide* or WH&O's *NATURAL Developers Handbook* do an excellent job of defining edit sessions and how the editor works. What I want to do here is to describe some aspects that may not be so obvious.

There are multiple objects one can edit within the three fundamental editors - program, data, and map. The Program Editor edits the largest number of different objects - COPYCODE, description, HELPROUTINE, program, subprogram and subroutine. Once the Program Editor is invoked, you can identify the desired object by entering 'SET TYPE ...'. Therefore, if you accidently entered the Program Editor and started to work and then realized you were creating a subprogram, you just enter 'SET TYPE SUBPROGRAM' or SET TYPE 'N' on the editor command line. The mode setting is also important.

The mode can be set for the session via the parameter SM or NATURAL Security if it is installed. You can alter the mode also by typing over the mode on the MAINMENU screen. Once NATURAL Security sets the mode to structured, you cannot alter it. There are a few other aspects of the editor worth describing.

On the top line in the editor is a direction indicator which is marked as '+' or '-'. The purpose of the indicator is to establish the direction for any copy, move, or insert line commands. It also, affects the scan and add operations. Lines can be blocked by marking them with the '.X' and '.Y' commands. The delete edit command and the copy and move line commands work with line markers. The scan capability is slightly improved. Once a scan operation is completed you can re-execute it by typing 'SC='. There are also EX and EY commands that will delete up to the line marked 'X' or from to the bottom for a line marked 'Y'.

A PF key can be defined in the Program Editor PROFILE to define a repeated scan operation as a key toggle. The scan will default from the current position forward or backward depending on the direction marker.

Other editor improvements include the '.J' and '.I' line commands. The '.J' command allows the joining of lines again the direction of the operation is affected by the direction indicator). The '.I' command has been expanded to: (1) define how many lines are to be inserted at what line position; (2) include a data area - cataloged local or parameter; (3) include a NATURAL map, which results in an INPUT USING MAP statement generated with the list of input fields. If you have Construct installed, there is a .G capability which is awesome. For example, if you develop a subprogram NAMED SUBXXX and you want to test it, you enter .G(driver,SUBXXX) and you achieve generating a program front end to which you pass parameters and SUBXX is executed. If your shop has Construct but uses it sparingly, the .G capability should be investigated and word spread to the staff.

If you enter '*' on the command line, you will get a display of the last command entered*. By entering '*=' you re-execute the last command entered (both '*' and '*=' may be assigned to PF keys for ease of use, see PROFILE section). Multiple commands can be entered on the command line by separating them with a comma. For example, 'B,-H' causes positioning to the bottom and backing up half a page, 'T,SC= ' repositions to the top and repeats the SCAN command.

The CHANGE command is a powerful command. It is used to search for a string of characters to replace with a replacement string. It is especially useful in the data editor where it can change characters behind the scenes. It is also the only way of changing a file name in the data editor since the data editor does not allow you to type over the name. However, use this command in the data editor with caution.

The STRUCT command is used to perform structural indentation on the program source. This is useful for readability. It does **NOT** mean that the program suddenly becomes “structured”. There are third party vendor tools to aid in this effort. STRUCT can be assigned to a PF key in the editor PROFILE making more convenient to use.

*The command LAST will display the last 9 commands executed from which a command can be selected or commands reorder and reexecuted.

Program Editor Profile.

The Program Editor allows the user to establish a custom edit profile, as seen in Figure 2.8. Included are such editor parameters as an auto-save capability and the PF designators. Once invoked these keys are good for the duration of the user session. On the editor command line type 'PROFILE' to get:

```
12:34:56          ***** NATURAL PROFILE MAINTENANCE *****           01/04/97
                        - Global Editor Profile -

Profile Name .. APDDBS__
PF and PA Keys
  PF1 ... HELP___________   PF2 ... ERRORTA________   PF3 ... EXIT___________
  PF4 ... LIST___________   PF5 ... SC=____________   PF6 ... _______________
  PF7 ... -______________   PF8 ... +______________   PF9 ... _______________
  PF10 .. SC=____________   PF11 .. t, sc=_________   PF12 .. _______________
  PF13 .. _______________   PF14 .. _______________   PF15 .. _______________
  PF16 .. _______________   PF17 .. _______________   PF18 .. _______________
  PF19 .. _______________   PF20 .. _______________   PF21 .. _______________
  PF22 .. _______________   PF23 .. _______________   PF24 .. _______________
  PA1 ... _______________   PA2 ... _______________   PA3 ... _______________

Automatic Functions
  Auto Renumber .. Y  Auto Save Numbers .. 5__  Source Save into .. APDDBS__

Additional Options .. N


Command ===>
Enter-PF1---PF2---FP3---PF4---PF5---PF6---PF7---PF8---PF9---PF10--PF11--PF12-
      Help        Exit  AddOp Save  Flip                          Del   Canc
```

Figure 2.8. Menu for the Editor PROFILE Command.

By choosing 'Additional options', a pop-up window gives the following options:

```
                        ADDITIONAL OPTIONS


    +  Editor Defaults ......... N
    +  General Defaults ........ N
    +  Colour Definitions ...... N
```

Figure 2.9. Additional options in the Editor PROFILE.

The next three windows illustrate the response to 'Y' to any of these three options:

```
                        EDITOR DEFAULTS

    Escape Character for Line Command .. .
    Empty Line Suppression ............. Y
    Source Size Information ............ Y
    Source Status Message .............. Y
    Absolute Mode for SCAN/CHANGE ...... Y
    Range Mode for SCAN/CHANGE ......... N
    Direction Indicator ................ +
```

Figure 2.10. Editor Defaut Settings Options in the Editor PROFILE.

```
                    GENERAL DEFAULTS

   Editing in Lower Case ............. Y
   Dynamic Conversion of Lower Case ... Y
   Position of Message Line ........... TOP
   Cursor Position in Command Line .... N
   Stay on Current Screen ............. N
   Prompt Window for Exit Function .... N
   ISPF Editor as Program Editor ...... N
```

Figure 2.11. General Defaults Options in the Editor PROFILE.

```
                    COLOUR DEFINITIONS

 Edit Work Area                   Split Screen Area
   Command Line ........ NE
   Label Indicator ..... RE          Label Indicator .... RE
   Line Numbers ........ TU          Line Numbers ....... TU
   Editor Lines ........ GR          Editor Lines ....... TU
   Scan and Error Line.. RE          Scan Line .......... RE
   Information Text .... TU          Information Text ... TU
   Information Value ... YE          Information Value .. YE
   Information Line .... TU
```

Figure II.12 Color Definitions from the Editor PROFILE.

AUTO RENUMBER is a useful feature included in the editor profile. This allows for renumbering to occur for RUN, SAVE, and STOW commands automatically. This is especially useful for people who copy or move lines immediately before issuing RUN from the editor.

The AUTO SAVE AFTER feature allows you to save your program into a member name provided in the profile (default is 'EDITWORK'). It is suggested to change the default member name to your own USERID. The number following 'AUTO SAVE AFTER' must be > 0. This counter specifies the number of e strokes before a save is done. Be careful of the count. A setting of 1 can cause some overhead since a save will be done after every ENTER key stroke.

If you are executing under NATURAL Security, a link is established between your User ID and your editor profile. This is controlled by the NATURAL editor which allows you to invoke the system default set by typing PROFILE for the first time on the EDIT Command Line. If you modify the system set and save the edited profile, only your INIT-USERID value is allowed as the profile name. This link forces the editor to invoke your profile upon entering any future edit session. However, you can save only the one set and modify it as required. Doing this changes the set of keys that are 'ON HOLD' for the editor session. Also, USR0070N in SYSEXT can be used to change the default editor profile.

The Data Editor

Once again, the Software AG *NATURAL Reference Manual* does an excellent job of describing the Data Editor as well as WH&O's *NATURAL Developers Handbook*. If you require external data areas, you must invoke the Data Editor.

Choosing fields from views is easy to do. The Data Editor command '.V' allows for the selection from a field list for that DDM. The field selection is terminated by entering '.' in the selection field. This can be done on the SAME screen as your last selected field. Secondly, issuing the '.I' line command in the Program Editor allows for any local/parameter data area to be included. Changes can be made here which cannot be made in the Data Editor (lengths, name of DVN for example).

Here are some hints on using the Data Editor that are not obvious. First, do not select PE elements at the group level. You can both specify the group name and select the components and which occurrences at the elementary level. Second, if you want the C* variable for a selected PE group, you can use the '.*' command. You can also enter the C* variable name by using the '.I' command. Remember that the '.D' on a group with subsidiary fields eliminates the entire group.

Third, you can invoke the Extended Edit Function for fields by typing '.E' in the first position. This function allows you to to initialize values &/or provide an edit mask. The edit mask and initial values allowed must obey NATURAL's syntax rules according to the format of the field. Beware, however, that the order in which you execute the sub-options of '.E' make a difference. An initial value must be done *before* defining an edit mask.

Fourth, if you want to add fields from a view in a later edit session, you can type the '.V' on the line marked 'V' in the Type column or type split screen mode (S V command). Lastly, creating external views eliminates source line entries. If the program you construct is very large, removing lines can be very difficult. A quick band-aid solution is to remove any local definitions of views and put them in an external local data definition, stow the local data area and re-compile the program referencing the data area as external. This approach is only for extreme emergencies; otherwise I would suggest rewriting the code and modularize the program into logical units.

Finally, the CHANGE command (abbreviated CH) is a powerful command. You can use it to change any literal, including literals behind the scenes, used by the Data Editor. A good technique for creating copy areas of view is to employ the MOVE BY NAME by marking and copying, then retyping the view name as a group name. Use the CHANGE command to replace the identifying characters used by the Data Editor to differentiate a view from a user group. This saves a lot of retyping time.

Data definition can be assigned values through the data area by including keywords INIT and CONST. INIT allows for an initial value to be assigned at startup. Also, this value is used when this variable is referenced via a RESET INITIAL. By using the keyword CONST, the variable is treated as a named constant and therefore cannot be altered during the program execution.

The security access definition is stored in views, ie. READ or UPDATE. A change in the current access after a SYSSEC change requires that the data area containing the view be reSTOWed.

III. Other NATURAL System Commands

The Software AG *NATURAL Reference Manual* and SAG Training Program presents materials on NATURAL System Commands. They are also found in the appendix of WH&O's *NATURAL Developers Handbook*. However, in my years of working with the NATURAL system, some of the commands were not so obvious to use. In addition, there are some new commands that may not be familiar to some users.

ADHOC

If a program ends with the END statement, then the program is both compiled and executed. However, if it ends with the ENDHOC statement, then ADHOC mode is terminated and command mode is entered. It is rumored this is going away in version 2.3.

BUS

The function Buffer Usage Statistics (or BUS) provides an overview for the allocation and used portions of certain storage areas in the NATURAL environment, as is displayed in Figure 2.14. You can use FETCH RETURN to the function BUS to get snapshot of the percentages of buffers used during execution. This technique is similar to the one discussed in Chapter 6 on TEST DBLOG.

IOB

Internal I/O area, serves as a buffer for NATURAL, e.g., for building messages.

PAGE

I/O Buffer, in this area NATURAL can build a 250 by 250 map.
Size is equal to LS * PS + 256

```
18:23:28              ***** NATURAL DBA UTILITY *****              97-04-01
COMPLETE               - Buffer Usage Statistics -                  MVS/ESA
```

No	Name	Offset	Size	Used	%	MaxUse	%
1	IOCB	00000000	2664	2664	100	2664	100
2	TIOB	00000A68	3968	788	19	788	19
3	CMPRTSZ	000019E8	2048	1152	56	1216	59
4	GETPHTAB	000021E8	1024	32	3	44	4
5	ISIZE	000025E8	4096	3104	75	3104	75
6	SYSFTAB	000035E8	1014	1014	100	1014	100
7	BPMWORK	000039E0	288	245	85	288	100
8	IOBV12	00003B00	2944	0	0	0	0
9	RUNSIZE	00004680	16384	6776	41	9764	59
10	GETMUB	00008680	243512	0	0	0	0
11	.USIZE	00008680	65528	2004	3	5051	7
12	.IOB	00018678	3072	2740	89	3072	100
13	.SAVE	00019278	2880	288	10	720	25
14	.ESIZE	00019DB8	79872	4224	5	29235	36
15	.DATSIZE	0002D5B8	92160	2520	2	24116	26
16	EPLTAB	00043DB8	2796	1392	49	1420	50
17	GETMIO	000448A8	44132	0	0	0	0
18	.PAGE	000448A8	8000	0	0	2500	31
19	.OVERLAY	000467E8	320	320	100	320	100
20	.SCREEN	00046928	1928	1928	100	1928	100
21	.PAGEATT	000470B0	19256	4020	20	4104	21
22	.OVLYATT	0004BBE8	768	58	7	768	100
23	.SCRNATT	0004BEE8	13860	4272	30	4272	30
24	FSIZE	0004F510	65568	32	0	32	0
25	DSIZE	0005F530	20480	32	0	32	0
26	EXTBUF	00064530	16384	816	4	816	4
27	PRNTWORK	00068530	6784	6330	93	6784	100
28	WORK07	00069FB0	1928	0	0	0	0
29	WORK29	0006A738	1928	0	0	0	0
30	WORK30	0006AEC0	1928	0	0	0	0
31	PRINT07	0006B648	1928	0	0	0	0
32	PRINT29	0006BDD0	1928	0	0	0	0
33	PRINT30	0006C558	1928	0	0	0	0
34	Total		445654	49931	11	104052	23
35			436K	49K		102K	

Figure 2.14. Buffer Usage Statistics Snapshot

OVERLAY

Overlay I/O buffer. This area can hold four lines, such as the message line, statistics and PF key lines.

SCREEN

I/O Buffer, used for V2.1 terminal I/O, window of what is displayed

PAGEATT

Page Attribute Buffer
This is equal to (SCREEN*10)/AVERIO.

SAVE (save area)

Here is where the save registers are stored. The register conventions are:

```
Register 0     Standard work register
Register 1     Parameter address register
Register 2     Internal BAL register
Register 3     Work / BAL register
Register 4     Work / Page buffer pointer
Register 5     Source / command pointer
Register 6     Object pointer
Register 7     Work register
Register 8     Command Block pointer
Register 9     Base register
Register 10    Base register / IOCB pointer
Register 11    Base register
Register 12    Benuzter Block (BB) pointer
Register 13    Save area pointer
Register 14    Return register
Register 15    Target register / return code
```

OVLYATT

Overlay Attribute Buffer

SCRNATT

Screen Attribute Buffer

DYNAM

Dynamic load area - CDYNAM*16 - area for dynamic names and addresses. This table maintains the addresses of dynamically loaded non-NATURAL programs in order to keep from reloading.

Real printers, such as PC3 printer designations, and work file buffers cost 1,928 bytes each.

CATALL

The CATALL command allows you to catalog any of the objects defined in a NATURAL library. The entire library can be catalogued or any part of it. The facility catalogs them in the order requested on the screen. The utility enters non-conversational mode after starting and reports the results of cataloging each object in the request. You should NOT try to press the ENTER key for it may interrupt the process. If you want to restart the process it automatically continues from the point of interruption or failure See Figure 2.15.

If you want to restart from the top you must delete a report in your library named CATALL. The facility uses the report to determine the starting point.

There are a few points about CATALL which make it effective. CATALL works for the library to which you are logged on. You can issue CATALL even if the source does not exist. It then aligns GDAs, reassigns object module dates, etc. CATALL works in batch as long as you remember to pass the correct number of parameters. If you wish to catalog a specific object or a group with the same prefix, you must remember to mark the type as well. Nothing is done if you do not do so; however, nothing is reported either.

Because your terminal is tied up until completion, it may be better to invoke the CATALL function in batch. For example:

```
CATALL DRNP000,,X,,,,,,,,,X (See Figure II.5.)
```

causes the facility to catalog the NATURAL.1 program `DRNP000`. If you accidently mark the wrong position, for example you entered the above as:

```
CATALL DRNP000,,X,,,,,,,,,,X
```

you do not get an incorrect report, but the facility also does NOT do what you intended.

```
CATALL *,X,,,,,,,,X,,X
```

Causes the facility to catalog all NATURAL programs and maps. The re-cataloging only takes place if the counterpart module exists.

```
CATALL *,X,,,,,,,,X,,X,STOW,KEEP
```

Causes the facility to stow all NATURAL programs and maps and retain the report of the CATALL operation. The report is retrievable by re-entering CATALL. The report is stored in an object named 0CATALL. To re-start a CATALL operation, you must delete the existing 0CATALL. Do not presss any keys during an online CATALL while the report is on the screen. This causes an interruption and produces errors which are not real.

EXECUTE REPEAT

This execute option allows a program with multiple screens to output them without any intervening prompts. This can be used to give an interesting display with a "cartoon" effect.

```
 12:34:56              ***** NATURAL CATALL Command *****              97-04-01
 User APDDBS                 - Start of Cataloging -              Library APDDBS

 Catalog Program  *_______

         Mark to  X Recatalog only if module exists
              or  _ Catalog ALL Source-programs

 Mark to catalog  X Global Data Areas
                  X Parameter   Areas
                  X Local  Data Areas
                  X Copycode (save)
                  X Text     (save)
                  X External Subroutines
                  X Subprograms (CALLNAT)
                  X Helproutines
                  X Version 2 Maps
                  X Version 1 Maps
                  X Version 2 Programs
                  X Version 1 Programs
Command ===>
```

Figure 2.15. Menu Presented When GLOBALS Command Executed

SCAN

The SCAN command allows you to scan and to report or replace a character string in an object, a range of names by object type, or all objects in a library. The improved version of this command provides a menu of expanded options and a new help facility. The SCAN functions can be invoked both online and in batch. To do this, you can specify a direct command using keywords FUNC, DVAL, TYPE, SVAL, LIB, ABSOL, RVAL, and OBJ. For example, the command

```
SCAN FUNC=S,SVAL=UPDATE,LIB=COURSES,OBJ=REG*,TYPE=N
```

allows for all subprogram objects with a prefix of 'REG' in library named COURSES to be scanned for the value 'UPDATE'. Beware of this - I have been told SCAN does not require a user to be linked to a library!

SETUP/RETURN

These commands provide a mechanism to establish control points for applications so that one can transfer between applications in one NATURAL session. If you issue SETUPwith no application name, no LOGON will be generated upon RETURN. This allows for multiple return points within the same application. If SETUP * is issued, NATURAL uses the current value of *LIBRARY-ID to construct a LOGON command when RETURN is issued. If RETURN * is used, you can choose the application which you wish to return.

UPDATE {ON | OFF}

This command is not new. Its purpose is to allow or prevent data base updating. If it is set to OFF, not only is no UPDATE, STORE, or DELETE NATURAL statement executed, but hold logic is also ignored.

KEY

This command allows for the establishments or changing existing PA/PF key assignments. These are the keys which are sensitive at session level. They are independent of key settings via the SET KEY statement in a NATURAL program.

SYSPROF

This command invokes a pop-up window which displays the current system file definitions for the NATURAL session.

DUMP (On-line Dump Facility)

Occasionally, a NATURAL program might experience an event which causes a system abend, e.g., a S0C7. This severe error causes NATURAL to abend. This doesn't help to debug the program on-line, and if you had not saved the program or its changes, they are lost. You can prevent this from happening by either logging onto NATURAL using the Dynamic Parameter Facility or asking the Systems/DBA staff to set the NATURAL parameter DU=OFF.

After that, a NATURAL error 954 - 956 will be generated and the DUMP command can be issued. The information displayed is very important in helping Software AG to diagnose the problem, see Figure 2.16.

```
Code       ILC  0000 PSW 000000 00 00000000 Disp  00000000 Csect  ?
R0-7 00000015 00000000 00000000 00000000 00000000 00000000 00000000 00000000
R8-F 00000000 00000000 00000000 00000000 00000000 00000000 00000000 00000000

066A5BF0 47F0F020 D5C1E3E2 E3E4C240 F2F2F040 .00.NATSTUB 220   Slot Name
066A5C00 F9F160F0 F760F0F4 40F1F34B F5F94040 91-07-04 13.59     N1
066A5C10 066BC3A0 066B6BD0 066B66F0 066BEE64 .,C..,,..,.0.,..   N2
066A5C20 066C0D90 06718418 0671C7C8 066B4EF8 .%....d...GH.,+8   N3
066A5C30 066B4B62 00000000 066C7454 06766FE0 .,.......%....?.   Cur. Nucleus
066A5C40 58F0C2A0 58F0F0E4 58F0F000 12FF072F .0B..00U.00.....   NATURAL
066A5C50 06F007FE 58F0C2A0 58F0F0E4 58F0F074 .0...0B..00U.00.   Load Point
066A5C60 12FF072F 06F007FE 58F0C2A0 58F0F0E4 .....0...0B..00U   066A5BF0
066A5C70 58F0F004 07FF58F0 C2A058F0 F0E458F0 .00....0B..00U.0   Entry Point
066A5C80 F00807FF 58F0C2A0 58F0F0E4 58F0F00C 0....0B..00U.00.   066A5BF0
066A5C90 12FF072F 06F007FE 58F0C2A0 58F0F0E4 .....0...0B..00U   Length
066A5CA0 58F0F010 12FF072F 06F007FE 58F0C2A0 .00......0...0B.   000C1410
066A5CB0 58F0F0E4 58F0F014 07FF58F0 C2A058F0 .00U.00....0B..0   Relocation
066A5CC0 F0E458F0 F01807FF 58F0C2A0 58F0F0E4 0U.00....0B..00U   ABSOLUTE
066A5CD0 58F0F01C 07FF58F0 C2A058F0 F0E458F0 .00....0B..00U.0   Cur. Location
066A5CE0 F02007FF 58F0C2A0 58F0F0E4 58F0F024 0....0B..00U.00.   NATSTUB
```

Figure 2.16. DUMP Facility Output

There are several commands that are useful to point to different areas of the dump. Here is the help pop-up revealing those commands:

```
HELP
  Choose one of the following
  THREAD positioning commands.

  REMEMBER: Addresses being in
  a thread will be relocated
  after a thread change.

  _ AFB    System File Table
  _ BB     User area
  _ CST    Command Stack
  _ DDR    ENTIRE Buffer
  _ DIR    Directories
  _ FUL    Editor Work Area
  _ GAA    Graph. Array area
  _ GDA    Graphics Data Area
  _ GLC    Global Area (Com.)
  _ GLS    Global Area (Sys.)
  Command ===> ________________
```

Figure 2.17. Dump Commands Help

Reading a NATURAL dump

From my close friend and NATURAL expert, Darrell Davenport, comes this more detailed dump reading description:

> In the dump, you can follow R12 to the BB. At offset hex 14 is the line number. It is packed 3 bytes (the fourth byte is the level number). It follows the NATURAL zap level constant (which is what I scan for). In our case the zap level is 2207 which stands for NATURAL.2 SM 7. Also, you can follow R10 to the IO control block where at offset hex 290 is the real register save area. In this RSA you can usually find the non-NATURAL info if needed (in case it died of non-NATURAL causes). Also in the IOCB is other valued information. like the last key pressed, current map name, last issued error message, error counter, etc.
>
> However, I have found the option for program information and PF18 or zaps to be of most use. You can also look at the buffer pool entries and look at the BB area pointing from register 12.

FAMOUS BB layout:

```
***********************************************************************
*  C O N T E N T                                                      *
***********************************************************************
```

- 1. BB-PREFIX AREA
- 2. VERSION-CONTROL
- 3. USER DEFINED AREA
- 4. COMMUNICATION AREA
- 5. SECURITY PART
- 6. DATE & TIME AREA
- 7. RELOCATION AREA
- 8. ADABAS PARM AREA
- 9. MISC
- 10. ROUTINES ADDRESSES
- 11. FLAGS
- 12. INLINE - INSTRUCTIONS
- 13. V1.2/V2.0 OVERLAY AREA
- 14. CONTROL BLOCKS POINTERS
- 15. NTPRM / DYNAMIC PARAMETERS
- 16. DISPATCHER & WORK AREAS
- 17. TABLE MANAGEMENT
- 18. EQUATES
-

At hex offset 0010:

```
BBZAPLV  DS    CL4                        ZAP LEVEL
ANWNR    DS    PL3                        STATEMENT NUMBER
SRLEVEL  DS    XL1                        SUBROUTINE LEVEL
CMSV1314 DS    0D                         SYSNONYM FOR NEXT TWO ADDR
CMSAVE13 DS    A                          SAVE R13
CMSAVE14 DS    A                          SAVE R14
CMIOSAVE DS    9D                         GENERAL SAVE AREA
```

At hex offset 0080:

```
***************  U S E R   I D E N T I F I C A T I O N   A R E A ***
USID     DS    CL8                        USERID FROM LOGON
APLLID   EQU   USID                       APPLICATION ID
USIDAC   DS    CL8                        CURRENT USER ID IN SINGLE CMD
USRID    DS    CL8                        INDIVIDUAL USER ID
IAPLLID  DS    CL8                        INTERNAL APPLIC ID
GRPID    DS    CL8                        GROUP-ID IF LOGGED ON BY GROUP
CLUSTER  DS    CL8                        APPLICATION CLUSTER
USNAM    DS    CL32                       USER NAME
APNAM    DS    CL32                       APPLICATION NAME
*                                         USER-ID IF LINKED VIA USER
*                                         BLANK IF APPL.NOT DEFINED
***************  P R O G R A M    C O M M U N I C A T I O N      ***
PRONAM   DS    CL8                        PROGRAM NAME
PRONAMM  DS    CL8                        SAVED PROGRAM NAME
ERRTA    DS    CL8                        ERROR RECOVERY TRANSACTION
PRSTRT   DS    CL8                        STARTUP TRANSACTION IN APPL.
```

(continued on the next page)

```
***************   U S E R   I D E N T I F I C A T I O N   A R E A ***
USID      DS    CL8                        USERID FROM LOGON
APLLID    EQU   USID                       APPLICATION ID
USIDAC    DS    CL8                        CURRENT USER ID IN SINGLE CMD
USRID     DS    CL8                        INDIVIDUAL USER ID
IAPLLID   DS    CL8                        INTERNAL APPLIC ID
GRPID     DS    CL8                        GROUP-ID IF LOGGED ON BY GROUP
CLUSTER   DS    CL8                        APPLICATION CLUSTER
USNAM     DS    CL32                       USER NAME
APNAM     DS    CL32                       APPLICATION NAME
*                                          USER-ID IF LINKED VIA USER
*                                          BLANK IF APPL.NOT DEFINED
***************   P R O G R A M   C O M M U N I C A T I O N       ***
PRONAM    DS    CL8                        PROGRAM NAME
PRONAMM   DS    CL8                        SAVED PROGRAM NAME
ERRTA     DS    CL8                        ERROR RECOVERY TRANSACTION
PRSTRT    DS    CL8                        STARTUP TRANSACTION IN APPL.
PRREST    DS    CL8                        RESTART TRANSACTION IN APPL.

At hex offset  04DA:
DISPLERR  DS     Y                         LAST ISSUED ERROR MSG
```

TECH

The TECH command displays the following information about your current session.

```
NATURAL TECH Command
12:34:56                                        97-04-01

Steplib  DBID FNR          User ............... APDDBS
-------- ---- ----         Library ............ APDDBS
BUSYSTEM 240  147          Version / SM Level . 2.2 / 0008
SYSTEM   240  147          Startup Transaction. MAINMENU
                           NATURAL SECURITY ... Yes
                           Operating System ... MVS/ESA
                           TP Monitor ......... COMPLETE
                           Device Type ........ PC
                           Terminal ID ........ 3    140

                           Last Error Number .. 974
                           Last Error Line .... 961
                           Last Error Type .... System
                           Error Transaction ..
Last Command .. DUMP
```

Figure 2.18. TECH Command Display.

STRUCT

This command is very useful to debug where NATURAL does not 'see' the necessary END constructs for much of its structured mode syntax. In the following example, I have deliberately commented out an 'END-IF' so you can see the output. The highlighted lines show the order of observations to make to find the problem.

```
 12:34:56             *** NATURAL STRUCT COMMAND ***            97-04-01
 User APDDBS                    - Menu -              Library    APDDBS

                     Code  Function
                     ----  -----------------------------------------
                       G   Generate structured source into work area
                       D   Display structure of source
                       P   Print structure of source
                       W   Write structure of source into work area
                       ?   Help
                       .   Exit
                     ----  -----------------------------------------

  Code ................ D
  Source name ......... ________ If blank, current source
  Display compressed .. N (Y/N)
  Shift value ......... 2 (1 - 9)
  Align comments ...... Y (Y/N/L)
  Display messages .... Y (Y/N)
  Return to STRUCT .... N (Y/N)

Command ==>
```

Figure 2.19. STRUCT Command menu.

```
2150      IF READ-WORK-FILE THEN                                 *RI
2160         READ WORK FILE 2 SEARCH-OBJ-NAME                    *RIW
2170           PERFORM FIND-OPERATION                            *RIW
2180      *    WRITE '=' SEARCH-OBJ-NAME                         *RIW
2190        SYSTEM-FILE-CHOICE (*) := SAVE-SYSTEM-FILE-CHOICE(   *RIW
2200        DEVL ENV :- SAVE-DEVL-ENV                            *RIW
2210        TEST-ENV := SAVE-TEST-ENV                            *RIW
2220        PROD-ENV := SAVE-PROD-ENV                            *RIW
2230 2160    END-WORK                                            *RIW
2240 2150 ELSE                                                   *RI
2250         PERFORM FIND-OPERATION                              *RI
2260      * D-IF                                                 *RI
2270      *                                                      *RI
2280      IF NOT ONLINE THEN                                     *RII
2290         ESCAPE BOTTOM                                       *RII
2300 2280 END-IF                                                 *RII
2310      *                                                      *RI
2320      IF DO-ONE-TIME THEN                                    *RII
2330         ESCAPE BOTTOM                                       *RII
2340 2320 END-IF                                                 *RII
2350      *                                                      *RI
2360      SYSTEM-FILE-CHOICE (*) := SAVE-SYSTEM-FILE-CHOICE(*)   *RI
2370      DEVL-ENV := SAVE-DEVL-ENV                              *RI
2380      TEST-ENV := SAVE-TEST-ENV                              *RI
2390      PROD-ENV := SAVE-PROD-ENV                              *RI
2400 2240 END-REPEAT                                             >END-IF EXPECTE
2410      STOP                                                   *R
```

Figure 2.20. STRUCT Command Example Output for Option 'D'

The NATPAGE Facility

The NATPAGE Facility, available under both COM-PLETE and CICS, provides the capability to record and replay screens. This is accomplished with a series of terminal control commands - %E, %I, %O, %P, %S. An explanation of these terminal commands is found in Chapter 4.

The space allowed for this facility is controlled by the NATURAL parameter PD. This parameter specifies the number of screens which can be saved in a session. The oldest screen saved is overwritten if more are saved than is set with this parameter. The recorded pages are saved on the NATURAL system file.

List XREF

The 'LIST *' command returns a list of objects in a library in alphabetical order. I have often seen the question "Is it possible to get the same list in user-id order?" or " in alphabetical order for one user id?". The answer is yes, by using the LIST XREF facility. With versions of Predict 3.2.3 and under, L X results are restricted to the current library. It is an absolute must for any organization. The resources required must be seen as acceptable by every organization.

To answer the above question, here's a scenario. The library GALAXY has objects which have been stowed by multiple users, and I am interested in getting a list of objects created by user 'APPDPH' and I can restrict a date range. I execute LIST XREF (or L X) followed by the X option (Figures 2.21 and 2.22 below).

```
NAT2467  1 set(s) exist.
12:34:56           *****  P R E D I C T  3.2.2  *****                97-04-01
Library: GALAXY              -   Xref Menu   -          DBnr:   240 Fnr:   147

                         Code  Object
                         ----  ------------------------------
                           I   Invoked programs
                           D   Data areas and variables
                           V   Views and fields
                           C   Copycode
                           E   Error numbers
                           P   Printers
                           W   Work files
                           S   Retained sets
                           R   Processing rules
                           X   Report programs with xref data
                           A   Verify application
                           N   Create new sets via selection
                           O   Operate on sets
                          ---  ------------------------------
                    Code:      ( ? Help   . Terminate)
Command ===>

Enter-PF1---PF2---PF3---PF4---PF5---PF6---PF7---PF8---PF9---PF10--PF11--PF12-
      Invp  GDAV  Quit  Sets  Rule  Copy  Xref  View  OSet  SPfk  Main  Exit
```

Figure 2.21. XREF Main Menu

```
NAT2435 Set nr 1 has been saved.
12:34:56           *****  P R E D I C T  3.2.2  *****                 97-04-01
Library: GALAXY              -  Report Program -          DBnr:   240 Fnr:   147


         Program: ________ (?) Program type: _ (?)
         User Id: APPDPH__      Terminal Id: ________
       from date: 1997-01-01        to date: 1997-05-08
      short list: Y (Y,N)          save set: Y (Y,N)

      Report                        Nr     Report                        Nr
      ----------------------------- --     ----------------------------- --
      Statistical data               1     PREDICT description            2
      NATURAL program list           3     Using/Referenced programs      4
      Views, Da-views and fields     5     Da-areas and variables         6
      Workfiles, printers, errors... 7     format extended description    N

      expand copycodes / rules       N     suppress empty reports         Y
      ----------------------------- --     ----------------------------- --


Command ===>
Enter-PF1---PF2---PF3---PF4---PF5---PF6---PF7---PF8---PF9---PF10--PF11--PF12-
      Invp  GDAV  Quit  Sets  Rule  Copy  Xref  View  OSet  SPfk  Main  Exit
```

Figure II.22. XREF option 'X'.

This facility, introduced in NATURAL.2, is based on PREDICT"s active references. Various retrieval functions are available. The data is built primarily depending on the XREF NATURAL parameter value set to ON/FORCE. Also, there is a NATURAL Security flag, which when set to 'Y' writes XREF data for start-up / re-start / error modules. The option 'X' (and others) can only retrieve XREF data ***if it exists***.

NOTE: This setting has a negative impact for SYSMAIN, NATUNLD and INPL. Particularly, for INPL, you cannot load with replace an object which has no XREF data into a library if the XREF flag is set to 'Y'. And yet, if it is set to 'N' you will not create XREF data no matter what the Natparm value is since the NATURAL Security value overrides it.

There are a few points of which one should be aware when working with LIST XREF and creating sets. First, various options off the main menu have suboptions of 'U' and 'R' as you can see in Fig.2I.23 and Fig. 2.25. It is the 'R' suboption which creates sets, not 'U'. Secondly, there is a built in limit of the size of sets. They cannot have more than 800 objects when created within L X.

```
12:34:56          *****  P R E D I C T  3.2.2  *****                97-04-01
Library: APDDBS         - Data Area and Variables -    DBnr:   240 Fnr:   147


        Code Function                          Ss Pg PT DA DT DB St Va Us
        ---- -------------------------------- -------------------------
          C  Count references to variable         *  O  *  O     *  *  O
          U  Program using variables              *  O  *  O     *  *  O
          R  Variable referenced in programs   O  *  O  *  O     *  *  O
          D  Program using data areas          O  *  O  *  O  O
          E  Data area referenced in programs  O  *  O  *  O  O
        ---- -------------------------------- -------------------------
              Code: E           Save set: Y   (Y,N)
           Program:          Program type:    (?)
         Data area: CONSTANT Data area type:  (?)
        Data block:
         Structure:
          Variable:
             Usage:   (?)

Command ===>
Enter-PF1---PF2---PF3---PF4---PF5---PF6---PF7---PF8---PF9---PF10--PF11--PF12-
      Invp  GDAV  Quit  Sets  Rule  Copy  Xref  View  OSet  SPfk  Main  Exit
```

Figure II.23. L X (XREF) screen for option 'D'

```
MORE
12:34:56          *****  P R E D I C T  3.2.2  *****                97-04-01
Library: APDDBS         - Data Area and Variables -    DBnr:   240 Fnr:   147
Command: DA-AREA CONSTANT (*) REF PROG * (*) BLOCK *               Page:     1

      Data area    ref. in program       with block
--------------------------------------------------------------------------
    1 L:CONSTANT           1 P:CALENDAR
                           2 P:CAL2
                           3 P:COUNTLIB
                           4 P:DBNP0057
                           5 P:DBNP0068
                           6 P:FETCHLST
                           7 P:FIND
                           8 P:SCANFILE
                           9 P:SCANFUSR
                          10 P:SCANT
                          11 P:SCANTEXT
                          12 P:SCANT2
                          13 P:SCANT3
                          14 P:SCAN2
                          15 P:SCAN240
                          16 P:SCAN240D
                          17 P:SCAN240X
```

Figure 2.24. XREF output generated by option 'D', suboption 'E'

```
12:34:56             *****  P R E D I C T  3.2.2  *****               97-04-01
Library: SYSDIC                  - Copycode -           DBnr:    240 Fnr:    143



                Code Function                          Ss Pg PT Co Us
                ---- ------------------------------- --------------
                  C  Count references to copycode         *  O  *
                  U  Program using copycodes              *  O  *  O
                  R  Copycode referenced in programs   O  *  O  *  O
                ---- ------------------------------- --------------

                  Code: R              Save set: R (Y,N)
               Program:            Program type:   (?)
              Copycode: DANC0001          Usage:   (?)




Command ===>
Enter-PF1---PF2---PF3---PF4---PF5---PF6---PF7---PF8---PF9---PF10--PF11--PF12-
      Invp  GDAV  Quit  Sets  Rule  Copy  Xref  View  OSet  SPfk  Main  Exit
```

Figure 2.25. XREF option 'C'

```
MORE
12:34:56             *****  P R E D I C T  3.2.2  *****               97-04-01
Library: SYSDIC                  - Copycode -           DBnr:    240 Fnr:    143
Command: PROG * (*) USING COPYCODE DANC0001 USAGE *                 Page:      1

      Program                Copycode    Usage
 -----------------------------------------------
    1 P:BUMNTFI            1 DANC0001    Direct
    2 P:BUMNTFI2           1 DANC0001    Direct
    3 P:BUMNTFP            1 DANC0001    Direct
    4 P:BURETDA            1 DANC0001    Direct
    5 P:BURETFI            1 DANC0001    Direct
    6 P:DANP0060           1 DANC0001    Direct
    7 P:DBNPFLMG           1 DANC0001    Direct
    8 P:DBNPMNT1           1 DANC0001    Direct
    9 P:DBNPMNT2           1 DANC0001    Direct
   10 N:DBNSFLSC           1 DANC0001    Direct

***** END OF LIST *****
```

Figure 2.26. XREF output generated by option 'C', suboption 'U'.

```
MORE _
12:34:56             *****  P R E D I C T  3.2.2  *****               97-04-01
Library: SYSDIC                  - Copycode -           DBnr:    240 Fnr:    143
Command: COPYCODE DANC0001 REF PROG * (*) USAGE *                   Page:      1

      Copycode             Program        Usage
-----------------------------------------------
    1 DANC0001           1 P:BUMNTFI      Direct
                         2 P:BUMNTFI2
                         3 P:BUMNTFP
                         4 P:BURETDA
                         5 P:BURETFI
                         6 P:DANP0060
                         7 P:DBNPFLMG
                         8 P:DBNPMNT1
                         9 P:DBNPMNT2
                        10 N:DBNSFLSC

***** END OF LIST *****
```

Figure 2.27. XREF output generated by option 'C', suboption 'R'.

Here you can see the results of requesting to create a set of objects which reference copycode named 'DANC0001' (Figure 2.28). There are other ways in which sets can be created, for example via option 'N', as you can see in Figures 2.29. and 2.30. Because SYSDIC delivers the DDMs for the various records types stored I the dictionary, you can write programs to read/write set records. More about this is available in a paper titled "Impact Analysis based on XREF and Sets" at web site **www.wizinc.com**.

```
NAT2436 Operations terminated normally.
19:18:27          ***** P R E D I C T  3.2.2  *****             97-05-08
Library: SYSDIC             - Operate on sets -      DBnr:   240 Fnr:   143

O  Nr  created with function                               Date   C/W Count
-  --  --------------------------------------------------  -------- --- -----
_  1   DA-AREA DBNAU001 (*) REF PROG * (*) BLOCK *        96-10-23  W      1
_  2   COPYCODE DANC0001 REF PROG * (*) USAGE *           97-05-08        10











-  --  --------------------------------------------------  -------- --- -----
O :   (I,U,X,Y      Intersection,Union,Difference X minus Y)
      (C,W,E,L      Cat,Stow,Edit,List)
      (D,P,S,T,N,?  Display,Purge,Send,Sort by type,Sort by name,Help)
Enter-PF1---PF2---PF3---PF4---PF5---PF6---PF7---PF8---PF9---PF10--PF11--PF12---
```

Figure 2.28. XREF Menu Presented for option 'O'.

```
12:34:56          ***** P R E D I C T  3.2.2  *****             97-04-01
Library: GALAXY              - Create  Set -         DBnr:   240 Fnr:   147


      Object name: *_______
   Mark to select: _ all stowed objects
                   _ all objects not yet catalogued
                   _ all catalogued objects without source

   Mark to select: _ Global Data Areas
                   _ Local  Data Areas
                   _ Parameter Data Areas
                   _ Subroutines
                   _ Subprograms (CALLNAT)
                   _ Helproutines
                   _ Version 2 Maps
                   _ Version 1 Maps
                   _ Version 2 Programs
                   _ Version 1 Programs

Command ===>
Enter-PF1---PF2---PF3---PF4---PF5---PF6---PF7---PF8---PF9---PF10--PF11--PF12-
      Invp  GDAV  Quit  Sets  Rule  Copy  Xref  View  OSet  SPfk  Main  Exit
```

Figure 2.29. XREF option 'N' screen.

```
12:34:56           *****  P R E D I C T  3.2.2  *****               97-04-01
Library: APDDBS            - Verify References  -       DBnr:   240 Fnr:   147


           Code Function                                Ss DA DT PT Pg
           ---- ---------------------------------------- --------------
              D Data areas not referenced                O     O
              V Variables in data area not referenced       *  O
              C Copycodes not referenced
              N Error numbers not referenced
              P Programs not referenced                  O        O
              I Programs not impl./ref. starting from one            R
           ---- ---------------------------------------- --------------
                  Code: ?        Save set: N (Y,N)
             Data area:          Data area type :   (?)
          Program type:   (?)
               Program:


Command ==>
Enter-PF1---PF2---PF3---PF4---PF5---PF6---PF7---PF8---PF9---PF10--PF11--PF12-
      Invp  GDAV  Quit  Sets  Rule  Copy  Xref  View  OSet  SPfk  Main  Exit
```
Figure 2.30. XREF option 'A' screen.

Sometimes, trying to combine sets may return an error stating the sets are in different orders. This can especially happen if you are programmatically creating sets. However, this is easy to overcome; simply use the 'T' or 'N' options, then try again.

Addendum

Calling NATURAL Subprograms from a Non-NATURAL Environment

If you are going to be calling NATURAL repeatedly from any non-NATURAL language environment that supports a BAL call interface, NATURAL provides under CICS, Com-plete batch, et. al., an interface referred to as the 3GL callnat interface. This interface eliminates the overhead incurred by starting up NATURAL repeatedly each time you need to process a NATURAL subprogram. You initialize the NATURAL session once and allow it to serve as the controlling program. It in turn is linked with the 3rd GL interface nodule (or can dynamically load it) and calls the Cobol program of choice. Once the NATURAL environment is initialized, any further CALLs to execute a CALLNAT doesn't require the re-initialization of NATURAL.

There are some considerations to making this work.

- The 3rd GL CALLNAT interface module is accessible from the NATURAL nucleus.
- The 3rd GL program executes a CALL 'interface module' USING obj-name (8 byte field) with parameter list.
- The NATURAL nucleus issues CALL '3rd GL program'.
- The start of NATURAL should include a LOGON to library where subprograms for execution can be resolved.
- Construct the appropriate interface module
- You will need a sufficient DATSIZE for holding the parameter list, et. al.
- Under CICS you will need to define to proper PPT entries.

A Debugging Tip

A SAG-L user asked:

> "Does anyone know of a way to see what the "COMMAND / PROGRAM NAME..." was when a NAT0080 is encountered? I believe this error is occurring from a STACK TOP COMMAND 'xxx' where 'xxx' is garbage. But after scanning the library I can't find the offending code. If we could capture what the invalid command or program was it might be easier to find and fix. You don't get a program name or line number from NATURAL with this error since the attempt to execute is what causes the error. This is a CONSTRUCT application and I suspect the stack is being

manipulated by a CONSTRUCT subprogram. I looked at USR1037N in SYSEXT but could not get this to return ANYTHING after an abend. Anyone had success using that subprogram?"

Darrell Davenport, one of the best NATURAL technicians in the business, wrote this tip to help in the following debugging scenario:

The first thing is to trace your steps backward to the last known executing program. You can use the TEST facility to do this, or use the USR0600N subroutine (from library SYSEXT) to show the last 10 modules executed.

Then, in the last module, just before the exit point, code the following:

```
IF *DATA = -1           /* If a command is on the top of the stack    */
  SET CONTROL '.S'      /* prepare for special-stack control mode     */
  INPUT #alpha-field    /* look (into the stack), but do not touch it */
  PRINT 'The top of the stack contains:' #alpha-field /* show it */
END-IF
```

Figure 2.25. Debugging tip using stack controls.

List Facility Tips.

Use the LIST Command LAYOUT option to list a MAP as the USER sees it.

```
LIST mapname FORMATTED
```

Yes it would... but have you tried the list expanded feature? It will list the object with all (or some, depending on what you specify) of the ldas, pdas, copycode, subprograms, and subroutines included. For example, L MYPGM EXPAND would expand everything within MYPGM. LIST MYPGM EXPAND PDA01 would only expand pda PDA01.The following is the syntax for the list command:

```
   General syntax for the LIST command:
   LIST ( ( SEQ ) (object-type) object-name ( ( WITH ) DIR )(expand option)
          / COUNT (object-name( < / > ))
          / DIR   (object-name)
          / XREF                            )

   Expand-option:
   EXPAND ( FORMATTED ) ( COMMENTS / n ) (expand-type...10) object-name

   SEQ   List the specified sources sequentially without selection
menu.
   COUNT List number and size of objects and sources sorted by object types.
   DIR   List directory of source in source area or specified sources.
   XREF  List cross references.
   EXPAND Expand specified sources. Optionally expand comments or nlines.
   FORMATTED Only relevant for maps. They are expanded formatted, if
required.
```

There is a program called CSUINCL that I believe comes with CONSTRUCT that you can use to expand copycode within your program. Just type in CSUINCL on the command line while editing your program and it will expand the copycode, commenting out the "include" statements.

Chapter 3

NATURAL Data Structures

I. Data Types

NATURAL supports a variety of data types. They are:

Type	Format
Alphanumeric	A
Binary	B
Numeric, unpacked	N
Numeric, packed	P
Date	D
Time	T
Attribute Control	C
Integer	I
Floating point	F
Logical	L

A data type can be assigned certain constants. Alphanumeric, binary data types treat data in its internal hexadecimal format. The numeric data types: packed, unpacked, integer, and floating take numeric constants, either signed or unsigned. A sample NATURAL local data definition for the various types is illustrated on the next page.

Data types are assigned to user-defined variables or database fields. I have a simple philosophy for naming variables. First, do not use the '&' character in any label; I have already seen trouble when they appear in the names of NATURAL objects created below 2.2. Second, ALWAYS qualify names when they appear in data structures.

```
0010 DEFINE DATA LOCAL
0020 1 #ALPHA       (A10)
0030 1 #PACKED1     (P5)
0040 1 #PACKED2     (P5.2)
0050 1 #UNPACKED1   (N5)
0060 1 #UNPACKED2   (N5.2)
0070 1 #BINARY      (B4)
0080 1 #ATTRIBUTE   (C)
0090 1 #LOGIC-VAR   (L)
0100 1 #DATE-VAR    (D)
0110 1 #TIME-VAR    (T)
0120 1 #INTEGER1    (I1)
0130 1 #INTEGER2    (I2)
0140 1 #INTEGER3    (I4)
0150 1 #SHORT-REAL  (F4)
0160 1 #LONG-REAL   (F8)
END-DEFINE
```

Alphanumeric

The alphanumeric data type supports alphanumeric constants of length 1 to 253 characters. Alphanumeric constants are enclosed in single quotes. Two single quotes allow for a quote to be included in the string. String constants can be concatenated with a hyphen. For example, 'ABC' - 'DEF' is a 6 byte string of 'ABCDEF'. Alphanumeric constants are left justified in the receiving field, but also are subject to MOVE RIGHT JUSIFIED.

Integer

The integer data type can be expressed in three lengths: one byte, half word, or full word. The type is defined as follows:

```
0010 1   #I        (I1)
0020 1   #J        (I2)
0030 1   #K        (I4)
```

The range of values for the I1 definition is -128 to 127, for the I2 definition the range is -32768 to 32767 and for the I4 definition it is - (2 to power 64) to + ((2 to power 63) - 1).

Numeric

The unpacked numeric data type is defined with the format N. Unpacked decimal can contain up to 29 digits and can be defined with a decimal position. Each numeric digit requires one byte of storage. A sign and/or decimal point can be included with the full 29 digit capacity. The decimal point is implied and the sign is maintained in the last (rightmost) byte. The fractional length must not be greater than 7; otherwise, a NAT0047 will occur at compile time. Computational errors of truncation are only recognized at run time. Figure 3.1 shows the results of certain assignments.

Variable	Value	INTERNAL VALUE	Reason For Error
#X (N8)	12345	F0F0F0F1F2F3F4F5	
#X (N5.2)	123.45	F0F0F1F2F3F4F5	
#X (N5.2)	123.456	F0F0F1F2F3F4F5	
#X (N6)	1234567	ERROR NAT1301	Value cuases truncation
#X (N5.0)	1234	F0F1F2F3F4	
#X (N6.8)	12345.123	ERROR NAT0047	Frac. Definition
#X (N21.7)	12345.678	ERROR NAT0047	Total bytes > 27
#X (N8)	-12345678	F1F2F3F4F5F6F7D8	
#X (N6.1)	1234567.1	ERROR NAT1305	Integer portion > definition
#X (N6.1)	-123456.7	F1F2F3F4F5F6D7	

Figure 3.1. Samples of Numeric Assignment Values

IBM System/360 and its children handle all six possible values of the sign nibble 'B' and 'D' are considered negative, the other four are considered positive. So NATURAL should accept as packed any byte string in which all but the last nibble are 0 - 9 and in which the last nibble is A to F.

Packed

The packed numeric data type is defined with the format P. Packed decimal is similar to unpacked, except for the number of bytes required to store which is determined by the formula, INT (number of digits / 2) + 1, where INT (integer) means truncate decimal.

For #X (P7.2), the length is 5 bytes. For #X (P10), the length is 6 bytes. The table in Figure 3.2 shows the results of certain assignments.

Variable	Value	INTERNAL VALUE	Reason For Error
#X (P8)	12345	000012345C	
#X (P5.2)	123.45	0012345C	
#X (P5.2)	123.456	0012345F	
#X (P6)	1234567	ERROR NAT1301	Value cuases truncation
#X (P5.0)	1234	01234C	
#X (P6.8)	12345.123	ERROR NAT0047	Frac definition >> definition
#X (P21.7)	12345.678	ERROR NAT0047	Total bytes > 27
#X (P7)	-12345678	1234567D	
#X (P6.1)	1234567.1	ERROR NAT1305	Integer portion > definition
#X (P6.1)	-123456.7	1234567D	
#X (P6.1)	12345.6	0123456C	

Figure 3.2. Samples of Packed Assignment Values

Testing Packed Field Data

A packed field may be tested for the sign bit, that is, a "C" or "F" which both mean a positive value, or 'D' for a negative. For a 3 byte (character field):

```
1 #PACKED-AS-ALPHA (A3)    /* Which is a 5 digit packed field
1 REDEFINE #PACKED-AS-ALPHA
   2 #PACKED         (P5)
```

A test of:

```
IF #PACKED-AS-ALPHA = MASK (NNNNZ) /* or #PACKED = MASK (NNNNZ)
     WRITE 'Field is packed' #PACKED
ELSE
     WRITE 'Not a valid packed field' #PACKED-AS-ALPHA (EM=HHHHHH)
END-IF
```

The 'Z' mask definition character is a check for valid signed digit, which, if any sign appears, is always in the rightmost character. The 'Z' (Zoned) mask definition can also be used to validate non-packed signed numeric fields. For instance:

```
1  #A (A3) INIT <'12J'>
1  REDEFINE #A
    2  #A-N (N3)
       :
WRITE #A-N          /* Shows -121
       :
IF #A = MASK (NNZ) /* True, is a valid signed number

IF #A = MASK (NNN) /* False, does not contain 3 numeric digits,
                   /* even though the field is a perfectly valid
                   /* numeric field.
```

To pack a number: Number is (as we would see it on a NATURAL display) -121. In a 3 character zoned decimal format: 12J. As a 3 character hex format: F1 F2 D1.

Number is (as we would see it on a NATURAL display) -121. As a 2 byte Packed Number - hex format:12 1D. Try this program:

```
DEFINE DATA LOCAL
1 #P   (P3)   INIT <-121>
END-DEFINE
WRITE #P #P (EM=HHHH)
END
then test #P = MASK (NNN), and MASK (NNZ)
```

- Always use the Packed field name in the MASK statement.
- Do NOT use the name of an Alpha field which is either a redefinition of the packed field, or the field the Packed field redefines. (NATURAL realizes this is a Packed field by it's definition, and DOES check things differently)

- In the MASK, specify the total number of digits the Packed field can contain, making the last one a "Z" in the MASK.

A violation of the above three points will result in an incorrect result.

Someone on SAG-L once asked the question of testing for potential overflow and an interesting response involving a NATURAL solution was presented:

```
The call program checks the length of the first parameter,
and returns in the second parameter the lowest value that
will give overflow. So if the first parameter is (N3), the
second one will contain 1000 after the call. Overflow can
now be checked against the returned value.

DEFINE DATA LOCAL
*
1 #N1                 (N2) INIT <99>
1 #N2                 (N2) INIT <2>
1 #N3                 (N2)
*
* Fields below should go in some common LDA *
1 #LARGE              (P22.7)
1 #OVERFLOW-VALUE     (P29)
1 #OVERFLOW           (L)
*
END-DEFINE
*
COMPUTE #LARGE = #N1 + #N2
INCLUDE OVERFLOW '#N3'
*
WRITE #N3 #OVERFLOW (EM='No overflow'/'Overflow') *
END
```

```
The copycode looks like this:

CALL 'GN001' &1& #OVERFLOW-VALUE
*
RESET #OVERFLOW
*
IF #LARGE GE #OVERFLOW-VALUE
  MOVE TRUE TO #OVERFLOW
  SUBTRACT #OVERFLOW-VALUE FROM #LARGE
  ELSE
    IF #LARGE LE #OVERFLOW-VALUE * -1
      MOVE TRUE TO #OVERFLOW
      ADD #OVERFLOW-VALUE TO #LARGE
    END-IF
END-IF
*
MOVE #LARGE TO #N3
```

It would be intriguing to write a general routine that isn't fixed to a specific length. Something like: `IF (#n1 + #n2)  > 10 ** Length(#N1).`

Binary

The binary data type is defined with the format B. This data type can contain from 1 to 126 bytes of data. Binary types can be assigned hexadecimal constants. Hexadecimal constants are preceded with 'H'. One way of getting characters which can not be entered from the keyboard is to use hex notation. For example, the tilde character is H'A1'. Hexadecimal constants can be assigned to both alphanumeric and binary data types.

Binary data types are useful in many ways. For one, they can used to store unsigned numbers in situations for which storage is an issue. The value 2,147,483,647 can be stored in a B4 formatted field. Any larger value causes a NAT1301 error. This is expected since the before mentioned value is equivalent to X'7FFFFFFF'. The next simple program shows the conversion between decimal, packed, and binary.

```
DEFINE DATA LOCAL
1 #N (N12)
1 #P (P12)
1 #B (B4)
END-DEFINE
*
#N := 344212256
MOVE #N TO #B
WRITE  NOTITLE '=' #N '=' #B
WRITE 'ASSUME VALUE:' #B (EM=H(8)) 'IS THE START VALUE.'
MOVE #B TO #P
ADD 10 TO #P
MOVE #P TO #B
WRITE 'NEW VALUES:' '=' #P '=' #B (EM=H(8))
END
```

```
MORE _
Page      1                                            97-04-01  12:34:56

#N:      344212256 #B: 14844320
       ASSUME VALUE: 14844320 IS THE START VALUE.
NEW VALUES: #P:      344212266 #B: 1484432A
```

What the above program shows is the savings on storage. The field #N requires 12 bytes, #P requires 7 bytes and #B requires only 4 bytes. I have seen a Production definition of an ADABAS file in which storage savings were required because records had reached the maximum compressed record length (the infanous response code NAT3049). Binary arrays will take much less storage that either packed or decimal arrays, but packed data is preferable for high performance arithmetic.

Logical

The logical data type is defined with the format L. The logical data type is one byte in length. Thus far, one can not assign a relational expression to a logical variable- only a logical constant or another logical variable. Logical constants are TRUE and FALSE. The constant TRUE has a binary value of 1; FALSE has binary value of 0.

You can write a logical variable value with an edit mask, as:

```
WRITE #LOGICAL (EM=FALSE/TRUE)
WRITE #LOGICAL (EM=NO/YES)
WRITE #LOGICAL (EM=NOWAY/OK)
```

These statements display ‘FALSE’, 'NO' and 'NOWAY', respectively, if #LOGICAL has a value of false; otherwise the string ‘TRUE’, 'YES' and 'OK' display. Without a specific EM, a FALSE value displays as blank and TRUE value displays as 'X'.

Floating Point

Real variables are equivalent to the floating point data type and are defined with the format F4 or F8. Exponential notation can be used with floating point constants. This notation is useful for dealing with very large or very small quantities, where magnitude is more important than precision. The value 3.5 E+9 is another way of signifying 3.5 * 10 to the power of 9, or 3,500,000,000. The value 8.992 E-8 is another way of specifying 8.992 * 10 to the power -8, or 0.00000008992. Fixed point constants cannot be stored with greater than 7 decimal place accuracy. However, one can attain extended precision with an F8 definition. Thus, one can 'see' an extended value (beyond the maximum 7 place decimal numbers of N or P formats) by using the FL option for output statements for any length of the mantissa, ranging from 1 to 16. The default seems to be 16. NATURAL only truncates beyond what is requested. Try the following example:

```
DEFINE DATA LOCAL
1 NUM1 (F4)
1 NUM2 (F8)
END-DEFINE
*
NUM1 := 1.1234567E-2
NUM2 := 1.123456789123E-4
WRITE '=' NUM1 / NUM2 (FL=9)
END
```

The floating point data type can be assigned either fixed point constants or floating point numbers using exponential notation. A fixed point constant implies sign, digits and decimal point. A fixed point constant cannot have more than 27 digits total and prints a result of (sign)n.nnnnnnE(sign)mm, where the range of magnitude is between 5.397605 E-79 and 1.684996 E+65 regardless of the format (F4 or F8). Any positive number smaller than 5.397605E-79 becomes zero. If one desires, one can see up to 16 characters in the mantissa of the floating point value by using FL= 16 with the output statement.

Date/Time

The date and time data types are defined with the format of D and T respectively. Format D contains date information only; format T contains both date and time information. Time is denoted in tenths of a second. The internal formats for date and time data types are P6 and P12 respectively. Date and time constants can be assigned to variables of type date and time respectively or by using the *DATX and *TIMX systems variables. Date constants are quoted strings preceded with the character 'D'. Similarly, time constants are quoted strings preceded with the character 'T'.

Internally, time variables contain both date and time information. When dealing with time constants however, eg. T'12:00:01', only the time component is allowed. If you want to provide a constant with both date and time components you must use the extended time notation:

```
IF *TIMX > E'03/01/1997 07:00:00'
  WRITE 'It is morning of March 1'
END-IF
```

With the use of edit masks used with formats D and T, you can review date / time formatted variables in different ways. The output of *DATX depends on the NATURAL parameter DTFORM (which if not set defaults to YY/MM/DD). With the MOVE EDITED statement, you can also manipulate date/time variables in some clever ways. Try these example programs:

```
0010 DEFINE DATA LOCAL
0020 1 #DAY   (A10)
0030 1 #MONTH (A10)
0040 1 #D     (D)
0050 END-DEFINE
0060 #D := *DATX
0070 MOVE EDITED #D (EM=N(9)) TO #DAY /* STRIP OFF DAY OF THE WEEK
0080 *
0090 IF #DAY = 'Wednesday' THEN
0100    WRITE 'THIS IS WEDNESDAY'
0110 ELSE
0120    WRITE 'THIS IS NOT WEDNESDAY'
0130 WRITE '=' #DAY
0140 *
0150 MOVE EDITED #D (EM=L(10)) TO #MONTH  /* STRIP OFF THE MONTH
0160 *
0170 IF #MONTH = 'May' THEN
0180    WRITE 'THIS IS THE MONTH OF MAY'
0190 ELSE
0200    WRITE 'THIS IS NOT THE MONTH OF MAY'
0210 WRITE '=' #MONTH
0220 END
```

```
THIS IS WEDNESDAY
#DAY: Wednesday
THIS IS THE MONTH OF MAY
#MONTH: May
```

In the event you do not want to worry about upper case versus lower case, you could append the following code to the above routine to remove this dependency.

```
EXAMINE #DAY TRANSLATE INTO UPPER CASE  or
EXAMINE SUBSTRING (#DAY,2,10) TRANSLATE INTO UPPER CASE
```

```
0010 **
0020 ** PROGRAM. DTFORM4
0030 ** AUTHOR.  JIM WISDOM
0040 **
0050 ** PROGRAM TO DISPLAY NOHDR THE DATE/TIME FORMATS WITH
0060 ** NEW NATURAL DEFINITIONS
0070 **
0080 ASSIGN
0100   #TIME                (A22) = '*TIME          (A10) = '
0110   #DATX                (A22) = '*DATX          (D)    = '
0120 **
0130   DISPLAY NOHDR #DATX      *DATX
0140   DISPLAY NOHDR #DATX      *DATX (EM=MM/DD/YY)
0150   DISPLAY NOHDR #DATX      *DATX (EM=ZM/ZD/YY)
0160   DISPLAY NOHDR #DATX      *DATX (EM=MM/DD/YYYY)
0170   DISPLAY NOHDR #DATX      *DATX (EM=N(9)^L(10),^DD,^YYYY)
0180   DISPLAY NOHDR #DATX      *DATX (EM=N(9)^L(10),^DD,^R)
0190   DISPLAY NOHDR #TIME      *TIMX (EM=HH:II:SS^AP)
0200 END
```

```
*DATX          (D)   =  89-05-30
*DATX          (D)   =  05/30/89
*DATX          (D)   =   5/30/89
*DATX          (D)   =  05/30/1989
*DATX          (D)   =  Tuesday May, 30, 1989
*DATX          (D)   =  Tuesday May, 30, MCMLXXXIX
*TIME          (A10) =  11:19:52 AM
```

```
0010 DEFINE DATA LOCAL
0020 1 #G-DATE  (A8)
0030 1 #J-DATE  (A5)
0040 1 #DATE    (D)
0050 END-DEFINE
0060 ASSIGN #G-DATE = '03/01/88'
0070 MOVE EDITED #G-DATE
0080   TO #DATE (EM=MM'/'DD'/'YY)   /* STRIP GREGORIAN FORMAT
0090 WRITE 'ORIGINAL GREGORIAN DATE:' #G-DATE
0100   /    'YYYYJJJ FORMAT:' #DATE (EM=YYYYJJJ)
0110   /    ' JULIAN FORMAT:' #DATE (EM=YYJJJ)
```

```
0120 MOVE EDITED #DATE (EM=YY'-'MM'-'DD)
0130   TO #G-DATE
0140 WRITE 'REFORMATTED GREGORIAN DATE:' #G-DATE
0150 **
0160 MOVE '89150' TO #J-DATE
0170 MOVE EDITED #J-DATE TO #DATE (EM=YYJJJ)
0180 WRITE 'ORIGINAL JULIAN DATE:' #J-DATE
0190 / 'REFORMATTED JULIAN DATE:' #DATE (EM=MM'-'DD'-'YY)
0200 / 'REFORMATTED JULIAN DATE:' #DATE (EM=MM'-'DD'-'YYYY)
0210 END
```

```
ORIGINAL GREGORIAN DATE: 03/01/88
YYYYJJJ FORMAT: 1988061
 JULIAN FORMAT: 88061
REFORMATTED GREGORIAN DATE: 88-03-01
ORIGINAL JULIAN DATE: 89150
REFORMATTED JULIAN DATE: 05-30-89
REFORMATTED JULIAN DATE: 05-30-1989
```

Here are some tips with in regards to working with date / time values:

Comparing Time Variables and Date Constants

```
IF #TIME = D'1995-05-20' THRU D'1995-05-21'
   WRITE 'YES'
ELSE
   WRITE 'NO'
END-IF
```

First, I have always believed NATURAL's handling of 'time' edit masks should be similar to a 'date' edit mask in relation to empty field values.

```
 WRITE #D(EM=MM/DD/YY)                  displays blanks

 WRITE #T(EM=MM/DD/YY^HH:II:SS.T)       displays 01/02/00 00:00:00.0,
```

If you need to do this here is a tip: **Printing blanks for null time**

```
IF #TIME = 0
   MOVE (AD=N) TO #TIME-CV
ELSE
   MOVE (AD=D) TO #TIME-CV
END-IF
DISPLAY #TIME (CV=#TIME-CV EM=...)
```

Date Checking

Several customers utilize code similar to the following and have requested a written explanation of the difference between the valid dates for MASK statements versus MOVE EDITED statements.

```
RESET field1(A8) field2(D)
INPUT field1
IF field1 = MASK(YYYY)
   THEN MOVE EDITED field1 TO field2 (EM=YYYY)
ELSE
   DISPLAY 'error in date field'
```

As documented in the NATURAL Reference Manual, the MASK (YYYY) checks for a valid date between 0000 and 2699. The MOVE EDITED statement, however, checks for a valid Gregorian date. Keep in mind the Gregorian year begins in 1582; therefore, the date in the MOVE EDITED statement must begin with the year 1582 or greater. If the date is not 1582 or greater, you will receive a NAT1143 error.

In order to avoid receiving the NAT1143 error, there are three possible coding methods:

1. Write a check in the program (or processing rule) for YEAR >=1582 and YEAR <=2699.
2. Add an ON ERROR routine which captures the NAT1143 error and issues back to the end-user a 'friendly' message stating the year (on the MOVE EDITED statement) must begin with the year 1582 or greater.
3. Place text on the INPUT statement stating that the year to be input must begin with the year 1582 or greater.

A lot of discussion ensued on SAG-L about whether to define date fields as N8 or A8. There was mixed reaction, but the conversation can be easily summarized as follows.

Reasons for defining as N8	Reasons for defining as A8	Reasons for D/T formats
Any format of YYYYMMDD is already numeric	Part of descriptor keys on the data base and did not like packed components	Take advantage of accurate date calculations in NATURAL programs
If stored as alpha, need to verify data integrity	DATX fields unsatisfactory as modifiable fields on maps	
No hassle with the possibility of entering non-numeric data on maps	No additional move required if you want to use MOVE EDITED for date calculations	
If you want dates descending keys, you must use date complements		

My own personal choice is A8. This serves the MOVED EDITED nicely, easily manageable as components to super/sub descriptor keys, and is better positioned in a client-server environment. As much as I like D/T formats, they are particular to NATURAL and should be used only for NATURAL's use. Therefore, I would definitely shy from defining such fields (or any encoded field) into a database.

At the 8th NATURAL Conference, I challenged my colleagues to write a brief subroutine that could check for the validity of a Julian date. The MASK operation had no legitimate check, and there is certainly no IF #DATE IS (JULIAN). Several responses appeared in public forums and I present my personal favorites as to the "best" postings.

Solution #1:

```
DEFINE DATA
 LOCAL
  1 #JULIAN(N7) INIT <1996366>
  1 REDEFINE #JULIAN
    2 #JULIAN-YEAR(N4)
    2 #JULIAN-DAYS(N3)
  1 VALID-JULIAN(L)
  1 #LEAPDATE(N8) INIT <00000229>
  1 REDEFINE #LEAPDATE
    2 #LEAP-YEAR(N4)
END-DEFINE
*
PERFORM IS-JULIAN-VALID
*
IF VALID-JULIAN
    WRITE 'VALID JULIAN DATE!'
  ELSE
    WRITE 'SORRY, TRY AGAIN...'
  END-IF
*
```

```
DEFINE SUBROUTINE IS-JULIAN-VALID
ASSIGN VALID-JULIAN = FALSE
IF #JULIAN EQ MASK (YYYY001:365)
    ASSIGN VALID-JULIAN = TRUE
    ESCAPE ROUTINE
  END-IF
IF #JULIAN-DAYS = 366
    ASSIGN #LEAP-YEAR = #JULIAN-YEAR
    IF #LEAPDATE EQ MASK(YYYYMMDD)
        ASSIGN VALID-JULIAN = TRUE
        ESCAPE ROUTINE
      END-IF
  END-IF
END-SUBROUTINE
END
```

Solution #2:

```
0010 DEFINE DATA PARAMETER
0020 1 #JULIAN-INPUT(N7)          /* Pass your suspect value to this
0030 1 REDEFINE #JULIAN-INPUT     /* subprogram from anywhere.
0040   2 #JULIAN-ALPHA(A7)
0050 1 #RESP-CODE(I1)
0060 LOCAL
0070 1 #JULIAN-CHECK(D)
0080 END-DEFINE
0090 IF NOT ( #JULIAN-INPUT = MASK (YYYY001-366) )
0100   #RESP-CODE := 1
0110   ESCAPE ROUTINE
0120 END-IF
0130 MOVE EDITED #JULIAN-ALPHA TO #JULIAN-CHECK(EM=YYYYJJJ)
0140 ON ERROR
0150   IF *ERROR-NR GT 0
0160     #RESP-CODE := 2
0170     ESCAPE ROUTINE
0180   END-IF
0190 END-ERROR
0200 END
```

Solution #3:

```
DEFINE DATA LOCAL
1 #IN(A7)
1 #FEB29(A8) INIT<'YYYY0229'>
1 #WRONG-DATE(L)
END-DEFINE
INPUT #IN
IF #IN NE MASK(YYYY001-366)
  MOVE TRUE TO #WRONG-DATE
ELSE
  IF #IN = MASK(....'366')
    MOVE SUBSTR(#IN,1,4) TO SUBSTR(#FEB29,1,4)
    IF #FEB29 NE MASK(YYYYMMDD)
      MOVE TRUE TO #WRONG-DATE
    END-IF
  END-IF
END-IF
WRITE 'Date is' WRONG-DATE(EM='correct'/'incorrect')
```

Note that if #IN is to be MOVE EDITED into a D field, you should make sure that a valid range of years is checked for in the first MASK check.

Another interesting question was asked on SAG-L: Does anyone know how to convert NATURAL's internal date format to a displayable date (number)? Here was a posted solution:

```
0010 * RGZESP driver/mainline
0020 * Convert P7 (D-format) field to A10 (yyyy/mm/dd) w/o edit mask
0030 *
0040 DEFINE DATA LOCAL
0050 1 #D(D)         INIT<*DATX>
0060                 1 REDEFINE #D
0070   2 #P(P7)
0080 1 #A(A10)
0090 END-DEFINE
0100 *
0110 SET KEY PF3
0120 *
0130 REPEAT
0140   CALLNAT 'RGZESP2' #P #A              /* Conversion
0150   INPUT(AD=O IP=F)
0160              'RGZESP'
0170     ///      'Packed:'   #P(AD=MDL'_') /* Field to be converted
0180     40T 'Internal:' #D(EM=YYYY/MM/DD) /* NATURAL's verification
0190     // 40T 'Computed:' #A              /* Virtual field
0200    ////      'Press' 'PF3'(YEI) 'to Stop'
0210 *
0220   IF  *PF-KEY = 'PF3'
0230     THEN
0240       STOP
0250   END-IF
0260   IF  #P < 577813
0270    OR #P > 986518
0280     THEN
0290       REINPUT 'Valid range is 577813 (1582/01/01) '
0300               - 'thru 986518 (2700/12/31)'
0310   END-IF
0320 END-REPEAT
0330 END
```

```
0010 * RGZESP2 conversion subprogram
0020 * Convert P7 (D-format) field to A10 (yyyy/mm/dd) w/o edit mask
0030 *
0040 DEFINE DATA PARAMETER
0050 1 #P(P7)
0060 1 #A(A10)      1 REDEFINE #A
0070   2 #YY(N4)
0080   2 FILLER 1X
0090   2 #MM(N2)
0100   2 FILLER 1X
0110   2 #DD(N2)
0120 LOCAL
0130 1 #DAYS(N7)
0140 1 #LEAP-YEAR(L)
0150 1 #CNT(N3/12)                       /* Days until end of month
0160      INIT<, 31, 59, 90, 120, 151, 181, 212, 243, 273, 304, 334>
0170 END-DEFINE
```

```
0180 *
0190 #A    := '1582/01/01'              /* First 'valid' date
0200 #DAYS := #P - 577812               /* Since 1582/01/01
0210 *
0220 REPEAT WHILE #DAYS > 365           /* Determine the year
0230   PERFORM LEAP
0240   IF  #LEAP-YEAR
0250     THEN
0260       SUBTRACT 366 FROM #DAYS
0270     ELSE
0280       SUBTRACT 365 FROM #DAYS
0290   END-IF
0300   ADD 1 TO #YY
0310 END-REPEAT
0320 *
0330 PERFORM LEAP                       /* Days until end of February
0340 IF  #LEAP-YEAR
0350   THEN
0360     #CNT(3) := 60
0370   ELSE
0380     #CNT(3) := 59
0390 END-IF
0400 *
0410 FOR #MM = 12 1 -1                  /* Determine month & day
0420   IF  #DAYS > #CNT(#MM)
0430     THEN
0440       #DD := #DAYS - #CNT(#MM)
0450       ESCAPE BOTTOM
0460   END-IF
0470 END-FOR
0480 *
0490 *
0500 DEFINE SUBROUTINE LEAP
0510 DECIDE FOR FIRST CONDITION
0520   WHEN #YY / 400 * 400 = #YY       /* Leap year
0530        #LEAP-YEAR := TRUE
0540   WHEN #YY / 100 * 100 = #YY       /* Century is not a leap year
0550        #LEAP-YEAR := FALSE
0560   WHEN #YY / 4 * 4 = #YY           /* Leap year
0570        #LEAP-YEAR := TRUE
0580   WHEN NONE
0590        #LEAP-YEAR := FALSE
0600 END-DECIDE
0610 END-SUBROUTINE
0620 END
```

Attribute Control

The attribute control data type is defined with the format C and uses two bytes of storage; however, in defining this type, length is not specified. Attribute and color definitions can be assigned to control variables. Color definitions obviously only make sense where hardware supports color display. Attribute constants are enclosed in parentheses.

Rule: You can only specify one from each group.

Attribute Definitions

Attribute	Explanation	Statements where used C	D	F	I	P	R	W
	Group 1							
B	Displayed blinking	o	o	o	o	o	o	o
C	Displayed italic	o	o	o	o	o	o	o
D**	Displayed normal intensity	o	o	o	o	o	o	o
I	Display intensified	o	o	o	o	o	o	o
N	Display no-echo	o	o	o	o	o	o	o
U	Display underlined	o	o	o	o	o	o	o
V	Display reverse video	o	o	o	o	o	o	o
Y	Display controlled dynamically via format C variable		o		o			o
	Group 2							
L	Display left justified	o	o	o	o	o	o	o
	Default for alphanumeric fields							
R	Display right justified	o	o	o	o	o	o	o
	Default for numeric fields							
Z	Display numeric right justified zero filled	o	o	o	o	o	o	o
	Group 3							
A	Input field only (def.)					o		
M	Display value and accepts input	o			o		o	
O	Display value only, non-modifiable	o			o		o	
P	Display protected used with control parameters , DY parameter				o		o	
	Group 4							
E	Mandatory non-pull input before processing				o			
F**	Mandatory, with nulls acceptable				o			
	Group 5							
G	Input number of characters same as field length				o			
H**	Input value may have fewer characters than field length				o			
	Group 6							
T	Translate value to upper case	o	o	o	o	o	o	o
W**	Lower case values acceptable	o	o	o	o	o	o	o
	Group 7							
'c'	Fill character for attributes A or M				o			

Figure 3.3. Attribute Control Definitions

C=CALLNAT,D=DISPLAY,F=FORMAT,I=INPUT,P=PRINT,R=REINPUT,W=WRITE

Color Definitions

Attribute	Explanation	C	D	F	I	P	R	W
		Statements where used						
	Color Explanation							
BL	Blue		o	o	o	o	o	o
GR	Green		o	o	o	o	o	o
NE	Neutral (white)		o	o	o	o	o	o
PI	Pink		o	o	o	o	o	o
RE	Red		o	o	o	o	o	o
TU	Turquoise		o	o	o	o	o	o
YE	Yellow		o	o	o	o	o	o

Figure 3.4. Color Attribute Definitions
C=CALLNAT,D=DISPLAY,F=FORMAT,I=INPUT,P=PRINT,R=REINPUT,W=WRITE

Attribute definitions and color definitions may be combined to form attribute constants which are assigned to attribute control variables. Currently, control variables can be related to variables and single dimension arrays.

Control variables are attached to input fields in the INPUT statement, either within the program or through the map editor. A control variable can be assigned at the INPUT statement level or the field level. Whenever a control variable is attached to a field as well as at the INPUT level, then 'AD=Y' must be specified for the field. The example program on the following page shows an INPUT statement properly constructed to point this out. A single dimensioned array of control variables can be associated to a one-dimensional data structure, but a two-dimensional association is not currently allowed.

The option MODIFIED can be checked in an IF statement to determine if a field which has been assigned attributes dynamically has been modified. This is possible for control variables referenced in an INPUT statement are assigned the status of 'NOT MODIFIED' when the map is transmitted.

Upon being overwritten, the control variable for that field is assigned the status of 'MODIFIED'. If a control variable is at the statement level, then any field overwritten causes the modified tag to be set to 'MODIFIED'. The check for the modified data tag at the field level does NOT override the status at the statement level. Both will have a status of 'MODIFIED'.

```
DEFINE DATA LOCAL
1 #NAME     (A20) INIT <'WISDOM, JAMES'>
1 #CITY     (A16)
1 #STATE    (A2)
1 #ZIP      (A5)
1 #AGE      (N3)  INIT <30>
1 #CV-CITY  (C)
1 #CV-NAME  (C)
1 #CV-AGE   (C)
1 #CV-MAP   (C)
END-DEFINE
```

```
MOVE (AD=IP) TO #CV-AGE
FORMAT KD=OFF IP=OFF
INPUT       (CV=#CV-MAP)   //////
010T '*' (060)             /
010T '*  NAME:'
019T #NAME  (AD=M CV=#CV-NAME)
046T 'CITY:'
052T #CITY  (AD=A CV=#CV-CITY)
069T '*'                   /
010T '* STATE:'
019T #STATE (AD=AD)
042T 'ZIP CODE:'
052T #ZIP   (AD=AD)
069T '*'                   /
010T '*   AGE:'
019T #AGE   (AD=M CV=#CV-AGE)
069T '*'                   /
010T '*' (060)
*
IF #CV-MAP MODIFIED
   WRITE 'MAP IS MODIFIED. UPDATE WILL OCCUR.'
END-IF
*
IF #CV-CITY MODIFIED
   WRITE 'CITY ENTRY RECORDED.'
END-IF
*
IF #CV-NAME NOT MODIFIED
   WRITE 'DEFAULT NAME IS USED'
END-IF
END
```

Understanding this, here is a sample subprogram which forces a control variable to be test as MODIFIED regardless of anything typed into the field.

```
* TITLE  : Force a control variable related to field to be modified
* DESC   : This routine receives a control variable and returns a
*          value which will test as modified even though no entry
*          was input on screen.
* WRITTEN: 04/14/96  BY: WISDOM, JIM
* CHANGED:   /  /        BY:
*
DEFINE DATA PARAMETER
  1 PASSED-CV (C)
LOCAL
  1 CV-FIELD (C)
  1 REDEFINE CV-FIELD
     2 CV-BIN (B2)
  1 CV-PACKED (P4)
END-DEFINE
*
MOVE PASSED-CV TO CV-FIELD
ADD 260 TO CV-PACKED
MOVE CV-PACKED TO CV-BIN
MOVE CV-FIELD  TO PASSED-CV
END
```

Remember that attribute control variables are passed as data through the parameter list like any other data field. Therefore, the error -
`NAT0247 'SPECIFICATION ERROR IN AUTOMATIC PARAMETER IN A FORM/MAP' -`
is as much a problem of having not defined your control variable in your program or not placing it in your variable list as much as any field physically visible on a map.

Another SAG-L expert shared some information on SAG-L about how one determines what attributes a Control Variable has (at least the MVS version):

A Control Variable - format (C) - takes two bytes.

First Half-byte determines the AD parameter:
0 - Default
3 - Blinking
5 - Intensified
7 - Non-display
9 - Reverse Video
B - Underlined
D - Cursive/Italic
F - Default

Second Half-byte determines the status:
0 - Modifiable, Modified
1 - Modifiable, Not Modified
9 - Protected (AD=P)

Third half-byte is not used. (always 0)

Fourth half-byte determines the color (CD):
0 - Default
1 - BL
2 - RE
3 - PI
4 - GR
5 - TU
6 - YE
7 - NE

For example, moving (AD=UP CD=RE) to a control variable, will result in the control variable having the value H'B802'.

II. Data Type Assignments

Here are examples of NATURAL statements that assign the constants to their appropriate data types. The user variables are the same as described at the beginning of the chapter.

Data Type	Assignment statement	
Alphanumeric	MOVE 'ABCD'	TO #ALPHA
	MOVE 'ABC'-'DEF'	TO #ALPHA
Binary	MOVE H'C1C2'	TO #BINARY
Unpacked decimal	MOVE 1234	TO #UNPACKED1
Packed decimal	MOVE 1234	TO #PACKED1
Integer	MOVE 12	TO #INTEGER3
Floating Point	MOVE 123.45	TO #SHORT-REAL
	MOVE 1.23E4	TO #LONG-REAL
Logical	MOVE TRUE	TO #LOGIC-VAR
Control	MOVE (AD=P)	TO #ATTRIBUTE
Date	MOVE D'12/25/1988'	TO #DATE-VAR
Time	MOVE T'12:00:00'	TO #TIME-VAR

Assignments can also be accomplished by using an ASSIGN statement, except for control variables. The word 'ASSIGN' is optional in Report Mode, but is mandatory in Structured Mode (more on modes in Chapter 4). Here are a few examples.

```
[ASSIGN] #LOGICAL = TRUE
[ASSIGN] #TIME    = D'12/25/1988'
[ASSIGN] #FLOAT   = 12.9E-4
[ASSIGN] #INTEGER = 45
```

III. Arithmetic Operators / Math Functions

Operation	Function	Comment
Addition	+	
Subtraction	-	
Multiplication	*	
Division	/	
Exponentiation	**	integer exponents only
Square root	SQRT	operand cannot be an expression,
NATURAL		evaluates negative operands as
positive		
Absolute value	ABS (X)	
Arctangent	ATN (X)	
Cosine	COS (X)	
Exponential function	EXP (X)	
Fractional part of field	FRAC (X)	
Integer part of field	INT (X)	
NATURAL log (base e)	LOG (X)	
Sign of field	SGN (X)	returned values are -1,0,+1
Sine	SIN (X)	
Tangent	TAN (X)	
Extract numeric value from alpha field	VAL (X)	
Returns B4 return code	RET (X)	'X' is a user program, invoked via CALL
Internal unique field representatiion	POS (X)	'X' is a field name

Figure 3.5. Arithmetic Operators & Mathematical Functions; (X) Argument Field

For those not familiar with the exponential function, it returns a value in base e. For example, #Y = EXP(#X), where #X = 2, returns a value of 7.38905 because it is approximately e, which the Calculus teaches us is approximately 2.71828, to the power of 2.

NATURAL logs can be converted to common logs (base 10) using the simple formula:

```
Common LOG(X) = NATURAL LOG (X) / NATURAL LOG 10
```

Currently, none of the functions can be referenced in a WRITE/DISPLAY statement without getting a NAT0044 error. Also, it is documented that the format and length of the resultant takes on the characteristics of the operand to the function. I do not believe this fact is documented correctly. In my attempt to conclude if the trig functions took radian or degree measure, I wrote the following example:

```
0010 #X (P4) = 45
0020 #Y (P2.6) = SIN (#X)
0030 WRITE #Y
0040 END
```

If the documentation were correct, I would expect the result to be of format/length P4, but in fact the result was recorded with 6 position fractional accuracy. This implies that the result is actually the "largest and most accurate" of any operand in the compute statement.

Lastly, the trig functions take the values of its operands as radian measure, not degree measure. The "best" conversion formula to use in NATURAL is

```
RAD = PI * DEG  /  180.0
```

where PI is defined as a CONST of value 3.14157, DEG is defined to contain the degree measure (e.g., P4 format).

Two basic rules to remember while writing arithmetic statements:

RULE 1 - *Avoid mixed mode expressions as much as possible.*

RULE 2 - *Use packed arithmetic as much as possible.*

It is Rule 2 which illustrates the timing results one gets from comparing packed, unpacked, and integer data types in arithmetic statements, as:

```
0100 RESET #Z
0160 FOR #X  = 1 TO 400
0170     RESET #Z
0180     FOR #Y = 1 TO 5000
0190         ADD 1 TO #Z
0200     END-FOR
0210 END-FOR
```

```
STARTING TIME:      22:05:12.4
ELAPSED TIME:       29.7  SECONDS.
FINISHING TIME:     22:05:42.1
```

```
0100 RESET #Z
0160 FOR #X = 1 TO 400
0170     RESET #Z
0180     FOR #Y = 1 TO 5000
0190         ADD 1 TO #Z
0200     END-FOR
0210 END-FOR
```

```
STARTING TIME:      22:11:24.7
ELAPSED TIME:       32.1  SECONDS.
FINISHING TIME:     22:11:56.8
```

```
0120 RESET #Z
0180 FOR #X = 1 TO 400
0190     RESET #Z
0200     FOR #Y = 1 TO 5000
0210         ADD 1 TO #Z
0220     END-FOR
0230 END-FOR
```

```
STARTING TIME:      22:13:56.6
ELAPSED TIME: 21.8  SECONDS.
FINISHING TIME:     22:14:18.4
```

IV. Date & Time Arithmetic

Add and subtract operations are the only arithmetic operations allowed with date / time formatted variables. There are three ideas which one needs to understand in order to treat date / time variables correctly. One should be aware of the results of doing arithmetic operations mixing date and time formats.

Operation				Result
1.	(D)	+	(D)	not allowed
2.	(T)	+	(T)	(T)
3.	(D)	+	integer	(D)
4.	(T)	+	integer	(T)
5.	(D)	+	(T)	(T)
6.	(D)	–	(D)	integer, in days
7.	(T)	–	(T)	integer, in tenths of seconds
8.	(D)	–	integer	(D)
9.	(T)	–	integer	(T)
10.	(D)	–	(T)	integer, in tenths of seconds
11.	(T)	–	(D)	integer, in tenths of seconds

Integer values can be of formats N, P, or I.

Here are some examples of programs which use date/time arithmetic.

```
0010 DEFINE DATA LOCAL
0020 1 #TIME    (T)
0030 1 #DATE    (D)
0040 1 #HOUR    (P8) INIT <36000> /* TENTHS SECONDS IN 1 HOUR
0050 END-DEFINE
0060 *
0070 MOVE *TIMX TO #DATE
0080 MOVE *TIMX TO #TIME
0090 WRITE 'CURRENT DATE / TIME          ' '=' #DATE '=' #TIME
0100 *
0110 ADD ( 36 * #HOUR ) TO #TIME
0120 MOVE #TIME TO #DATE
0130 WRITE 'DATE / TIME 36 HOURS PLUS    ' '=' #DATE '=' #TIME
0140 *
0150 END
```

```
CURRENT DATE / TIME         #DATE: 89-12-10 #TIME: 19:00:20
DATE / TIME 36 HOURS PLUS   #DATE: 89-12-12 #TIME: 07:00:20
```

```
0010 DEFINE DATA LOCAL
0020 1 #TIME    (T)
0030 1 #SECOND  (P8) INIT <10>
0040 1 #MINUTE  (P8) INIT <600>
0050 1 #HOUR    (P8) INIT <36000>
0060 END-DEFINE
0070 *
0080 #TIME = *TIMX
0090 WRITE 'START TIME:' #TIME
0100 ADD #SECOND TO #TIME
0110 WRITE '1 SECOND LATER' #TIME
0120 *
0130 #TIME = *TIMX
0140 ADD #MINUTE TO #TIME
0150 WRITE '1 MINUTE LATER' #TIME
0160 *
0170 #TIME = *TIMX
0180 ADD #HOUR TO #TIME
0190 WRITE '1 HOUR LATER' #TIME
0200 *
0210 #TIME = *TIMX
0220 ADD ( 20 * #HOUR ) TO #TIME
0230 WRITE '20 HOURS LATER' #TIME
0240 END
```

```
START TIME: 11:02:19
1 SECOND LATER 11:02:20
1 MINUTE LATER 11:03:19
1 HOUR LATER 12:02:19
20 HOURS LATER 07:02:19
```

```
0010 DEFINE DATA LOCAL
0020 1 #TODAY              (D)
0030 1 #FUTURE-DATE        (D)
0040 1 #FUTURE-GREG-DATE   (A8) INIT <'89-12-30'>
0050 1 #CURRENT-DATE       (A8)
0060 1 #DAYS-DIFFERENTIAL  (P4)
0070 END-DEFINE
0080 #TODAY = *DATX
0090 MOVE EDITED #FUTURE-GREG-DATE TO #FUTURE-DATE (EM=YY'-'MM'-'DD)
0100 MOVE EDITED #TODAY (EM=YY'-'MM'-'DD) TO #CURRENT-DATE
0110 #DAYS-DIFFERENTIAL = #FUTURE-DATE - #TODAY
0120 WRITE 'THE NUMBER OF DAYS BETWEEN' #FUTURE-GREG-DATE
0130       'AND TODAY' #TODAY 'IS' #DAYS-DIFFERENTIAL
0140 END
```

```
THE NUMBER OF DAYS BETWEEN 89-12-30 AND TODAY 89-12-11 IS   19
```

```
0010 DEFINE DATA LOCAL
0020 1 #DAY1   (A10)
0030 1 #MONTH1 (A10)
0040 1 #D      (D)
0050 1 #DAY2   (A10)
0060 1 #MONTH2 (A10)
0070 END-DEFINE
0080 #D = *DATX
0090 * STRIP OFF CURRENT MONTH
0100 MOVE EDITED #D (EM=L(10)) TO #MONTH1
0110 * STRIP OFF CURRENT DAY
0120 MOVE EDITED #D (EM=N(9))  TO #DAY1
0130 WRITE 'THIS MONTH IS:' #MONTH1 'THE DAY IS:' #DAY1
0140 /      'THIS DATE IS:' #D
0150 *
0160 ADD 75 TO #D
0170 * STRIP OFF LATER MONTH
0180 MOVE EDITED #D (EM=L(10)) TO #MONTH2
0190 * STRIP OFF LATER DAY
0200 MOVE EDITED #D (EM=N(9))  TO #DAY2
0210 WRITE '75 DAYS LATER'
0220 / 'THE  MONTH IS:' #MONTH2 'THE DAY IS:' #DAY2
0230 / 'THE  DATE  IS:'   #D
0240 *
0250 END
```

```
THIS MONTH IS: June       THE DAY IS: Monday
THIS DATE IS: 89-06-19
75 DAYS LATER
THE  MONTH IS: September  THE DAY IS: Saturday
THE  DATE  IS: 89-09-02
```

```
0010 DEFINE DATA LOCAL
0020   1 #FINISH-DATE        (D)
0030   1 #FINISH-TIME        (T)
0040   1 #CURRENT-DATE       (D)
0050   1 #CURRENT-TIME       (T)
0060   1 #ONLINE-UP          (T)     INIT <T'06:00:00'>
0070 * 6 AM ANY DAY
0080   1 #ONLINE-DOWN        (T)     INIT <T'20:00:00'>
0090 * 8 PM ANY DAY
0100   1 #NEXT-ONLINE-UP     (T)     INIT <T'06:00:00'>
0110   1 #AVERAGE-RUN-TIME   (P2.2) INIT <8.5>
0120   1 #SECONDS-PER-HR     (P5)    INIT <36000>
0130   1 #NEXT-DAY           (A9)
0140 END-DEFINE
0150 #CURRENT-DATE = #CURRENT-TIME = *TIMX
0160 *
0170 * CHECK TO SEE IF THE BATCH WINDOW IS OPEN OR CLOSED.
0180 *
0190 ADD *DATX TO #ONLINE-UP
0200 ADD *DATX TO #ONLINE-DOWN
```

```
0210 *
0220 IF #CURRENT-TIME > #ONLINE-UP
0230   AND #CURRENT-TIME < #ONLINE-DOWN THEN
0240       WRITE 'NO BATCH JOB CAN RUN NOW.'
0250       'THE BATCH WINDOW IS CLOSED.'
0260 ELSE
0270       WRITE 'THE BATCH WINDOW IS NOW OPEN.'
0280 END-IF
0290 *
0300 *
0310 * QUESTION: IS THERE ENOUGH TIME / APPROPRIATE
0320 * CONDITIONS TO RERUN A JOB?
0330 *
0340 #FINISH-TIME := (#SECONDS-PER-HR * #AVERAGE-RUN-TIME)
0350              + #CURRENT-TIME
0360 MOVE #FINISH-TIME TO #FINISH-DATE
0370 *
0380 IF #CURRENT-DATE <= #FINISH-DATE THEN
0390    ADD ( *DATX + 1 ) TO #NEXT-ONLINE-UP
0410 ELSE
0420    ADD *DATX  TO #NEXT-ONLINE-UP
0430 END-IF
0440 *
0450 MOVE EDITED #FINISH-DATE (EM=N(9)) TO #NEXT-DAY
0460 *
0470 IF #AVERAGE-RUN-TIME > 34 THEN
0480    WRITE 'JOB NEEDS SPECIAL CONSIDERATION.'
0490            'IT EXCEEDS WEEKEND WINDOW.'
0510 ELSE
0520   IF #CURRENT-TIME > #ONLINE-UP
0530      AND #CURRENT-TIME < #ONLINE-DOWN THEN
0540        WRITE 'JOB CANNOT STARTUP NOW. ONLINES ARE UP.'
0560   ELSE   /* ONLINES ARE NOT UP ON SUNDAY
0570      IF #FINISH-TIME > #NEXT-ONLINE-UP
0580         AND #NEXT-DAY NE 'Sunday' THEN
0590         WRITE 'THIS JOB CANNOT RUN.'
0600        'ONLINES ARE DUE BEFORE FINISH TIME.'
0620      ELSE
0630        WRITE 'THIS JOB CAN RUN.'
0640        'IT WILL FINISH BEFORE ONLINE BEGINS.'
0650      END-IF
0660   END-IF
0680 END-IF
0690 END
```

```
PAGE      1                                    97-04-01  12:34:56

NO BATCH JOB CAN RUN NOW. THE BATCH WINDOW IS CLOSED.
JOB CANNOT STARTUP NOW. ONLINES ARE UP.
```

V. Mixed Mode Operations

Whenever doing assignments, computations, or comparisons you always do yourself a favor by making all the operands the *same data type*. This prevents unnecessary conversions internally and unexpected results.

Arithmetic expressions can be written involving multiple operations. Two questions come to mind in terms of performance. Does it make a difference to write an expression in its simpler components? Does it make a difference to use an arithmetic statement versus the NATURAL arithmetic operation?

Tests I have conducted indicate that not much is gained to separate a complex arithmetic expression into its atomic operations. Therefore, if you like to write calculations in the form of formulas do so. Remember, however, that NATURAL converts all arithmetic types to packed and uses packed instructions.

Tests indicate that performance is only slightly improved by using the arithmetic operators instead of the NATURAL arithmetic statement, e.g.:

```
0100 #P = #P + 1
```

or

```
0100 COMPUTE #P = #P + 1
```

instead of

```
0100 ADD 1 to #P
```

VI. User Defined Array Data Structures

NATURAL supports one, two and three dimensional arrays. Arrays can be based on any of the fundamental data types. First, let's see the definition of a user-defined array.

```
0010 DEFINE DATA LOCAL
0020   1 #STATES            (A2/1:50)
0030   1 REDEFINE #STATES
0040     2 #REDF-STATES     (A4/1:25)
0050   1 #ZIPCODES          (A5/0:49,0:9)
0060   1 REDEFINE #ZIPCODES
0070     2 #ZIP-LIST        (A5/1:500)
0080 END-DEFINE
```

Each of these represents a single named variable with multiple locations which are referenced with a subscript. The variable #STATES allows for 50 locations, #STATES(1), #STATES(2),... ,#STATES(50). An explicit subscript reference for #STATES over 50 at compile time generates NAT0280. At run time, any subscript reference for #STATE over 50 generates NAT1316.

The variable #ZIPCODES defines a two-dimensional structure of five bytes per table entry, a total of 2500 bytes. This structure is used to provide a row/column orientation. #ZIPCODES is defined to relate ten zip codes per state. Zero and negative subscripts are allowed as long as it is provided for in the range notation.

Redefinition of array structures is also allowed. The two examples above show how this is possible. Notice that #ZIPCODES, a two-dimensional array, can be redefined as a one-dimensional object. This could be useful for routines which are written to work with lists rather than matrices e.g., sorting. The redefinition of arrays is possible as long as the length does not exceed the base length; otherwise, a NAT0039 is generated. The following redefinition is also possible, allowing for an alphanumeric variable to be manipulated as a character array as well as a single buffer.

```
0010 DEFINE DATA LOCAL
0020   1 #WORD              (A80)
0030   1 REDEFINE #WORD
0040     2 #CHAR            (A1/1:80)
0050 END-DEFINE
```

Arrays can be initialized in the DEFINE DATA BLOCK as well. An array can be initialized to the same value throughout (other than 0 or blanks which are the default) by using the ALL keyword following either INIT or CONST:

```
0040   1 #ARR-P (P2/1:4,1:4) INIT ALL <10>
```

You can initialize any specific entry or any entire dimension by using the character 'V':

```
0040   1 #ARR-P (P2/1:4,1:4) INIT (1,V) (10,20,30,40)
```

initializes the entire first row with values 10, 20, 30, and 40. You should supply as many values as entries. Too many cause a NAT0094 at compile time, whereas too few will only fill as many slots as the running index will allow from the lower bound to the upper bound.

You can initialize a specific entry:

```
1 #ARR-P (P2/1:4,1:4) INIT (1,4) <10> (2,4) <20> (4,4) <40>
```

You can define an array of a variable number of elements as follows:

```
1 #MAX-ELEMENTS (N1) CONST <3>
1 #A-A1(A1/#MAX-ELEMENTS)
```

This checks fine but not if #MAX-ELEMENTS is defined in a GDA. In an LDA, 2.2.8 on the mainframe accepts either the CONST or INIT definition. Darrell Davenport of Washington State University gave an excellent explanation as follows:

The compiler is a single pass compiler. The GDA has a symbol table of its own that is not associated to the symbol table for LDAs. When building the data area segment of the program the compiler has to have a base and offset for each variable. While building the LDA segment, it cannot dynamically change the pointer to the GDA symbol table without losing its place.

Case Study: Defining 2 or 3-dimensional tables to handle database table unloads

Here is a scenario I have developed in concert with consultant Lew Levy about loading an ADABAS Table.

PROJECT ID	MEMBER-NAME	FUNCTION-ID
DB01	DBAJTW	10
DB01	DBASXW	10
DB01	DBASXW	12
DB05	DBALAA	08
DB05	DBALAA	11
DB05	DBALAA	14
DB05	DBASXW	10
DB05	DBASXW	12
DB06	DBAPPR	01
DB06	DBAPPR	02
DB06	DBAPPR	03
DB06	DBASXW	11
DB06	DBASXW	12
DB06	DBASXW	13
DB06	DBAJTW	09
DB06	DBAJTW	10

If one were to define a table to hold up to 200 entries on the ADABAS table, you might define the following structure:

```
0010 DEFINE DATA LOCAL
0020  1 PRJT-T (1:200)
0030    2 PROJ-ID    (A4)
0040    2 USER-ID    (A6)
0050    2 FUNC-CD    (A2)
0060 END-DEFINE
```

The space requirements for this table is 200 * 12 = 2,400 bytes. However, it turns out that there is a better way to define this structure. First, one must see that the relationships of the data is as following:

These relationships can be depicted as a 3-dimensional table!

```
0010 DEFINE DATA LOCAL
0020  1 PRJT-T       (1:10)
0030    2 PROJ-ID            (A4)
0040    2 USER-T     (1:10)
0050      3 USER-ID          (A6)
0060      3 FUNC-T  (1:10)
0070        4 FUNC-C         (A2)
0080 END-DEFINE
```

With such a structure, two questions arise - how do you load the data structure and how do you use it to observe certain relationships. The following two programs show the answer to both questions.

The input could come from an OS work file, a VSAM file, an ADABAS file, even a DB2 table. If the data is input in sort sequence by project, it makes life easier.

Let's assume the "best" scenario,
where the input is as follows:

```
DB01 DBAJTW 10
DB01 DBASXW 10
DB01 DBASXW 12
DB05 DBALAA 08
DB05 DBALAA 11
DB05 DBALAA 14
DB05 DBASXW 10
DB06 DBAPPR 02
DB06 DBASXW 12
DB06 DBAJTW 14
DB06 DBAJTW 09
DB06 DBAJTW 10
DB07 DBALAA 10
DB07 DBASXW 10
DB07 DBASXW 12
DB07 DBAJTW 08
DD07 DBAJTW 11
DB07 DBAPPR 02
DB11 DBAPPR 08
DB11 DBASXW 09
DB11 DBASXW 12
DB11 DBAJTW 12
DB11 DBAJTW 14
```

```
0010 DEFINE DATA LOCAL
0020 1 #WORK-RECORD
0030   2 #PROJ-NAME (A4)
0040   2 #FILLER1   (A1)
0050   2 #PROJ-MEM  (A6)
0060   2 #FILLER2   (A1)
0070   2 #FUNCTION-CD (A2)
0080 1 #IND1    (P3)      INIT <1>
0090 1 #IND2    (P3)      INIT <1>
0100 1 #IND3    (P3)      INIT <1>
0110 1 PRJT-T (1:10)
0120   2 PROJ-ID  (A4)
0130   2 USER-T  (1:10)
0140     3 USER-ID (A6)
0150     3 FUNC-T (1:10)
0160       4 FUNC-CD (A2)
0170 END-DEFINE
```

```
0180 *
0190 * INITIALIZE ARRAY BY READING WORK FILE
0200 * SORTED BY PROJECT, PROJECT MEMBER
0210 *
0220 READ WORK FILE 1 RECORD #WORK-RECORD
0230 *
0240   IF BREAK OF #PROJ-MEM
0250      DO ADD 1 TO #IND2
0260         RESET INITIAL #IND3
0270      DOEND
0280 *
0290   IF BREAK OF #PROJ-NAME
0300      DO ADD 1 TO #IND1
0310         RESET INITIAL #IND2 #IND3
0320      DOEND
0330 *
0340   PROJ-ID (#IND1) = #PROJ-NAME
0350   USER-ID (#IND1,#IND2) = #PROJ-MEM
0360   FUNC-CD (#IND1,#IND2,#IND3) = #FUNCTION-CD
0370   ADD 1 TO #IND3
0380 LOOP
0390 *
0400 FOR #IND1 = 1 TO 10
0410     IF PROJ-ID (#IND1) = ' ' THEN ESCAPE
0420     WRITE NOTITLE 'PROJECT:' PROJ-ID (#IND1)
0430     FOR #IND2 = 1 TO 10
0440         IF USER-ID (#IND1,#IND2) = ' ' THEN ESCAPE
0450         WRITE 10T 'USER:' USER-ID (#IND1,#IND2)
0460         FOR #IND3 = 1 TO 10
0470             IF FUNC-CD (#IND1,#IND2,#IND3) = ' ' THEN ESCAPE
0480             WRITE 15T 'CODES:' FUNC-CD (#IND1,#IND2,#IND3)
0490         LOOP
0500     LOOP
0510 LOOP
0520 END
```

```
PROJECT: DB01 USER ID: DBAJTW CODES: 10
PROJECT:      USER ID: DBASXW CODES: 10
PROJECT:      USER ID:        CODES: 12
PROJECT: DB05 USER ID: DBALAA CODES: 08
PROJECT:      USER ID:        CODES: 11
PROJECT:      USER ID:        CODES: 14
PROJECT:      USER ID: DBASXW CODES: 10
PROJECT: DB06 USER ID: DBAPPR CODES: 02
PROJECT:      USER ID: DBASXW CODES: 12
PROJECT:      USER ID: DBAJTW CODES: 14
PROJECT:      USER ID:        CODES: 09
PROJECT:      USER ID:        CODES: 10
PROJECT: DB07 USER ID: DBALAA CODES: 10
PROJECT:      USER ID: DBASXW CODES: 10
PROJECT:      USER ID:        CODES: 12
PROJECT:      USER ID: DBAJTW CODES: 08
PROJECT:      USER ID:        CODES: 11
PROJECT:      USER ID: DBAPPR CODES: 02
PROJECT: DB11 USER ID:        CODES: 08
PROJECT:      USER ID: DBASXW CODES: 09
PROJECT:      USER ID:        CODES: 12
PROJECT:      USER ID: DBAJTW CODES: 12
PROJECT: DB11 USER ID: DBAJTW CODES: 14
```

No matter what the source of the data, the input is best handled if it is in project / project member order. An interesting exercise for the enterprising programmer is to program the loading of the array without any order.

The space requirements for this table is 2,640 bytes, which is more than the previous definition. However, it allows for the removal of repetitive data and it can hold 1,000 combinations of 10 project IDs, 10 member names and 10 function codes. In the previous definition it would require 12,000 bytes of storage to hold the same number of combinations. However it should be noted that the first solution makes fewer assumptions, and places fewer restrictions on the data to be stored.

```
0010 DEFINE DATA LOCAL
0020 1 #IND1    (P2)
0030 1 #IND2    (P2)
0040 1 #IND3    (P2)
0050 1 J        (P2)
0060 1 K        (P2)
0070 1 PRJT-T (1:10)
0080   2 PROJ-ID  (A4)
0090   2 USER-T  (1:10)
0100     3 USER-ID (A6)
0110     3 FUNC-T (1:10)
0120       4 FUNC-CD (A2)
0130 END-DEFINE
0140 *
0150 * INITIALIZE ARRAY FOR SEARCHING
0160 *
0170 PROJ-ID (1) = 'DB01'
0180 PROJ-ID (2) = 'DB05'
0190 PROJ-ID (3) = 'DB06'
0200 PROJ-ID (4) = 'DB07'
0210 PROJ-ID (5) = 'DB11'
0220 *
0230 USER-ID (1,1) = 'DBAJTW'     FUNC-CD (1,1,1) = '10'
0240 USER-ID (1,2) = 'DBASXW'     FUNC-CD (1,2,1) = '10'
0250 USER-ID (2,1) = 'DBALAA'     FUNC-CD (1,2,2) = '12'
0260 USER-ID (2,2) = 'DBASXW'     FUNC-CD (2,1,1) = '08'
0270 USER-ID (3,1) = 'DBAPPR'     FUNC-CD (2,1,2) = '11'
0280 USER-ID (3,2) = 'DBASXW'     FUNC-CD (2,1,3) = '14'
0290 USER-ID (3,3) = 'DBAJTW'     FUNC-CD (3,3,1) = '09'
0300 USER-ID (4,1) = 'DBALAA'     FUNC-CD (3,3,2) = '10'
0310 USER-ID (4,2) = 'DBASXW'     FUNC-CD (4,1,1) = '10'
0320 USER-ID (4,3) = 'DBAJTW'     FUNC-CD (4,2,1) = '10'
0330 USER-ID (4,4) = 'DBAPPR'     FUNC-CD (4,2,2) = '12'
0340 USER-ID (5,1) = 'DBAPPR'     FUNC-CD (4,3,1) = '08'
0350 USER-ID (5,2) = 'DBASXW'     FUNC-CD (4,3,2) = '11'
0360 USER-ID (5,3) = 'DBAJTW'     FUNC-CD (5,1,1) = '14'
0370 *                            FUNC-CD (5,1,2) = '09'
0380                              FUNC-CD (5,2,2) = '10'
```

```
0390 *
0400 * ANSWER QUESTION: IS PROJECT 'DB05' IN THE MATRIX?
0410 * IF YES, WHO IS ON THE PROJECT?
0420 *
0430 EXAMINE PROJ-ID (*) FOR 'DB05' GIVING INDEX #IND1
0440 *
0450 IF #IND1 = 0 THEN
0460    WRITE 'PROJECT NOT IN MATRIX'
0470 ELSE
0480    DO WRITE 'PERSONS ON PROJECT DB05'
0490       FOR J = 1 TO 10
0500           IF USER-ID (#IND1,J) = ' ' THEN ESCAPE
0510           WRITE USER-ID (#IND1,J)
0520       LOOP
0530    DOEND
0540 *
0550 * QUESTION: IF 'DBAJTW' IS ON PROJECT 'DB06', WHAT
0560 * FUNCTIONAL RESPONSIBILITIES DOES THIS PERSON HAVE?
0570 *
0580 EXAMINE PROJ-ID (*) FOR 'DB06' GIVING INDEX #IND1
0590 *
0600 IF #IND1 = 0 THEN
0610    WRITE 'PROJECT NOT IN MATRIX'
0620 ELSE
0630    DO
0640       EXAMINE USER-ID (#IND1,*) FOR 'DBAJTW'
0650         GIVING INDEX #IND1 #IND2
0660       IF #IND2 = 0 THEN
0670          WRITE 'DBAJTW NOT ON PROJECT'
0680       ELSE
0690          DO WRITE 'FUNCTION CODES FOR DBAJTW ON PROJECT DB06'
0700             FOR K = 1 TO 10
0710                IF FUNC-CD (#IND1,#IND2,K) = ' ' THEN ESCAPE
0720                   WRITE FUNC-CD (#IND1,#IND2,K)
0730             LOOP
0740          DOEND
0750    DOEND
0760 *
0770 * QUESTION: ON WHAT PROJECTS IS DBAJTW? EXAMINE FOR
0780 * FIRST ROW, COLUMN ENTRY THEN REMOVE TO CONTINUE SEARCH
0790 *
0800 WRITE 'PROJECTS FOR DBAJTW'
0810 REPEAT
0820    EXAMINE USER-ID (*,*) FOR 'DBAJTW'
0830      GIVING INDEX #IND1 #IND2
0840    IF #IND1 = 0 OR #IND2 = 0 THEN ESCAPE BOTTOM
0850    WRITE PROJ-ID (#IND1)
0860    USER-ID (#IND1,#IND2) = '$$'   /* REPLACE LATER
0870 LOOP
0880 EXAMINE USER-ID (*,*) FOR '$$'
0890   REPLACE WITH 'DBAJTW'           /* CHANGE BACK
0870 END
```

```
PAGE      1
PERSONS ON PROJECT DB05
DBALAA
DBASXW
FUNCTION CODES FOR DBAJTW ON PROJECT DB06
09
10
PROJECTS FOR DBAJTW
DB01
DB06
DB07
DB11
```

Multi Dimensions > 4

Although you are limited to 3 dimensional arrays in NATURAL, you can simulate higher order arrays. Any multi-dimensional array can be treated like a one dimensional array. For example, suppose you have GRADES (25,10), that is, 25 students with 10 grades each. If you want to access the third grade of the fifth student, you have two choices. If you also define GRADES (250), compute (which student - 1) * 10 + grade position. Therefore GRADES (3,5) in the first definition is equivalent to GRADES (25) in the one dimensional array. This techniques can extend to a four dimensional system. GRADES (I,J,K,L) reference in a four dimensional system is equivalent to GRADES (X), where if

> I index has Lim1, J index has Lim2, K index has Lim3
> X := (I - 1)*Lim1*lim2*Lim3 + (J - 1)*Lim2*Lim3 + (K-1)*Lim 3 + L

Therefore, in this example GRADES has the dimensional limits (4,3,2,3). To refer to GRADES (2,2,1,1) in a one dimensional system. The formula states

```
X = (2 - 1)*4*3*2 + (2 -1)*3*2 + (1 - 1)*2 + 3 = 24 + 6 + 0 + 3 = 33
```

A second technique is to think of a 4 dimensional system as a one dimensional plus a three dimensional system. Allow each reference of the one dimensional system to run through all the index ranges for the three dimensional system

VII. Block Data Structures

A Global Data Area can be subdivided into logical data blocks. In this way, data storage requirements are reduced by allowing data blocks to be overlaid during program execution. A large GDA may be read into ESIZE with a particular program only requiring a fraction of the space. Data blocking can provide a means to better manage the space allocated for programs within an application.

Elements can be identically named within data structures in different data blocks. Fundamentally, I disagree with this practice; however, it is possible since only one block is available at one time. In the example below you can see that #Z is defined in both
SUB-BLOCK1 and in SUB-BLOCK-2. Again, I repeat, this is not a good idea.

```
B    MASTER-BLOCK
  1 MAIN-DATA-AREA
  2 #ARR1                              A    70 (1:10)
B    SUB-BLOCK-1                               MASTER-BLOCK
  1 DATA-BLOCK-1
  2 #X                                 A    20
  2 #Y                                 A    50
R 2 #Y                                         /* REDEF. BEGIN : #Y
  3 #R1                                A    10
  3 #Z                                 A    40
  2 #ARR2                              A    80 (1:100)
  2 #ARR3                              P     4 (1:100)
B    SUB-BLOCK-2                               MASTER-BLOCK
  1 DATA-BLOCK-2
  2 #X                                 A    25
  2 #Y                                 A    25
  2 #ARR4                              A    20 (1:100,1:5)
  2 #ARR5                              P     4 (1:100)
  2 #Z                                 A     4
```

The master block contains 700 bytes of storage, whereas the block SUB-BLOCK-1 contains 9,074 bytes and SUB-BLOCK-2 contains 11,058 bytes of storage.

A similar large GDA defined to take care of all storage requirements for all programs would look like the following and occupy 19,424 bytes of storage.

```
  1 MAIN-DATA-AREA
  2 #ARR1                              A    70 (1:10)
  2 #X1                                A    20
  2 #X2                                A    25
  2 #Y1                                A    50
R 2 #Y1                                        /* REDEF. BEGIN : #Y
  3 #R1                                A    10
  3 #Z1                                A    40
  2 #Y2                                A    25
  2 #Z2                                A     4
  2 #ARR2                              A    80 (1:100)
  2 #ARR3                              P     4 (1:100)
  2 #ARR4                              A    20 (1:100,1:5)
  2 #ARR5                              P     4 (1:100)
```

Below are three simple programs linked via the FETCH mechanism. MAINPGM, PGM01, and PGM02 have smaller ESIZE requirement than if they were linked to the second GDA listed above.

```
0010 * MAINPGM
0020 DEFINE DATA GLOBAL
0030  USING BLOCKGDA
0040  WITH MASTER-BLOCK
0050            LOCAL
0060 *
0070 END-DEFINE
0080 * PROGRAM CODE
0090 FETCH 'PGM01'
0100 END
```

```
0010 * PGM01
0020 DEFINE DATA GLOBAL
0030  USING BLOCKGDA
0040  WITH MASTER-BLOCK.SUB-BLOCK-1
0050            LOCAL
0060 END-DEFINE
0070 *
0080 FETCH 'PGM02'
0090 END
```

```
0010 * PGM02
0020 DEFINE DATA GLOBAL
0030  USING BLOCKGDA
0040  WITH MASTER-BLOCK.SUB-BLOCK-2
0050            LOCAL
0060 *
0070 END-DEFINE
0080 *
0090 END
```

Any program referencing the same GDA block will have the values retained from the first references to them. This way, an application can be developed to "manage" the ESIZE (maximum of 64K managed in 32K blocks) more effectively than defining one large area.

VIII. MU/PE Data Base Structures

The fundamental principle for how to handle MU/PE structures is to remember to describe in the format buffer what you want from ADABAS. This is what the OBTAIN provided in NATURAL 1.2. There are, however, some considerations of which one must be aware to handle them correctly in NATURAL.

Scenario #1: A program requires a determined amount of MU/PE occurrences in report mode (1.2 style code).

```
0010 *
0020 * THIS PROGRAM IS A 2.1 REPORT MODE PROGRAM WRITTEN IN
0030 * 1.2 STYLE THE FIELD LK CONTAINS SOURCE CODE
0040 * OCCURRENCES ON THE NATURAL SYSTEM FILE.
0050 *
0060 SET GLOBALS LS=100
0070 *
```

```
0080 READ NAT-SYS-TEST-47 BY SOURCE-CD-ID
0090   OBTAIN SOURCE-TEXT (1:60)
0100   FOR #I (P2) = 1 TO C*SOURCE-TEXT
0110       WRITE SOURCE-TEXT (1.#I)
0120   LOOP (0110)
0130 LOOP (0080)
0140 END
```

```
0010 *
0020 * THIS PROGRAM IS A 2.1 REPORT MODE PROGRAM WRITTEN IN
0030 * 1.2 STYLE WHILE USING STRUCTURED DATA. THE FIELD
0040 * SOURCE-TEXT (LK) CONTAINS SOURCE CODE OCCURRENCES ON
0050 * THE NATURAL SYSTEM FILE. THE FORMAT BUFFER IS BUILT TO
0060 * CONTAIN 60 OCCURRENCES OF SOURCE-TEXT.
0070 *
0080 DEFINE DATA LOCAL
0090 1 #I (P2)
0100 1 FUSER VIEW OF NAT-SYS-TEST-47
0110   2 SOURCE-CD-ID
0120   2 C*SOURCE-TEXT
0130   2 SOURCE-TEXT (1:60)
0140 END-DEFINE
0150 *
0160 SET GLOBALS LS=100
0170 *
0180 READ FUSER BY SOURCE-CD-ID
0190 *
0200   FOR #I = 1 TO C*SOURCE-TEXT
0210       WRITE SOURCE-TEXT (#I)
0220   LOOP (0200)
0230 LOOP (0180)
0240 END
```

Scenario #2: A program requires an undetermined amount of MU/PE occurrences in report mode using GET SAME logic.

```
0010 DEFINE DATA LOCAL
0020  1 #ALIAS (A8)
0030  1 I (P2)  INIT<1>
0040  1 REQUEST-FILE VIEW OF ACPT-TEST-PROD-REQ
0050    2 REQ-ID
0060    2 C*REQ-CHECKPT-GRP
0070    2 REQ-CHECKPT-USER-ID (I)
0080    2 REDEFINE REQ-CHECKPT-USER-ID
0090      3 #USER (A6)
0100 END-DEFINE
0110 READ REQUEST-FILE BY SP-REQ-ID-SEQ STARTING FROM 'DRM'
0120   FOR I = 1 TO C*REQ-CHECKPT-GRP
0130       GET SAME (0120)
0140       IF REQ-CHECKPT-USER-ID (0140/I.1) = 'DRMJTW' THEN
0150          ASSIGN #ALIAS = 'WISDOM'
0160       ELSE
0170          ASSIGN #ALIAS = 'IMPOSTER'
0180       END-IF
0190       WRITE I REQ-ID REQ-CHECKPT-USER-ID (0140/I.1)
0200             #USER #ALIAS
0210   END-FOR
0220   NEWPAGE
0230 END-READ
0240 END
```

Scenario #3: A program reads an average count of PE occurrences and uses conditional logic to reread the remainder using GET SAME logic.

```
0010 DEFINE DATA LOCAL
0020 1 I (P2)
0030 1 REQUEST-FILE VIEW OF ACPT-TEST-PROD-REQ
0040   2 REQ-ID
0050   2 C*REQ-CHECKPT-GRP
0060   2 REQ-CHECKPT-USER-ID (1:7) /* MOST LIKELY NUMBER OF
0070   2 REQ-CHECKPT-USER-ID (I)   /* OCCURRENCES FOR GET SAME
0080 END-DEFINE                    /* LOGIC (6 TO WHATEVER)
0090 *
0100 ASSIGN I = 8
0110 READ REQUEST-FILE BY SP-REQ-ID-SEQ
0120   REJECT IF C*REQ-CHECKPT-GRP = 0
0130   DISPLAY REQ-ID REQ-CHECKPT-USER-ID (*) C*REQ-CHECKPT-GRP
0140   NEWPAGE
0150   IF C*REQ-CHECKPT-GRP > 7 THEN
0160     FOR I = 8 TO C*REQ-CHECKPT-GRP
0170       GET SAME (0110)
0180       WRITE I REQ-ID REQ-CHECKPT-USER-ID (0180/I.1)
0190     END-FOR
0200   END-IF
0210   NEWPAGE
0220 END-READ
0230 END
```

Ye Ole MOVE INDEXED Syntax.

Software AG no longer documents a syntax that as of NATURAL.2.8 continues to syntax check. Fewer and fewer NATURAL programmers are familiar with the syntax as the older generation passes the baton to the next group. I though it prudent to document the syntax since many 1.2 / 2.1 NATURAL programs in the world still contain the syntax. My belief is that an organization should invest in one of several products which are available which can convert the older style of MU/PE handling using structured arrays.

```
MOVE INDEXED {field[(r/i)]}  [<{field[(r/i)]}>]
             {constant    }  [<{constant    }>]

          TO field [(r,fl/i)]  [<{field[(r/i)]}>]
                               [<{constant    }>]
```

The MOVE INDEXED statement is used to perform a move operation in which an offset is applied to the sending and/or receiving fields.

For the sending field, the byte offset used is equal to the number of bytes defined for the field or constant specified after the keyword INDEXED multiplied by the value of the constant or field specified within the delimiters < > minus 1.

For example, MOVE INDEXED A <5> results in an offset of 8 if the length of A is 2.

For the receiving field, the byte offset used is equal to the number of bytes defined for the field or constant specified after the keyword TO multiplied by the value of the constant or field specified within the delimiters < > minus 1.

For example, TO F <2> results in an offset of 2 if the length of F is 2.

The offset specified for the receiving field is checked to ensure that it lies within the NATURAL program's data area. The specification of an offset which references beyond the defined length of the receiving field may produce unpredictable results, including a possible abnormal termination of the program.

IX. Mask Option

Often, a programmer is required to analyze or ask questions about the contents of alphanumeric quantities. There are a few techniques NATURAL provides to do this.

First, masking is a powerful tool for determining specific contents of fields. This can also be done positionally. Here is an example of a report mode program which shows various checking that can be done using the mask definition.

.	ignore position
*	Wild character, indicates any number of positions to ignore
A	alpha character check (upper or lower case)
N	numeric check 0 - 9
Z	checks for ASCII/EBCDIC appropriate nibbles
C	check for alphanumeric (upper or lower case, numerics)
H	check for A-F, 0-9
P	check for printable character (upper or lower case, special char, numeric)
S	check for special characters
U	check for upper case A-Z
L	check for lower case a-z
'c'	check for character(s) bounded by apostrophes
DD	check for value of day 1 thru 31
MM	check for value of month 1 - 12
YY	check for valid year
YYYY	check for valid year
MMDDYY	check for valid combination of month-day-year
N1-N2	check for inclusive range of numeric value N1 - N2 (also N1:N2)
X	relative position check

Figure 3.6. MASK Option Definitions

```
0010 DEFINE DATA LOCAL
0020 1 #FOREIGN-ID (A9)
0030 1 #IN-VALUE   (A6)
0040 1 #IN-NUM-VAL (A6)
0050 1 #PROGRAM    (A8)
0060 1 #SYS-ID     (A2)
0070 END-DEFINE
```

```
0080 *
0090 INPUT 'INPUT VALUE> ' #IN-VALUE
0100 IF #IN-VALUE = MASK(AAAAAA) THEN
0110    WRITE 'DATA IS STRICTLY ALPHABETIC'
0120 ELSE
0130 IF #IN-VALUE = MASK (NNNNNN) THEN
0140    WRITE 'DATA IS NUMERIC'
0150 ELSE
0160 IF #IN-VALUE = MASK (CCCCCC) THEN
0170    WRITE 'DATA IS STRICTLY ALPHANUMERIC'
0180 ELSE
0190    WRITE 'DATA CONTAINS SPECIAL CHARACTERS'
0200 *
0210 * POSITION CHECKING
0220 *
0230 IF #IN-VALUE = MASK (SSS) THEN
0240    WRITE 'DATA CONTAINS SPECIAL CHARS IN FIRST 3 POSITIONS'
0250 *
0260 IF #IN-VALUE = MASK (...NNN) THEN
0270    WRITE 'LAST 3 DIGITS ARE NUMERIC'
0280 *
0290 IF #IN-VALUE = MASK ('J'.'W') THEN
0300    WRITE 'THE FIRST AND THIRD CHARACTERS ARE J AND W'
0310 *
0320 * RANGE CHECKING
0330 *
0340 INPUT 'TYPE IN FOREIGN STUDENT ID>' #FOREIGN-ID
0350 IF #FOREIGN-ID NOT = MASK(800-899NNNNNN) THEN
0360    REINPUT 'NOT A FOREIGN STUDENT! ID STARTS 800 - 899 '
0370 WRITE 'FOREIGN STUDENT IS ACCEPTED.'
0380 *
0390 INPUT 'TYPE 6 DIGIT NUMBER>' #IN-NUM-VAL
0400 *
0410 IF #IN-NUM-VAL = MASK(...500) THEN
0420    WRITE 'IN VALUE HAS VALUES <= 500 IN LAST 3 DIGITS'
0430 *
0440 * CHECKING AGAINST CONTENTS OF CONSTANTS OR VARIABLES
0450 *
0460 INPUT 'TYPE IN PROGRAM NAME>' #PROGRAM
0470 #SYS-ID = 'FI'
0480 *
0490 IF #PROGRAM = MASK (XX) 'SA' THEN
0500    WRITE 'PROGRAM BELONGS TO THE STUDENT ACCOUNTS SYSTEM'
0510 ELSE
0520 IF #PROGRAM = MASK (XX) #SYS-ID THEN
0530    WRITE 'PROGRAM BELONGS TO THE FINANCIAL AID SYSTEM'
0540 END
```

Below is another example which shows how masking can be used to validate month / day relationships in a date field. This leads to the controversy often discussed concerning validating a legal year. It also points out the difference between MASK (YYYYMMDD), MASK (1582-2699MMDD) and MOVED EDITED. The range for MASK (YYYYMMDD) is outside that for the MOVED EDITED which is based on the Gregorian calendar. Therefore, if you want to date validate and use the MOVED EDITED feature, you can only accept dates in the range. Simple enough. The following program illuminates the differences in approaches.

```
DEFINE DATA LOCAL
   1 IN-DATE (A8)
   1 REDEFINE IN-DATE
      2 CENTURY (N4)
   1 #DATE   (D)
END-DEFINE
*
IN-DATE := '19960229'  /* writes legal date
IN-DATE := '00960229'  /* fails MOVED EDITED - NAT1143
IN-DATE := '20000229'  /* writes legal date
IN-DATE := '19001232'  /* writes Illegal date
*
IF IN-DATE = MASK (YYYYMMDD) THEN
    MOVE EDITED IN-DATE TO #DATE (EM=YYYYMMDD)
   WRITE 'Legal date'
ELSE
   WRITE 'Illegal date'
END-IF
*
IN-DATE := '19960229'  /* writes Illegal date
IN-DATE := '00960229'  /* writes Illegal date
IN-DATE := '20000229'  /* writes Illegal date
IN-DATE := '19001232'  /* writes Illegal date
*
IF IN-DATE = MASK (1582-2699MMDD)
     MOVE EDITED IN-DATE TO #DATE (EM=YYYYMMDD)
     WRITE 'Legal date'
ELSE
     WRITE 'Illegal date'
END-IF
*
* For the next code segment, I will never get a NAT1143 in MOVE EDITED
*
IN-DATE := '19960229'  /* writes Legal date
IN-DATE := '00960229'  /* writes Illegal date
IN-DATE := '20000229'  /* writes Legal date
IN-DATE := '19001232'  /* writes Illegal date
*
IF IN-DATE = MASK (YYYYMMDD) AND CENTURY > 1581 AND CENTURY < 2700
   MOVE EDITED IN-DATE TO #DATE (EM=YYYYMMDD)
   WRITE 'Legal date'
ELSE
  WRITE 'Illegal date'
END-IF
END
```

Here is another example of a phone number edit.

```
DEFINE DATA LOCAL
1 PHONE (N10) INIT <6172675875>
1 EDIT-PHONE (A14)
END-DEFINE
*
MOVE EDITED PHONE (EM=' ('999') '999'-'9999)
   TO EDIT-PHONE
WRITE '=' EDIT-PHONE
END
```

X. The IS Operator / VAL Function

This is an excellent addition to the NATURAL language constructs. Consider the need to check the contents of an operand before converting it to another format, for example, convert a numeric field contained in an alphanumeric field using the VAL function. You wouldn't want a needless runtime error because conversion is not possible. For example, an alphanumeric field contains an integer of unspecified length. Therefore, the MASK function is not useful. However, you need to know if the field is numeric in order to use the VAL function. Here is a sample program showing this scenario.

```
0010 DEFINE DATA LOCAL
0020 1 #DATE-FIELD  (A8)
0030 1 #INPUT-FIELD (A5)
0040 1 #NUMF(N5.5)
0050 END-DEFINE
0060 *
0070 #INPUT-FIELD = '123.4'
0080 #DATE-FIELD  = '07/02/89'
0090 *
0100 IF #INPUT-FIELD = MASK (N5) THEN /* MASK NOT HELPFUL HERE
0110    WRITE 'NUMERIC'
0120 ELSE
0130    WRITE 'WHOLE FIELD NOT NUMERIC'
0140 *
0150 IF #INPUT-FIELD IS (N5) THEN /* IS OPERATOR PROVIDES ANSWER
0160    DO WRITE 'CONTENTS OF FIELD IS NUMERIC'
0170       #NUMF = VAL (#INPUT-FIELD)
0180       WRITE 'NUMERIC VALUE =' #NUMF
0190    DOEND
0200 ELSE
0210    DO WRITE 'NOT NUMERIC'
0220    DOEND
0230 *
0240 IF #DATE-FIELD IS (D) THEN   /* CHECK DEFINITION ACCORDING
0245 *                            /* TO DTFORM PARAMETER
0250    WRITE 'LEGAL DATE FORMAT'
0260 ELSE
0270    WRITE 'ILLEGAL DATE FORMAT. CORRECT FORMAT IS' *DATX
0280 END
```

```
WHOLE FIELD NOT NUMERIC
CONTENTS OF FIELD IS NUMERIC
NUMERIC VALUE =    123
ILLEGAL DATE FORMAT. CORRECT FORMAT IS 89-06-07
```

There is no longer a problem with the IS operator returning the correct result when testing for negative data.

Packed, Numeric Data - IS/VAL function

```
DEFINE DATA LOCAL
1 #P-A (A3)
END-DEFINE
MOVE 123 TO #P-A
DISPLAY #P-A (EM=HHH)
IF #P-A IS (P5)
  WRITE 'PACKED'
ELSE WRITE 'NOT PACKED'
END-IF
END
```

In the above program, it seems that the numeric is being stored as packed, being re-formatted for the display and then being compared for the 'IS' operator in it's original packed format and is showing as a valid packed.

I believe that the IS option acts identically for N and P. It answers the question "is it legal to convert the given value to the given format?", or at least that is how I have always interpreted it. Since it is legal to convert '123' to P5, it returns true. In other words, the following is legal:

```
COMPUTE #P-N = VAL (#P-A)
```

where #P-N is defined as (P5) for the given value of #P-A. The IS test does not test if field is packed, but tests to see if the VAL(field) function can successfully convert the alphabetic string to a packed field. It is used to avoid a NAT error as follows:

```
1 #DFIELD (A5)
2 #PFIELD (P5)

IF #DFIELD IS (P5)
  ASSIGN #PFIELD = VAL(#DFIELD)
  ELSE REINPUT "FIELD IS NOT NUMERIC"
END IF
```

A tip on numeric checking: I would not rely on MASK function. First, it requires that the entire field is filled. A left-justified alpha field not fully filled will have H'40' in the right most bytes. This will not pass MASK (NNNNN) and it does pass MASK (PPPPP). This is a problem, but you must understand what is means to be 'packable'. If you enter '1234@' into an A5 field, it will fail the MASK numeric test and pass the MASK packed test. It seems that MASK does exactly that, it tests the bytes, whereas the IS operator checks to see if it can be converted successfully. Also, what about decimal numbers? An input of '123.4' fails the mask but passes the IS operator.

XI. Scan Option

The SCAN option can check for a specific value anywhere in a field. The option can be invoked as part of a logical condition criteria. It is NOT positionally oriented. In the following program segment, a "table" of acceptable function values is defined and the SCAN operates against it.

```
0010 DEFINE DATA LOCAL
0020 1 #KEYWORD-LIST(A80) INIT <'.CMD.OPT.EVAL.UTIL.MACRO.'>
0030 1 #FUNCTION-REQUEST (A50)
0040 1 #TOKENS (A10/1:20)
0050 1 SPACE   (A1)
0060 1 I(P2)
0070 END-DEFINE
0080 #FUNCTION-REQUEST = 'CMD OPEN EMPLOYEES GET RECORD 2 DIS AB AC'
0090 *
0100 SEPARATE #FUNCTION-REQUEST INTO #TOKENS (*)
0110  WITH DELIMITER SPACE
0120 *
0130 IF #KEYWORD-LIST = SCAN #TOKENS (1) THEN
0140    WRITE 'LEGAL FUNCTION'
0150 END
```

XII. String Analysis

The statement provides a powerful vehicle for analyzing contents of strings, which are really character arrays, and searching lists. Substring capability is also available by using this function. Chapter 4 discusses in more detail the substring capability and other options for EXAMINE that were new with version 2.2.

Example: Locate a substring within a string.

```
0100 EXAMINE #STR1 FOR #STR2 GIVING POSITION #POS
```

Example: Replace a substring with another string.

```
0100 EXAMINE #STR1 FOR #STR2 REPLACE WITH #STR3
```

Example: Count the number of occurrences of a substring within a string and also determine the length of string 1.

```
0100 EXAMINE #STR1 FOR #STR2 GIVING NUMBER #POS LENGTH #LEN
```

Example: Delete the first occurrence of string 2 within string 1.

```
0100 EXAMINE #STR1 FOR #STR2 DELETE FIRST
```

XIII. Searching Lists

An array can be searched for a particular value or to determine the index position of the value. Here is an array of program names which is searched for an existing name and its index determined.

```
0010 DEFINE DATA LOCAL
0020  1 #PGM-ARRAY (A8/500)
0030  1 #INDEX     (P4)
0040  1 #SEARCH-FIELD (A8)
0050 END-DEFINE

0100 #SEARCH-FIELD = 'TAL0530N'
0110 EXAMINE #PGM-ARRAY(*) FOR #SEARCH-FIELD
0120   GIVING POSITION #INDEX
```

There is one issue for which care must be taken in searching lists. If the search string is shorter than the occurrences, the wrong EXAMINE statement can return unexpected results. In the following example an attempt is made to determine which position in the array contains the word 'A'. The answer that should be returned is four. Only the last EXAMINE statement with the FULL option for both the search field and the array returns the correct count.

```
0010 DEFINE DATA LOCAL
0020 1 #ARRAY (A8/5)
0030 1 #INDEX (P4)
0040 1 #SEARCH-FIELD (A8) INIT<'A'>
0050 END-DEFINE
0060 *
0070 #ARRAY (1) = 'THAT'
0080 #ARRAY (2) = 'MAN'
0090 #ARRAY (3) = 'IS'
0100 #ARRAY (4) = 'A'
0110 #ARRAY (5) = 'FAKER.'
0120 EXAMINE #ARRAY(*) FOR #SEARCH-FIELD
0130   GIVING INDEX #INDEX
0140 WRITE '=' #INDEX
0150 EXAMINE #ARRAY(*) FOR FULL #SEARCH-FIELD
0160   GIVING INDEX #INDEX
0170 WRITE '=' #INDEX
0180 EXAMINE FULL #ARRAY(*) FOR #SEARCH-FIELD
0190   GIVING INDEX #INDEX
0200 WRITE '=' #INDEX
0210 EXAMINE FULL #ARRAY(*) FOR FULL #SEARCH-FIELD
0220   GIVING INDEX #INDEX
0230 WRITE '=' #INDEX
0240 END
```

```
#INDEX:      1
#INDEX:      0
#INDEX:      1
#INDEX:      4
```

Make sure that the search string is defined with a format the same length as the array occurrences, unless you mean to do substring analysis.

Sometimes a question arises as to whether or not every position of an array contains the same value. This is possible to do. Assume an array has length #ARRAY-LEN. Here is the technique for answering this question.

```
0010 DEFINE DATA LOCAL
0020 1 #ARR (A4/1:10)
0030 1 #IND (P2)
0040 1 #NUM (P2)
0050 1 #LEN (P2) INIT <10>
0060 1 #STR (A4) INIT <'**'>
0070 END-DEFINE
0080 *
0090 MOVE  '**'  TO #ARR (*)
0100 MOVE  '***' TO #ARR (1)
0110 EXAMINE FULL #ARR (*) FOR FULL #STR
0120   GIVING NUMBER IN #NUM INDEX #IND
0130 IF #NUM = #LEN THEN DO
0140    WRITE 'ALL OCCURRENCES MATCH'
0150    DOEND
0160 ELSE DO
0170    WRITE 'ALL OCCURRENCES DO NOT MATCH'
0180    DOEND
0190 WRITE '-' #LEN '=' #NUM 'INDEX OF FIRST MATCH:' #IND
0200 END
```

```
ALL OCCURRENCES DO NOT MATCH
#LEN:  10 #NUM:   9 INDEX OF FIRST MATCH:   2
```

I compared the performance of the EXAMINE statement for arrays versus a binary search routine written in NATURAL. The EXAMINE statement outperformed the binary search algorithm. However, it remains to compare the EXAMINE to an assembler routine.

XIV. Sorting Lists

Sorting represents one of the most common programming processes. You may often have the requirement to put a table into sorted order. There are many different sorting algorithms that can be used. The decision to use a particular method depends on both people and machine resources. For people, it is a function of time. For the machine, it is a function of time, resources and space. I will present a few methods here and describe their advantages or disadvantages.

A few words are in order about the code which follows.

First, the routines are coded as subroutines within a program and could have also been coded as subprograms. I leave this as an exercise for the inquisitive programmer. This is important to understand. You can write subroutines with their own local data area block, as long as they are external subroutines and not inline. For inline subroutines, it is my own habit to define data by employing prefixes for variables needed within a subroutine. A hotly contested practice is to "pass" a parameter to a subroutine through the stack. This is certainly not necessary and has a negative affect issuing a page eject for subsequent I/O.

It is also inefficient and is limited to ESIZE minus GDA allocation and PF keys. I have discontinued this practice especially now since NATURAL provides subprogram units.

Comparison of Sort Methods

Tables or files are a list of records, or table entries, with key k(i) which is associated to record r(i). For this list or file to be ordered is to say that for i<j implies that k(i) precedes k(j) in some ordering on these keys.

A sort is internal when the records are sorted in main memory and external when auxiliary storage is used. I am interested in discussing algorithms for internal sorting.

You can either sort the table entries directly or sort an auxiliary table of pointers (i.e., sorting by address). This is important to understand because the overhead in moving a large amount of data can be prohibitive.

To sort or not to sort, that is the question. Circumstances will guide the programmer to make the decision. If the decision is yes, then the method of sorting is also important since no one method is completely superior to all others. There are interrelated efficiency issues for each algorithm.

Whereas some sort algorithms are easy to code, this is no longer necessary with the NATURAL SORT statement. You can view various sort routines in Appendix XX. A sort routine should be both efficient for the situation and correct. There are two basic operations any sort algorithm must perform in order to do what is expected: (1) ask questions about comparative order and; (2) move data. Needless to say, minimizing these operations is what qualifies one sort algorithm as "better" than another. Here is an example which compares the NATURAL SORT and a simple bubble SORT.

```
0010 *
0020 * COMPARISON OF NATURAL SORT VS. IN-CORE BUBBLE SORT USING TAGS (KEYS)
0030 * AND RECORDS
0040 *
0050 DEFINE DATA LOCAL
0060 1 #START-TIME  (A10)
0070 1 #FINISH-TIME (A10)
0080 1 #ET (N7)
0090 1 REDEFINE #ET
0100   2 #ELAPSED-TIME (N6.1)    /* DIFFERENTIAL TO SETTIME
0110 1 #J (P4)
0120 1 LIST-SORTED (L)
0130 1 #MAX-QTY (P4) CONST <98>
0140 1 #ARRAY (#MAX-QTY)
0150   2 #ARRAY-KEY (A10)
0160   2 #ARRAY-RESULT (A2)
0170 1 #INITIAL-ARRAY (#MAX-QTY)
0180   2 #INITIAL-ARRAY-KEY (A10) INIT <
0190 'HOUSTON','NEW YORK', 'BOSTON','CHICAGO', 'MIAMI' ,'DALLAS','PHOENIX',
0200 'HOUSTON','NEW YORK', 'BOSTON','CHICAGO', 'MIAMI' ,'DALLAS','PHOENIX',
0210 'HOUSTON','NEW YORK', 'BOSTON','CHICAGO', 'MIAMI' ,'DALLAS', 'PHOENIX',
0220 'HOUSTON','NEW YORK', 'BOSTON','CHICAGO', 'MIAMI' ,'DALLAS', 'PHOENIX',
0230 'HOUSTON','NEW YORK', 'BOSTON','CHICAGO', 'MIAMI' ,'DALLAS', 'PHOENIX',
0240 'HOUSTON','NEW YORK', 'BOSTON','CHICAGO', 'MIAMI' ,'DALLAS', 'PHOENIX',
0250 'HOUSTON','NEW YORK', 'BOSTON','CHICAGO', 'MIAMI' ,'DALLAS', 'PHOENIX',
0260 'HOUSTON','NEW YORK', 'BOSTON','CHICAGO', 'MIAMI' ,'DALLAS', 'PHOENIX',
0270 'HOUSTON','NEW YORK', 'BOSTON','CHICAGO', 'MIAMI' ,'DALLAS', 'PHOENIX',
0280 'HOUSTON','NEW YORK', 'BOSTON','CHICAGO', 'MIAMI' ,'DALLAS', 'PHOENIX',
0290 'HOUSTON','NEW YORK', 'BOSTON','CHICAGO', 'MIAMI' ,'DALLAS', 'PHOENIX',
0300 'HOUSTON','NEW YORK', 'BOSTON','CHICAGO', 'MIAMI' ,'DALLAS', 'PHOENIX',
0310 'HOUSTON','NEW YORK', 'BOSTON','CHICAGO', 'MIAMI' ,'DALLAS', 'PHOENIX',
0320 'HOUSTON','NEW YORK', 'BOSTON','CHICAGO', 'MIAMI' ,'DALLAS', 'PHOENIX'
0330 >
0340   2 #INITIAL-ARRAY-RESULT(A2) INIT <
0350 'TX', 'NY', 'MA', 'IL', 'FL', 'TX', 'AZ',
0360 'TX', 'NY', 'MA', 'IL', 'FL', 'TX', 'AZ',
0370 'TX', 'NY', 'MA', 'IL', 'FL', 'TX', 'AZ',
0380 'TX', 'NY', 'MA', 'IL', 'FL', 'TX', 'AZ',
0390 'TX', 'NY', 'MA', 'IL', 'FL', 'TX', 'AZ',
0400 'TX', 'NY', 'MA', 'IL', 'FL', 'TX', 'AZ',
0410 'TX', 'NY', 'MA', 'IL', 'FL', 'TX', 'AZ',
0420 'TX', 'NY', 'MA', 'IL', 'FL', 'TX', 'AZ',
0430 'TX', 'NY', 'MA', 'IL', 'FL', 'TX', 'AZ',
0440 'TX', 'NY', 'MA', 'IL', 'FL', 'TX', 'AZ',
0450 'TX', 'NY', 'MA', 'IL', 'FL', 'TX', 'AZ',
0460 'TX', 'NY', 'MA', 'IL', 'FL', 'TX', 'AZ',
0470 'TX', 'NY', 'MA', 'IL', 'FL', 'TX', 'AZ',
0480 'TX', 'NY', 'MA', 'IL', 'FL', 'TX', 'AZ'
```

```
0490 >
0500 1 #TEMP-KEY    (A10)
0510 1 #TEMP-RESULT (A2)
0520 1 #KEY         (A10)
0530 1 #RESULT      (A2)
0540 1 #EL-QTY      (P4)
0550 1 #TEMP-CTR    (P4)
0560 1 #OUTER-CTR   (P4)
0570 1 #INNER-CTR   (P4)
0580 1 #SECOND-LAST (P4)
0590 END-DEFINE
0610 PERFORM MODIFIED-BUBBLE-SORT
0620 *
0630 MOVE #INITIAL-ARRAY-KEY(*) TO #ARRAY-KEY(*)
0640 MOVE #INITIAL-ARRAY-RESULT(*) TO #ARRAY-RESULT(*)
0650 *
0660 USING. SETTIME
0670 ASSIGN #START-TIME = *TIME
0680 FOR #TEMP-CTR = 1 TO #EL-QTY
0690   MOVE #ARRAY-KEY(#TEMP-CTR) TO #KEY
0700   MOVE #ARRAY-RESULT(#TEMP-CTR) TO #RESULT
0710 END-ALL
0720   SORT #KEY ASC
0730       USING #RESULT
0740     ADD 1 TO #J
0750     MOVE  #KEY TO #ARRAY-KEY (#J)
0760     MOVE  #RESULT TO #ARRAY-RESULT (#J)
0770   END-SORT
0780 ASSIGN #ET = *TIMD (USING.)            /* NATURAL'S TIME DIFFERENTIAL
0790 ASSIGN #FINISH-TIME = *TIME
0800 WRITE   'TIMING FOR NATURAL SORT'
0810 WRITE   '======================'
0820 WRITE   'STARTING TIME:  ' #START-TIME
0830       / 'ELAPSED TIME:   ' #ELAPSED-TIME (EM=ZZZZZ.99) 'SECONDS.'
0840       / 'FINISHING TIME: ' #FINISH-TIME
0850 *
0860 FOR #J = 1 TO #EL-QTY
0870   WRITE #ARRAY-KEY (#J) #ARRAY-RESULT (#J)
0880 END-FOR
0890 STOP
0900 *
0910 DEFINE SUBROUTINE MODIFIED-BUBBLE-SORT
0920 *                  ********************
0930 MOVE #MAX-QTY TO #EL-QTY
0940 MOVE #INITIAL-ARRAY-KEY(*) TO #ARRAY-KEY(*)
0950 MOVE #INITIAL-ARRAY-RESULT(*) TO #ARRAY-RESULT(*)
0960 *
0970 ASSIGN #START-TIME = *TIME
0980 COMPUTE #SECOND-LAST = #EL-QTY - 1
0990 BUBBLE. SETTIME
1000 FOR #OUTER-CTR = #SECOND-LAST TO 1 STEP -1
1010   ASSIGN LIST-SORTED = TRUE
1020   FOR #INNER-CTR = 1 TO #OUTER-CTR
1030     IF #ARRAY-KEY(#INNER-CTR) > #ARRAY-KEY(#INNER-CTR + 1)
1040       MOVE #ARRAY-KEY(#INNER-CTR) TO #TEMP-KEY
1050       MOVE #ARRAY-RESULT(#INNER-CTR) TO #TEMP-RESULT
1060       MOVE #ARRAY-KEY(#INNER-CTR+1) TO #ARRAY-KEY(#INNER-CTR)
1070       MOVE #ARRAY-RESULT(#INNER-CTR+1) TO #ARRAY-RESULT(#INNER-CTR)
1080       MOVE #TEMP-KEY TO #ARRAY-KEY(#INNER-CTR + 1)
1090       MOVE #TEMP-RESULT TO #ARRAY-RESULT(#INNER-CTR + 1)
```

```
1100        ASSIGN LIST-SORTED = FALSE
1110      END-IF
1120    END-FOR
1130    IF LIST-SORTED
1140       ESCAPE BOTTOM
1150    END-IF
1160 END-FOR
1170 ASSIGN #ET = *TIMD (BUBBLE.)           /* NATURAL'S TIME DIFFERENTIAL
1180 ASSIGN #FINISH-TIME = *TIME
1190 WRITE   'TIMING FOR IN-CORE MODIFIED BUBBLE SORT'
1200 WRITE   '======================================='
1210 WRITE   'STARTING TIME:   ' #START-TIME
1220       / 'ELAPSED TIME:    ' #ELAPSED-TIME (EM=ZZZZZ.99) 'SECONDS.'
1230       / 'FINISHING TIME: ' #FINISH-TIME /
1240 FOR #J = 1 TO #EL-QTY
1250   WRITE #ARRAY-KEY (#J) #ARRAY-RESULT (#J)
1260 END-FOR
1270 RESET #J
1280 END-SUBROUTINE
1290 END
```

The amount of time necessary to complete the sort is proportional to the number of key comparisons and the number of moves required. The methods presented here fall into two groups. If a list is of length n, the time required by these methods is proportional to n ** 2 (i.e., n squared) or n log(n). This criteria is not a sole reason for choosing a particular sort method. Also, the nature of the list itself (e.g., is it partially sorted, inversely sorted , etc.) affects the decision.

Chapter 4

NATURAL 2 Enhancements

I. New / Enhanced Statements

There are a few new statements as well as extensions to existing statements. The Software AG manuals superbly explain these statements. I will however to list them and very briefly describe them. Also, for some I will describe when and how to use them.

A. Program Communication / Interaction

CALL (enhanced)

This statement invokes a non-NATURAL external routine. The parameter list may have up to forty items. Standard linkage conventions are used. Any group item in the parameter list is expanded into its individual items. This can have an undesirable side effect upon entering the called routine and returning. Alignment within the data area is the programmer's responsibility and follows the rules of NATURAL data alignment. After successfully returning from the called routine, the calling program returns control to the next statement after the CALL statement.

The return code is available through register 15 and is obtainable using the NATURAL system function RET. NATURAL under CICS expects the return code to be posted in CICS TWA (transaction work area) or CICS COMMAREA. If you expect to load the return code into register 15, forget it. CICS destroys register 15 (thank you Darrell Davenport of Washington State University).

A SAG-L user asked if one can write an Assembler program to be called by NATURAL programs in both a batch and CICS environment. The example they had in mind involved creating a 8 byte unique token based on the STCK time. An international SAG-L participant suggested the following:

"Use "SET CONTROL 'P=S'". Coding this before each call to the subroutine causes NATURAL to pass control to the subroutine with a BALR 14,15 branch instead of the normal EXEC CICS LINK. Additionally, register conventions will be identical to those of MVS/Batch or MVS/TSO (parameter list pointed to by R1, return register R14, savearea pointed to by R13, etc.). Register 12 still points to the BB (NATURAL User Buffer) if you need that. You don't have to use the DFHEIENT prolog either. Of course, this works only for a routine that contains no EXEC CICS statements. Also, there is still the catch that the subroutine must be defined in the PPT and contained in a DFHRPL library (it is loaded by NATURAL via an EXEC CICS LOAD before branching to it via BALR).

In short, use SET CONTROL 'P=S' and you can use the identical routine in CICS, TSO, and batch, without statically linking it to the NATURAL nucleus. By the way, NATURAL/TSO and NATURAL/Batch will treat SET CONTROL 'P=S' as a NO OP; no need to code IF *TPSYS NOT = 'CICS' ".

CALLNAT

This statement provides the capability to invoke a NATURAL subprogram. Parameters can be described as read-only (AD=O) or can be modified by the subprogram (AD=M). The position of the parameters is important because the corresponding elements in the parameter data area of the subprogram must agree in type, format and length. If you pass a group name, NATURAL replaces it with the group items. In example two below, you see that an element of an array and an entire array is passed. NATURAL passes by address, but it also checks type, format, and length. The subprogram is in either the application library or the steplib.

```
0240 CALLNAT 'HONS006'  #HALL  #ROOM  #YR-SEM
0250 CALLNAT 'MBUBSRT5' #L #STATE (1) #IND (*)
0260 CALLNAT 'SUBR001'  #PARM1 (AD=O) #PARM2 (AD=M)
```

A subprogram (or map) object name ending with an ampersand (&) may be dynamically specified when invoked. The last character of the object name in the invoking module is replaced by the value of the NATURAL System Variable *LANGUAGE at execution (English = 1, German = 2). This means that the statement - CALLNAT 'SUBTEST&' - invokes SUBTEST1 or SUBTEST2 or SUBTEST3 or whatever the value of *LANGUAGE has ben set to prior to execution of the CALLNAT.

A variable passed with the attribute 'AD=O' can be changed at lower levels. This is required in NATURAL 2.2.8 for variables with CONST definition (it is my personal belief that CONST variables are not too useful). If PGM1 CALLNATs SUBR1 and passes a parameter with 'AD=O', and SUBR1 CALLNATs SUBR2 without 'AD=O' specification, then SUBR2 can modify it and the new value appears in the PGM1 block. You must, therefore, specify the attribute at every CALLNAT level if you do not want the variable modified outside the parent block.

The statement "CALLNAT #SUBR-NAME #PARM1 #PARM2" causes cross reference to refer to this as a dynamic call. The can present problems during impact analysis. However, my good friend Al Blakey from the State of Vermont uses the following technique to get the best of both worlds:

```
DEFINE SUBROUTINE #DYNAMIC-CALLS
DECIDE FOR FIRST CONDITION
  WHEN #WHICH = 'ABC'
    CALLNAT 'ABC' PARM1 PARM2
  WHEN #WHICH = 'PQR'
    CALLNAT 'PQR' PARM1 PARM2
  WHEN #WHICH = 'XYZ'
    CALLNAT 'XYZ' PARM1 PARM2
  WHEN NONE
    IGNORE
END-DECIDE
END-SUBROUTINE
```

Arrays may be passed to a subprogram for a varying or dynamic index. This is described in more detail in section C in this chapter.

DEFINE SUBROUTINE

DEFINE SUBROUTINE allows you to define a program block internal/external to a NATURAL program. These statements are executed via the PERFORM statement. The data available to an external subroutine is the global data area of the invoking program module or data defined in a local DEFINE DATA block or stacked data.

FETCH / FETCH RETURN

This statement allows for a NATURAL program to be loaded and executed. The program is written like a main program;however, the RETURN option allows for the program to be handled like a combination of a subprogram and a subroutine - it can be passed parameters and it has access to the GDA. The FETCH RETURN mechanism increments the level in which you are processing. Control is released to the fetching program when an END or ESCAPE ROUTINEis encountered.

A FETCHed object does not change the level at which programs execute. Data is available to the FETCHed module either via the NATURAL stack or the global data area. Any parameters listed with the FETCH statement are placed on top of the stack. They are accepted by the first INPUT statement encountered in the fetched module.

INCLUDE

This statement allows you to include source from an external member of type COPYCODE. The object is included at compile time. However, you can not use copy code for entries in DEFINE DATA. A change in an INCLUDE segment requires a recompile. This statement provides a means to standardizing in-line code.

INCLUDE may specify up to 99 insertion values to be inserted into COPYCODE object dynamically at compile time. Each value is enclosed in a set of apostrophes when specified. The COPYCODE invoked by the INCLUDE must have an equal number of substitution target variables. COPYCODE accepts values into "&n&" variables where "n" corresponds to variables from INCLUDE list: "&1&"references the first variable in the list, "&6&" references the sixth and so forth.

```
INCLUDE PRINTJOB 'JOB-TITLE' 'DEPT'
```

requires a target variable for each parameter:

```
WRITE &1& &2&
```

which is translated at compile into:

```
WRITE JOB-TITLE DEPT
```

Literals and user-defined variables may also be included:

```
INCLUDE PRINTJOB '''ACCOUNTANT''' '#DEPT'
```

Any variable may be used more than once if needed. The following sample COPYCODE requires the invoking CALLNAT to pass a minimum of 6 argument variables

```
READ EMPLOYEE BY &1&
  STARTING FROM &2&
  IF &1& GT &3&
    ESCAPE BOTTOM
  END-IF
  DISPLAY &1& &4& &5& &6&
END-READ
```

The following CALLNAT:

```
CALLNAT PRINTJOB 'JOB-TITLE' '''ACCOUNTANT'''
  '''AZ''' 'DEPT' 'SALARY(1)' 'CITY'
```

translates the COPYCODE at compile into:

```
READ EMPLOYEE BY JOB-TITLE
STARTING FROM 'ACCOUNTANT'
  IF JOB-TITLE GT 'AZ'
    ESCAPE BOTTOM
  END-IF
  DISPLAY JOB-TITLE DEPT SALARY(1) CITY
END-READ
```

PERFORM

NATURAL allows for a subroutine (not to be confused with a subprogram) to be either inline or external. If it is inline, all of the data available to the same program module is available to the subroutine. If it is external, only the data in the GDA of the program module is available to the subroutine or any stacked data. However, local data can be defined to an external subroutine by writing the DEFINE DATA statement at the top of the subroutine.

Subroutines can be nested, i.e., a PERFORM statement can be executed from within a subroutine. NATURAL supports up to 15 deep; however, the number is restricted by the amount of memory required. I try not to design with more than three to five.

STACK

This statement allows for data, commands, and the names of NATURAL programs to be executed to be placed upon the NATURAL stack. However, not all alphanumeric data can be placed on the stack.

If you write a standard STACK TOP DATA statement, then you must be careful about data which includes the ID character (usually defaults to ','), the CF character - '%', or the IA character - '='. In fact, low values - H'00' - and H'3F' are changed by the stack facility to H'40', or blanks, if they appear as trailing bytes in a string which is placed onto the stack. More problematic is that if the string has H'3F' anywhere in the string, all bytes to the right of it will also be converted to H'40'.

You can make NATURAL ignore the translation of bytes of data on the stack by using the FORMATTED option in the STACK statement. However, even the FORMATTED option does not prevent the stack from modifying trailing nulls. The situation where I saw this to be a problem was in a program which stacked ISN values as binary data redefined to be alphanumeric data. Needless to say the receiving program mistakenly referenced another record since the value is modified. This situation is corrected, however, by converting the value to be numeric before stacking or simply do not redefine the data.

*DATA is a variable through which NATURAL can provide a program valuable information about the stack. *DATA reveals what is actually on top of the stack. If *DATA < 0, a command is next on top of the stack. If *DATA = 0, the stack is empty. If *DATA > 0, the number reflects the number of items that are on the stack. Unfortunately, you can not find out exactly what is on the stack positionally nor remove or execute positionally except top or bottom.

Another SAG-L participant noticed that when a NATURAL program unexpectedly crashes - division by zero works - the command stack gets cleared.

If NATURAL detects an error, it first looks for an ON ERROR clause to handle the problem. If that fails, it clears the stack and puts the error information in the stack (prog-name, err-no, err-line...). Then it looks for the name of an error handler program in the *ERROR-TA variable. If that fails too it returns to the NEXT prompt to report the error.

As an example, a program can look like this:

```
DEFINE DATA
  LOCAL
  01 #N (N1)
END-DEFINE

ON ERROR                                /* detect before stack is cleared
  IF *ERROR = 1302                      /* If we should continue w/stack
     STOP                               /*    do so
  ELSE
     WRITE 'ERROR' *ERROR 'DETECTED'    /* otherwise handle error

WRITE NOTITLE *PROGRAM 'testing div by zero'
ASSIGN #N = 5 / #N
END
```

B. Loop Processing (non-ADABAS)

ESCAPE

This statement allows you to exit the current loop using option keywords TOP, BOTTOM, BOTTOM IMMEDIATE, ROUTINE, ROUTINE IMMEDIATE. The IMMEDIATE option bypasses any end of loop processing. You can provide a line number or label reference.

REPEAT

This statement has two mechanisms for testing a logical condition for terminating the loop - UNTIL and WHILE. These keywords can be placed before or after the statements performed in the loop. Placement of the UNTIL / WHILE clause after the statement(s) in the loop causes their mandatory execution. This is a good feature since some may have programming experiences with other languages which have the (until) option at the end. The difference is slight, but subtle. The loop terminates with the UNTIL option when the logical condition is true. For the WHILE option the loop terminates when the condition is false.

```
0010 DEFINE DATA
0020   LOCAL
0030 1 I       (P2)
0040 1 #BUFFER (A80) INIT
0050  <'This is a Test of the EFFicienCY of UPPER/lower translation'>
0060 1 REDEFINE #BUFFER
0070   2 #CHAR (A1/1:80)
0080 1 #UPPER-LETTERS  (A26) INIT <'ABCDEFGHIJKLMNOPQRSTUVWXYZ'>
0090 1 REDEFINE #UPPER-LETTERS
0100   2 #UPPER (A1/1:26)
```

```
0110 1 #LOWER-LETTERS  (A26) INIT <'abcdefghijklmnopqrstuvwxyz'>
0120 1 REDEFINE #LOWER-LETTERS
0130   2 #LOWER (A1/1:26)
0140 END-DEFINE
0150 *
0160 WRITE (LS=82) 'STRING BEFORE CONVERSION:' / #BUFFER
0170 REPEAT
0180  WHILE #CHAR (*) = MASK (L)
0190    ADD 1 TO I
0200    EXAMINE #BUFFER FOR #LOWER (I) REPLACE WITH #UPPER (I)
0210 LOOP
0220 WRITE (LS=82) 'CONVERTED STRING:' / #BUFFER
0230 END
```

```
Page      1                                              97-04-01  12:34:56
STRING BEFORE CONVERSION:
This is a Test of the EFFicienCY of UPPER/loWer translation
CONVERTED STRING:
THIS IS A TEST OF THE EFFICIENCY OF UPPER/LOWER TRANSLATION
```

At this point, I want to share an interesting observation. After conversing with friend Lew Levy, a past chairperson for the NATURAL Tips & Techniques SIG, I was convinced this routine would perform best. However, to my surprise, I tested the REPEAT...WHILE against my favorite and after 100,000 iterations it took 56% more time than the one listed below.

```
0200 EXAMINE #BUFFER FOR 'a' REPLACE WITH 'A'
0210 EXAMINE #BUFFER FOR 'b' REPLACE WITH 'B'
0220 EXAMINE #BUFFER FOR 'c' REPLACE WITH 'C'
                          ...
0440 EXAMINE #BUFFER FOR 'y' REPLACE WITH 'Y'
0450 EXAMINE #BUFFER FOR 'z' REPLACE WITH 'Z'
```

However, if you use a REPEAT WHILE or REPEAT UNTIL you must assure an exit point for the loop. Real programmers don't create infinite loops. Here is a harmless example to show the point.

```
0010 DEFINE DATA LOCAL
0020 1 I     (P4) INIT <1>
0030 1 #ARR (N4/1:500)
0040 END-DEFINE
0050 *
0060 #ARR (102) := 50
0070 REPEAT WHILE #ARR (I) = 0  /* FIND FIRST NON-ZERO ENTRY
0080   ADD 1 TO I
0090 END-REPEAT
0100 *
0110 WRITE '=' I
0120 END
```

```
Page      1                                              97-04-01  12:34:56
I:   102
```

Now remove line 0060 and a situation exists where the loop is endless. Modifying the REPEAT loop to check the index is possible,

```
0060 REPEAT WHILE #ARR (I) = 0 AND I < 500
```

C. Data Definition / Manipulation

COMPRESS

One can think of COMPRESS as an opposite function of SEPARATE. The contents of two or more fields can be combined into one target field.The 'LEAVING NO SPACE' option removes all trailing blanks or leading zeroes. Otherwise, only one space is included unless the new 'WITH DELIMITERS' option is used.

In this case, NATURAL provides either the INPUT DELIMITER defined for the session or any delimiter provided in the statement. The source field and target field can be the same.

```
0060 COMPRESS CITY STATE ZIP INTO #ADDR-LINE4
0070 COMPRESS #A #B INTO #A USING ':'
```

What if you want to compress alpha fields for downloading as a comma delimiter file, and some alpha fields may be blank? What if you wish to compress numbers with decimal points and negative signs?

COMPRESS FULL will certainly give you commas when one of the fields being compressed is blank. Example:

```
COMPRESS FULL #A #B #C #D INTO #E WITH DELIMITER ','
```

To include numeric signs and decimal points, try this:

```
    0010 DEFINE DATA
  0020 LOCAL
  0030 01 BUFFER (A6)
  0040 01 #NUM (N5) INIT <-100>
  0050 01 #SIGN-1 (A1)
  0060 END-DEFINE
  0070 IF #NUM < 0
  0080    MOVE '-' TO #SIGN-1
  0090    COMPRESS #SIGN-1 #NUM INTO BUFFER LEAVING NO SPACE
  0100 END-IF
0110 WRITE '=' BUFFER
  0120 END
```

In general, if you want to Compress with a numeric field you might want to also consider a separate MOVE EDITED to take care of minus signs (and plus signs if the user wants) and decimal points.

DEFINE DATA

Other than comments, this statement must be the first in a structured mode program. If it is optionally used in report mode, then it must also be the first statement.

```
0010 DEFINE DATA
0020    GLOBAL    USING GDA-NAME
0030    PARAMETER USING PDA-NAME
0040    PARAMETER
0050 1 #P-EMP-ID       (A8)
0060    LOCAL     USING LDA-NAME
0070    LOCAL
0080 1 #FIELD1         (A10)
0090 1 REDEFINE #FIELD1
0100   2 #FILLER1      (A8)
0110   2 #SUB-PART     (A2)
0120 1 #FIELD2         (L) INIT <FALSE>
0130 END-DEFINE
```

The Global Data Area can only be defined in the Data Editor, whereas the others can be defined either inline or in the Data Editor. The Global Data Area must be stowed in the same application library for which it is invoked. A more complete description of each data area - global, parameter, local - is in the next section.

Level 1 data elements are stored on a word boundary. They can also serve as a group name. Two different groups can have elements with the same name. However, references to these names must be qualified by prefixing the name with an index of 'group-name'. References to doubly defined database names may be qualified with line numbers or labels.

Keep group level names reasonably short, since they take up room in the symbol table for the program. At compile time, the symbol table resides in FSIZE.

```
LISTING OF LOCAL01
1 GROUP-ONE-LEVEL-NAME
2 #NAME1       A   10
2 #NAME2       P    4
2 #NAME3       A   10 (1:100)
2 #GROUP-TWO-LEVEL-NAME
3 #NAME5       A    4
3 #NAME6       A   20 (1:10,1:10)
2 #NAME7       A    1
1 #NAME8       A  250
1 #NAME9       A  100
```

Size in source: 429 BYTES
Size in USIZE buffer: 3368 BYTES
Size in buffer pool: 210 BYTES

```
LISTING OF LOCAL02
1 #G1
2 #NAME1       A   10
2 #NAME2       P    4
2 #NAME3       A   10 (1:100)
2 #G2
3 #NAME5       A    4
3 #NAME6       A   20 (1:10,1:10)
2 #NAME7       A    1
1 #NAME8       A  250
1 #NAME9       A  100
```

Size in source: 394 BYTES
Size in USIZE buffer: 3368 BYTES
Size in buffer pool: 175 BYTES

The INIT keyword assigns an initial value. The CONST keyword not only initializes but establishes a value which cannot be changed programmatically. The ALL keyword assigns the same value to all occurrences of an array.

```
1 #FIELD3        (P2/2,2) INIT (2,V)  <10,200>
1 #FIELD4        (N2/2,2) INIT (2,V)  <10,200>
1 #FIELD5        (N2/2,2) INIT ALL    <200>
```

In the above example, there are three intentional errors. First, the definitions for #FIELD4 and #FIELD5 will CHECK/STOW in NATURAL 2, but a run-time error of 1305 occurs. On the other hand, the definition of #FIELD3 will have a compile time error of NAT0094

Three additional parameters definable in this block are EM (edit mask), HD (column header defined in apostrophes), and PM (print mode).

FILLER nX provides filler bytes in a DEFINE DATA statement. This is valid for both database and user-defined variables:

```
DEFINE DATA LOCAL
1 STAFF VIEW OF EMPLOYEES
  2 PERSONNEL-ID
  2 NAME
  2 BIRTH
  2 REDEFINE BIRTH
    3 FILLER 2X
    3 #MONTH(N2)
1 #TODAYS-DATE(D) INIT <*DATX>
1 #A(A100)
1 REDEFINE #A
  2 #A1(A10/10)
  2 REDEFINE #A1
    3 #A1A (10)
     4 FILLER 1X
     4 #A1B(A5)
     4 FILLER 4X
END-DEFINE
MOVE 'ABCEDFGHIJKLMNOPQRSTUVWXYZ0123456789' TO #A
DISPLAY #TODAYS-DATE #A1(*) #A1B(*)
END
```

NATURAL System Variables may now be used for initialization (#TODAYS-DATE set to the System Variable *DATX).

Field #A1 is redefined twice with filler options specified to ignore the first five bytes of the field in the redefinition. Also, position 11, 12,and 13 are ignored. The second redefine sets up the data as an array and attaches an edit mask to each element.

FULL LENGTH will fill an alphanumeric field with the character(s)specified. A single character provides the same results as the ALL clause. When multiple characters are specified, the operation functions like the MOVE ALL statement.

LENGTH *n* fills the first *n* number of positions of the field with the character(s) specified.

```
DEFINE DATA LOCAL
1 #MYDATE(D) INIT <*DATX>
1 #F1(A25) INIT FULL LENGTH <'*'>
1 #F2(A25) INIT FULL LENGTH <'{..} '>
1 #F3(A25) INIT LENGTH 5 <'*'>
1 #F4(A25) INIT LENGTH 5 <'{.}'>
WRITE #MYDATE (EM=N(9),ÂL(9)ÂZD) /
DISPLAY #F1 #F2
DISPLAY T*#F1 #F3 #F4
END
```

Arrays can be passed to a subprogram with a variable number of occurrences using the index notation "1:V".

```
DEFINE DATA
LOCAL
1 #ARRAY(A5/5,10) INIT ALL <'9:99'>
1 #I(P2)
1 #J(P2)
END-DEFINE
MOVE 3 TO #I
MOVE 5 TO #J
CALLNAT 'SUBPGM' #ARRAY (1:#I,1:#J) #I #J
* DISPLAY ARRAY AFTER BEING MODIFIED IN THE SUBPROGRAM
DISPLAY #ARRAY(1,*) #ARRAY(2,*) #ARRAY(3,*)
  #ARRAY(4,*) #ARRAY(5,*)
END
```

You are allowed to specify variable ranges in NATURAL 2.2 as shown in the program example above. The Parameter Data Area needs to be coded using specific syntax to accept the passed array element values.

The first program in this example passes subscripts for each level of array. The DISPLAY statement dumps the entire array after execution of the subprogram illustrating the syntax necessary for this technique and the actual data returned to the program.

```
* SUBPROGRAM: SUBPGM
DEFINE DATA PARAMETER
1 #A(A5/1:V,1:V)
1 #I(P2)
1 #J(P2)
LOCAL
1 #1(P2)
1 #2(P2)
END-DEFINE
DISPLAY #A(1:#I,1:#J) /* DISPLAY EVERYTHING PASSED
FOR #1 1 #I
  FOR #2 1 #J
    COMPRESS #1 ':' #2 INTO #A (#1,#2) LEAVING NO
  END-FOR
END-FOR
NEWPAGE
END
```

The key to the solution are the values passed to #I and #J. When the array is passed back and displayed in the calling program, only the specified subset of occurrences are modified.

DIVIDE

The DIVIDE statement has a ROUNDED option which allows for the result to be stored to the precision of the receiving field. Also, the new REMAINDER option allow for the remainder to be stored. If ZD=ON is set, then an error message will appear upon attempting to divide by zero; if ZD=OFF, a result of 0 is returned.

```
0240 DIVIDE 5 INTO 47 GIVING #ANS REMAINDER #REM
```

EXAMINE

This statement has the flexibility of scanning target fields for strings and replacing, deleting and counting them. It is extremely useful for searching arrays.

A detailed explanation of this is in the section under searching strings. Remember that the GIVING options are positional. The order for the GIVING clause is NUMBER, POSITION, LENGTH and INDEX. Therefore, if you want the length of a string and the number of occurrences of a substring (more ob substring in the discussion of MOVE) in the same examine operation, you must write

```
0240 EXAMINE #BUFFER FOR #SUBSTRING
0250   GIVING NUMBER #SUB-COUNT LENGTH #LEN
```

and not

```
0240 EXAMINE #BUFFER FOR #SUBSTRING
0250   GIVING LENGTH #LEN NUMBER #SUB-COUNT
```

Another SAG-L user asks, when using :

```
EXAMINE FULL #STRING FOR #SEARCH REPLACE #REPLACE
```

with

```
#STRING(A75) , #SEARCH(A15), #REPLACE(A15)
```

NATURAL 2.2.6 (MVS) will replace #SEARCH with #REPLACE only if #SEARCH is greater than or equal to #REPLACE in length. If #REPLACE is longer than #SEARCH one gets a NAT1308 error. Is there an elegant solution to this problem?

Yes there is, but not quite like many SAG users on SAG-L suggested.

First, the EXAMINE ... REPLACE stops at the first occurrence unless you specify the FIRST option, or improperly use a FULL option. Secondly, the NAT1308 error will occur under a specific situation. NATURAL does not allow variable length fields and the EXAMINE statement cannot force a replacement that will allow the resultant string to be longer than its original length. An EXAMINE ... REPLACE where the actual length of #REPLACE is longer than #SEARCH is possible, but you must understand how the FULL option is employed in the context of EXAMINE...REPLACE. Study the following example and I think anyone will understand the inner workings of the EXAMINE statement.

```
DEFINE DATA LOCAL
1 LINE (A24)
1 SRCH-STR (A3)
1 REPL-STR (A3)
END-DEFINE
*
SRCH-STR := 'AB'
REPL-STR := 'DEF'
MOVE ALL 'AB ' TO LINE
WRITE '=' LINE / LINE (EM=H(48))
*
EXAMINE FULL LINE FOR FULL
   SRCH-STR
     WITH DELIMITERS
        REPLACE WITH REPL-STR
WRITE '=' LINE
*
MOVE ALL 'AB ' TO LINE
EXAMINE LINE FOR FULL
SRCH-STR
  WITH DELIMITERS
     REPLACE WITH REPL-STR
WRITE '=' LINE
*
MOVE ALL 'AB ' TO LINE
EXAMINE LINE FOR FULL
SRCH-STR
  REPLACE WITH REPL-STR
WRITE '=' LINE
*
MOVE ALL 'AB ' TO LINE
EXAMINE FULL LINE FOR FULL
SRCH-STR
  REPLACE WITH REPL-STR
WRITE '=' LINE
END
```

```
Page      1                                          97-04-01  12:34:56

 LINE: AB AB AB AB AB AB AB AB
 C1C240C1C240C1C240C1C240C1C240C1C240C1C240C1C240
 LINE: AB AB AB AB AB AB AB DEF
 LINE: AB AB AB AB AB AB AB AB
 LINE: DEFDEFDEFDEFDEFDEFDEFAB
 LINE: DEFDEFDEFDEFDEFDEFDEFDEF
```

Here is another application of EXAMINE - answering the question if something is on a list or not.

```
0010 DEFINE DATA LOCAL
0020 1 OK (P2)
0030 1 USER-ID (A8)
0040 1 SPECIAL-USER (L)
0050 1 PROCESS-ERR-MSG (A60) INIT
0060   <'You are not authorized for this function.'>
0070 1 SPECIAL-USERS-LIST (A8/1:12)
0080   INIT <'APDDAD','APDDBA','DSPSPH','APDDB2','APPDRO',
0090          'APPPHB','APPDS1','APPASB','ISDMUV','ISDJSC'>
0100 END-DEFINE
0110 PERFORM INITIALIZE-FIELDS
0120 *
0130 IF SPECIAL-USER THEN
0140    IGNORE
0150 ELSE
0160   WRITE USER-ID PROCESS-ERR-MSG
0170   STOP
0180 END-IF
0190 *
0200 STOP
0210 * EJECT
0220 DEFINE SUBROUTINE INITIALIZE-FIELDS
0230 USER-ID := *INIT-USER
0240 EXAMINE SPECIAL-USERS-LIST (*) FOR USER-ID
0250  GIVING NUMBER OK
0260 *
0270 IF OK > 0 THEN
0280    SPECIAL-USER := TRUE
0290 ELSE
0300    SPECIAL-USER := FALSE
0310 END-IF
0320 *
0330 END-SUBROUTINE
0340 *
0350 END
```

EXAMINE has been expanded with the addition of substring capabilities and the ability to search for patterns of characters within the field value. SUBSTRING examincs a portion of afield beginning at the first position of the field or a user-definedstarting position and scans left to right for the defined length.

```
DEFINE DATA LOCAL
1 #VARIABLE (A40) INIT <'COMMANDCOMPOSECOMPRESS'>
1 #NUMBER(I2)
1 #POSITION(I2)
END-DEFINE
EXAMINE SUBSTRING(#VARIABLE,1,3) FOR 'COM'
  GIVING NUMBER #NUMBER POSITION #POSITION
WRITE #VARIABLE / #NUMBER / #POSITION
END
```

Here is how to scan the argument field beginning at the first position for three positions for the value "COM". Processing elements of an array are handled in the same manner.

```
DEFINE DATA LOCAL
1 #VARIABLE (A10/12)
    INIT <'COMMAND','COMPOSE','COMPRESS','COMPUTE',
      'CONDITION','CONSTANT','CONTROL','COPIES',
      'COPY','COUNT','COUPLED'>
1 #NUMBER(I2)
END-DEFINE
EXAMINE SUBSTRING(#VARIABLE(*),1,3) FOR 'COM'
  GIVING NUMBER #NUMBER
WRITE #VARIABLE(*) / '=' #NUMBER
END
```

PATTERN provides the capability to test the target of an EXAMINE and exclude selected positions from the test. Its properties are analogous to the MASK option. A period, a question mark, or an underscore indicate a position is to be excluded from the evaluation.

```
EXAMINE #VARIABLE FOR PATTERN 'CO.P'
  GIVING NUMBER #COUNT /* COUNT SHOULD EQUAL 4
```

An asterisk or a percent sign indicate any number ofpositions are to be ignored.

```
EXAMINE #VARIABLE FOR PATTERN 'CO*P'
  GIVING NUMBER #COUNT /* COUNT SHOULD EQUAL 6
```

Pattern Symbols

%	Ignore multiple positions
*	Ignore multiple positions
.	Ignore a position
?	Ignore a position
_	Ignore a position

PATTERN characters may be combined into several configurations.

```
EXAMINE SUBSTRING(JOB-TITLE,1,4) FOR 'PROG' GIVING #COUNT
```

This EXAMINE example utilizes the SUBSTRING function to evaluate the first 4 positions of JOB-TITLE for the character string"PROG".

```
EXAMINE SUBSTR(JOB-TITLE,1,4) FOR 'PROG'
  REPLACE 'ANAG' GIVING NUMBER #NUMBER
```

This example program builds on the previous example adding a REPLACE clause that changes each JOB-TITLE with "PROG" in the first 4 positions to "ANAG", first 4 positions.

```
EXAMINE JOB-TITLE FOR PATTERN 'PROG*ER'
  GIVING NUMBER #NUMBER
```

The PATTERN capability provides a wider flexibility in searches for mixed values in the target field. This example scans the value of each JOB-TITLE for "PROG" followed by "ER" with any number of characters in between the two search values.

```
EXAMINE JOB-TITLE FOR PATTERN 'PROG......%ER'
  GIVING NUMBER #NUMBER
```

In the above, program JOB-TITLE is being tested for"PROG" followed by any six characters followed by"ER" with any number of character in between the two search values (the first is"PROG......"; the second is "ER").The "ER" must be found anywhere after the first 10 characters in JOB-TITLE.

```
EXAMINE JOB-TITLE FOR PATTERN 'PROG%ER......'
  GIVING NUMBER #NUMBER
```

In thus program example, the search criteria tests for "PROG" followed by any number of characters and then "ER......". The "ER" in this example must be followed by at least six characters for the condition to return a value in the GIVING clause.

```
EXAMINE JOB-TITLE FOR PATTERN '....PROG%ER'
  GIVING NUMBER #NUMBER
```

This PATTERN option now scans JOB-TITLE for any four characters followed by "PROG" then any number of characters followed finally by "ER". In other words, there has to be four characters preceding the characters"PROG" in addition to the other criteria.

```
EXAMINE JOB-TITLE FOR PATTERN '....PROG'
  GIVING NUMBER #NUMBER
```

The ignored positions again come into play during this EXAMINE PATTERN test. "PROG" preceded by a minimum of four characters are the only JOB-TITLEs accepted by this criteria.

EXAMINE TRANSLATE is used to translate the characters in a field into upper case, lower case, or into other characters in your own table of values.

The entire field is translated as specified unless SUBSTRING function has been specified translating only of a portion of afield. The INTO UPPER CASE option translates the argument field in its entirety into upper case characters; the INTO LOWER CASE option translates the argument field in its entirety into lower case.

The translation table can be used for one-to-one character translation. Each table element must have a format and length of A2 or B2. The first half of the element contains the target character while the second half contains the translation value. Execution of EXAMINE TRANSLATE USING will replace each occurrence of the argument field found in the table with the translation value from the table.

Periodically, users ask the question about case translation, especially in environments that do not yet support EXAMINE TRANSLATE. Efficiency is at times because of the need to execute it many times.

```
DEFINE DATA LOCAL
1 #I(P3)
1 #TEXT(Awww/hhh)
1 #LC(A26) INIT <'abc...xyz'>
1 REDEFINE #LC
  2 #LC-CHAR(A1/26)
1 #UC(A26) INIT <'ABC...XYZ'>
1 REDEFINE #UC
  2 #UC-CHAR(A1/26)
END-DEFINE
...
/* load text into #TEXT and perform LOWER-TO-UPPER
...

DEFINE SUBROUTINE LOWER-TO-UPPER
  FOR #I = 1 TO 26
    EXAMINE #TEXT(*) FOR #LC-CHAR(#I)
      AND REPLACE WITH   #UC-CHAR(#I)
  END-FOR
END-SUBROUTINE
```

That algorithm has the advantage of being platform independent; it converts 'a' to 'A' regardless of their representation. A stack of 26 EXAMINE statements is even more efficient.

Working with arrays has just as many EXAMINE TRANSLATE rewards as working with elementary fields.

```
DEFINE DATA
LOCAL
1 #A(A12/16) INIT (V) <'HAPPY','IS','THE','MAN','WITH','A','HOBBY',
         'FOR','HE','HAS','TWO','WORLDS','IN','WHICH','TO','LIVE.'>
END-DEFINE
PRINT #A(*)
EXAMINE #A(*) TRANSLATE INTO LOWER CASE
PRINT #A(*)
EXAMINE SUBSTRING(#A(1),1) TRANSLATE INTO UPPER CASE
PRINT #A(*)
EXAMINE SUBSTRING(#A(*),1) TRANSLATE INTO UPPER CASE
PRINT #A(*)
EXAMINE #A(*) TRANSLATE INTO UPPER CASE
PRINT #A(*)
END
```

```
MORE
Page      1                                                97-04-01  12:34:56

 HAPPY IS THE MAN WITH A HOBBY FOR HE HAS TWO WORLDS IN WHICH TO LIVE.
 happy is the man with a hobby for he has two worlds in which to live.
 Happy is the man with a hobby for he has two worlds in which to live.
 Happy Is The Man With A Hobby For He Has Two Worlds In Which To Live.
 HAPPY IS THE MAN WITH A HOBBY FOR HE HAS TWO WORLDS IN WHICH TO LIVE.
```

Here is an example of translation using s user-defined table:

```
*
* EXAMINE...TRANSLATE USING USER-DEFINED TABLE
*
DEFINE DATA
LOCAL USING XXEMPLDA
LOCAL
1 #TT(A52) INIT
  <'A1B2C3D4E5F6G7H8I9J1K2L3M4N5O6P7Q8R9S2T3U4V5W6X7Y8Z9'>
1 REDEFINE #TT
  2 #TRANSLATE-TABLE(A2/26)
1 #1(P2)
1 #2(P2)
END-DEFINE
READ EMPLOYEE BY CITY = 'D' THRU 'DETROIT'
  DISPLAY JOB-TITLE FIRST-NAME(AL=10)
    NAME (AL=10) CITY (AL=10)
  EXAMINE JOB-TITLE TRANSLATE USING #TRANSLATE-TABLE(*)
  WRITE JOB-TITLE /
  WRITE TITLE 'EXAMINE...TRANSLATE...USER TABLE - PAGE'
    *PAGE-NUMBER
END-READ
END
```

```
DEFINE DATA LOCAL
1 #VAR(A10/6) INIT <'ONE','TWO','THREE','FOUR','FIVE','SIX'>
1 #TABLE(A16) INIT <'A1M2E3R4I1C2A3N4'>
1 REDEFINE #TABLE
  2 #ELEMENT(A2/8)
END-DEFINE
EXAMINE #VAR(*) TRANSLATE USING #ELEMENT(*)
DISPLAY #VAR(*) #ELEMENT(*)
EXAMINE #VAR(*) TRANSLATE USING INVERTED #ELEMENT(*)
DISPLAY #VAR(*) #ELEMENT(*)
END
```

You must be careful with your table elements.Duplicates will cause serious and unexpected [re]translation problems. The INVERTED option will translate in reverse of the translation table. value will be returned to the previous value. This is true as long as there are no duplicate values in the translation portion of the table elements, the second half. Construction of the table is most important.

The SUBSTRING option permits applying the EXAMINE TRANSLATE option against a portion of the argument field. The SUBSTRING option will examine the argument field for a specified number of positions. Starting position and length or number of positions should be supplied.

```
EXAMINE JOB-TITLE TRANSLATE INTO LOWER CASE
```

The program simply translates all characters in JOB-TITLE into lower case characters.

```
EXAMINE SUBSTRING(JOB-TITLE,2,24) TRANSLATE INTO LOWER CASE
```

To look for blanks:

```
EXAMINE JOB-TITLE FOR H'40' GIVING POSITION #1
```

MOVE / MOVE ALL / MOVE EDITED / MOVE BY NAME/ LEFT / RIGHT

This statement has been enhanced a lot from the 1.2 version. First, you can move BY NAME, similar to COBOL's MOVE CORRESPONDING, except with more flexibility. Data elements defined in groups can be moved as long as the names are the same and the MOVE follows NATURAL's conversion rules. This means the formats do not have to be the same.

Secondly, right / left justification can be accomplished with MOVE. Thirdly, the MOVE EDITED is a great new option. This allows for the attachment of an edit mask upon the MOVE. The key to using it is to remember when the mask is applied. In the first statement, the mask is applied to #FIELD1 then the move is performed. In the second statement, the move is performed and then the mask is applied. The contents of #FIELD1 remains unchanged.

```
0240 MOVE EDITED #FIELD1 (EM=...) TO #FIELD2
0240 MOVE EDITED #FIELD1          TO #FIELD2 (EM=...)
```

New is the SUBSTRING capability. SUBSTRING permits moving a value to a portion of a target field.

```
MOVE #SENDING-FIELD TO SUBSTRING(#TRECEIVER,1,5)
```

And MOVE SUBSTRING permits moving a portion of the sending field to a portion of a receiving field.

```
MOVE SUBSTRING(#SENDER,6,5) TO SUBSTRING(#RECEIVER,1,5)

DEFINE DATA LOCAL USING EMPLDA
LOCAL
1 #J1(A25)
1 #J2(A25)
END-DEFINE
READ EMPLOYEE BY CITY = 'D' THRU 'DETROIT'
  DISPLAY JOB-TITLE FIRST-NAME(AL=10) NAME (AL=10) CITY (AL=10)
  MOVE SUBSTRING(JOB-TITLE,1,6) TO #J1
  MOVE SUBSTRING(JOB-TITLE,6,20) TO #J2
  WRITE '=' #J1 / '=' #J2 /
  WRITE TITLE 'MOVE SUBSTRING TEST - PAGE' *PAGE-NUMBER
END-READ
```

The MOVE SUBSTRING statement below unloads positions one through six into positions 19 through 24 of the same field.

```
DEFINE DATA LOCAL USING EMPLDA
END-DEFINE
READ EMPLOYEE BY CITY = 'D' THRU 'DETROIT'
  DISPLAY JOB-TITLE FIRST-NAME(AL=10)
    NAME (AL=15) CITY (AL=10)
  MOVE SUBSTRING(JOB-TITLE,1,6) TO SUBSTRING(JOB-TITLE,19,6)
  WRITE JOB-TITLE /
  WRITE TITLE 'MOVE SUBSTRING TEST - PAGE' *PAGE-NUMBER
END-READ
END
```

This program is manipulating the data to produce names for a mailing label application. The first substring operation clears all but the first position of FIRST-NAME. The next substring moves the first character of middle name into the third position of FIRST-NAME. The final substring fills the remaining positions of FIRST-NAME with the first fourteen characters of NAME. The results print the first and middle initials with the lastname.

```
DEFINE DATA LOCAL USING EMPLDA
END-DEFINE
READ EMPLOYEE BY CITY = 'D' THRU 'DETROIT'
  DISPLAY JOB-TITLE FIRST-NAME(AL=10)
    NAME (AL=15) CITY (AL=10)
  MOVE ' ' TO SUBSTRING(FIRST-NAME,2,19)
  MOVE SUBSTRING(MIDDLE-NAME,1,1)
    TO SUBSTRING(FIRST-NAME,3,1)
  MOVE SUBSTRING(NAME,1,14)
    TO SUBSTRING(FIRST-NAME,5,14)
  WRITE T*FIRST-NAME FIRST-NAME /
  WRITE TITLE 'MOVE SUBSTRING TEST - PAGE' *PAGE-NUMBER
END-READ
END
```

You can use this statement to reorganize the display of a date edit mask, especially if it is stored on an ADABAS file one way and you want to display it differently.

```
0010 DEFINE DATA LOCAL
0020 1 #DATE1 (A8)  INIT <'08-30-89'>
0030 1 #DATE2 (A8)
0040 1 #D     (D)
0050 END-DEFINE
0060 *
0070 IF #DATE1 = MASK (MM'-'DD'-'YY) THEN
0080    MOVE EDITED #DATE1                  TO #D (EM=MM'-'DD'-'YY)
0090    MOVE EDITED #D     (EM=YY'/'MM'/'DD) TO #DATE2
0100 END-IF
0110 *
0120 WRITE 'ORIGINAL DATE:' #DATE1 'MODIFIED DATE:' #DATE2
0130 END
```

```
ORIGINAL DATE: 08-30-89 MODIFIED DATE: 89/08/30
```

Here is an example of two MOVE options - MOVE EDITED and MOVE LEFT

```
0010 DEFINE DATA LOCAL
0020 1 #NUM                  (N8)  INIT <5000>
0030 1 #EDIT-NUM             (A15)
0040 1 #LEFT-JUST-DISP-FLD (A15)
0050 END-DEFINE
0060 *
0070 * PURPOSE IS TO DISPLAY A NUMERIC QUANTITY
0080 * WITH A COMMA LEFT JUSTIFIED
0090 *
0100 MOVE EDITED #NUM (EM=ZZZ,ZZZ)
0110  TO #EDIT-NUM
0120 MOVE LEFT JUSTIFIED #EDIT-NUM
0130  TO #LEFT-JUST-DISP-FLD
```

```
0140 WRITE '=' #NUM
0150       '=' #EDIT-NUM
0160       '=' #LEFT-JUST-DISP-FLD
0170 MOVE EDITED #NUM (EM=$ZZ,ZZZ)
0180  TO #EDIT-NUM
0190 WRITE '=' #NUM
0200       '=' #EDIT-NUM
0210 END
```

```
#NUM:      5000 #EDIT-NUM:   5,000        #LEFT-JUST-DISP-FLD: 5,000
#NUM:      5000 #EDIT-NUM: $5,000
```

Here is an example of MOVE BY NAME

```
0010 DEFINE DATA LOCAL
0020 1 #STRUCT-TABLE (0:15)  /* TABLE TO HOLD RECORDS
0030 2 DEPT           (A6)
0040 2 JOB-TITLE      (A25)
0050 2 FIRST-NAME     (A20)
0060 2 MIDDLE-I       (A1)
0070 2 NAME           (A20)
0080 1 REDEFINE #STRUCT-TABLE
0090   2 #TABLE-ENTRY  (A72/0:15)
0100 1 #SCR-TABLE (1:15)      /* TABLE FOR SCREEN DISPLAY
0110 2 DEPT           (A6)
0120 2 JOB-TITLE      (A25)
0130 2 FIRST-NAME     (A20)
0140 2 MIDDLE-I       (A1)
0150 2 NAME           (A20)
0160 1 REDEFINE #SCR-TABLE
0170   2 #SCR-LINE        (A72/1:15)
0180 1 #ISN           (P9/1:15)
0190 1 #CV            (C/1:15)  INIT <(AD=D)>
0200 1 #CT            (P3)
0210 1 #I             (P3)
0220 ****************************
0230 *    DATABASE   V I E W
0240 ****************************
0250 1 EMP VIEW OF EMPLOYEES
0260 2 DEPT
0270 2 JOB-TITLE
0280 2 FIRST-NAME
0290 2 MIDDLE-I
0300 2 NAME
0310 END-DEFINE
0320 *
0330 RD. READ EMP BY NAME
0340     ADD 1 TO #CT
0350     MOVE BY NAME EMP TO #STRUCT-TABLE (#CT)
0360     #ISN (#CT) := *ISN (RD.)
0370     IF #CT = 15 THEN
0380        PERFORM SCR-PRINT
0390     END-IF
0400 END-READ
0410 *
0420 IF #CT > 0 THEN            /* LAST SCREEN
0430    MOVE (AD=PN) TO #CV (#CT+1:15)
0440    PERFORM SCR-PRINT
0450    END-IF
0460 STOP
```

```
0470 DEFINE SUBROUTINE SCR-PRINT
0480 #SCR-LINE (1:#CT) := #TABLE-ENTRY (1:#CT)
0490 * INPUT USING MAP    /* SCREEN DISPLAY
0500 FR-LP.
0510  FOR #I = 1 TO #CT
0520    IF #SCR-LINE (#I) = #TABLE-ENTRY (#I) THEN
0530       ESCAPE TOP
0540    END-IF
0550    GT. GET EMP #ISN (#I)  /* CHANGE ON SCREEN OCCURRED REREAD
0560        MOVE BY NAME EMP TO #STRUCT-TABLE (0)
0570        IF #TABLE-ENTRY (0) NE #TABLE-ENTRY (#I) THEN /* CHANGE?
0580             WRITE 'DATA ON RECORD WITH ISN:'
0590               #ISN (#I) 'IS DIFFERENT.'
0600             ESCAPE TOP
0610        END-IF
0620        MOVE BY NAME #SCR-TABLE (#I) TO EMP
0630        UPDATE (GT.)
0640        END TRANSACTION
0650  END-FOR
0660 RESET #CT
0670 RETURN
0680 END
```

The RESET is now designed to work with arrays and groups. For arrays, you can use specific index references, range notation, or the wildcard character '*'. The INITIAL option allows for the assigning of the value as defined in the GDA or LDA. Executing the following results in:

```
0010 DEFINE DATA LOCAL
0020 1 #NAME  (A15) INIT <'MASTER OBIWAN'>
0030 END-DEFINE
0040 *
0050 WRITE '=' #NAME
0060 MOVE 'JIM WISDOM' TO #NAME
0070 WRITE '=' #NAME
0080 RESET INITIAL #NAME
0090 WRITE '=' #NAME
0100 END
```

```
#NAME: MASTER OBIWAN
#NAME: JIM WISDOM
#NAME: MASTER OBIWAN
```

SEPARATE

This statement allows you to split a field into two or more fields. A delimiter option allows for the examination for any special character or the INPUT DELIMITER character (the ID parameter). The GIVING clause allows for the count of non-null fields.

```
DEFINE DATA LOCAL
  1 #COLON                    (A1)  CONST <':'>
  1 #IN-TIME                  (A10)
  1 #HOURS                    (A2)
  1 REDEFINE #HOURS
    2 #NUM-HOURS              (N2)
```

```
  1 #MINUTES                  (A2)
  1 REDEFINE #MINUTES
    2 #NUM-MINS               (N2)
  1 #SECONDS                  (A4)
  1 REDEFINE #SECONDS
    2 #NUM-SECS               (N2)
    2 #TIME-OF-DAY-SIGNAL     (A2)
*
  INPUT #IN-TIME       /* TIME INPUT AS HH:MM:SSAM OR HH:MM:SSPM
  SEPARATE #IN-TIME INTO #HOURS #MINUTES #SECONDS
    WITH DELIMITER #COLON
*
  IF #TIME-OF-DAY-SIGNAL = 'PM' THEN
     ADD 12 TO #NUM-HOURS
  END-IF
```

The SEPARATE WITH DELIMITER option can lead to some confusing results. Since SM04, trailing blanks in the operand of the delimiter string are ignored. If you want a blank to be part of the delimiter string, place it other than last. The following program shows differences that can occur if you do not specify the DELIMITER option correctly.

```
0010 DEFINE DATA LOCAL
0020 1 BUFF (A60)
0030 1 WORD (A60/1:10)
0040 1 I (P2)
0050 END-DEFINE
0060 *
0070 BUFF := 'HELLO, MR. WISDOM. NEVER BELIEVE EVERYTHING-U-HEAR.'
0080 SEPARATE BUFF INTO WORD (1:10) WITH DELIMITER ',. '
0090 FOR I = 1 TO 10
0100   WRITE '=' I WORD (I)
0110 END-FOR
0120 *
0130 SEPARATE BUFF INTO WORD (1:10) WITH DELIMITER ', .'
0140 FOR I = 1 TO 10
0150   WRITE '=' I WORD (I)
0160 END-FOR
0170 *
0180 SEPARATE BUFF LEFT JUSTIFIED INTO WORD (1:10)
0190   WITH DELIMITER ', .'
0200 FOR I = 1 TO 10
0210   WRITE '=' I WORD (I)
0220 END-FOR
0230 END
```

```
I:   1 HELLO
I:   2  MR
I:   3  WISDOM
I:   4  NEVER BELIEVE EVERYTHING-U-HEAR
I:   5
I:   6
I:   7
I:   8
I:   9
I:  10
```

```
I:   1 HELLO
I:   2
I:   3 MR
I:   4
I:   5 WISDOM
I:   6
I:   7 NEVER
I:   8 BELIEVE
I:   9 EVERYTHING-U-HEAR
I:  10
```

```
I:   1 HELLO
I:   2 MR
I:   3 WISDOM
I:   4 NEVER
I:   5 BELIEVE
I:   6 EVERYTHING-U-HEAR
I:   7
I:   8
I:   9
I:  10
```

Here is another example:

```
0010 DEFINE DATA LOCAL
0020 1 JCL (A60)
0030 1 WORDS (A20/1:10)
0040 1 REM (N2)
0050 1 I (P2)
0060 END-DEFINE
0070 *
0080 JCL := '//DD1 DISP=SHR,DSN=APDDBS.UTIL.JCL'
0090 *
0100 SEPARATE JCL INTO
0110    WORDS (*) WITH DELIMITER ', ='
0120      GIVING NUMBER REM
0130 *
0140 FOR I 1 10
0150 WRITE '=' I WORDS (I) REM
0160 END-FOR
0170 *
0180 END
```

```
Page      1                                                97-04-01  12:34:56

I:   1 //DD1                   5
I:   2 DISP                    5
I:   3 SHR                     5
I:   4 DSN                     5
I:   5 APDDBS.UTIL.JCL         5
I:   6                         5
I:   7                         5
I:   8                         5
I:   9                         5
I:  10                         5
```

Use the REMAINDER option of the SEPARATE to see what would not fit into your target array fields. If you don't care about what ends up in the REMAINDER field, you can use SEPARATE...IGNORE to ignore values that were not separated.

SORT

The SORT statement is greatly enhanced. You can only have one SORT statement per program. You can provide a buffer area with the USING clause. You will have no difficulties with SORT as long as you specify the USING clause. You cannot specify USING view-group-name; you must mention each field. This way you know exactly what gets passed to the sort routine. The sort keys cannot be part of the data specified in the USING clause; so you may need to move the key values to another user field and specify those as the key fields. The USING clause is required in structured mode and is not required in report mode. If you use the DEFINE DATA as the means to define the data requirements, all the fields are passed to the SORT as intermediate storage if a USING clause is not coded.

A misunderstanding is how often records are written to intermediate storage. the number of records written is the product of the number of data base loops and the number of records encountered in each loop. This eliminates the possibility of creating a record per MU/PE processed in a view.

SORT vs. SORT USING

As a preliminary discussion to the SORT vs.SORT USING, one needs to understand how SORT works. The SORT invokes the system utility in batch but uses an internal routine on-line. On-line the record length restriction equals 4K. The sort field(s) length cannot exceed 254 bytes. There are some differences if you code your sort requirements in a report mode 1.2 styled program vs. 2.1/2.2 structured mode style.

The major difference between SORT vs. SORT USING is the rule NATURAL uses to create the sort record. Either you can explicitly define the contents of the record or NATURAL will construct a record with everything but the kitchen sink. This huge amount of data can cause other system resources to be invoked, such as ESA hiperspaces and dataspaces. This is fine if they are available; if not, the job may die.

Here are two SORT statements from the same program and you can see the difference in the requirements needed to allow the program to successfully sort. The program referenced 55 user fields and 10 database fields. The amount of data constructed in the sort buffer is very different dependent upon whether or not one uses SORT USING or not. Your choice will also reduce the amount of CPU is consumed.

```
1180 SORT #COL #CLASS USING #GPI943 #CR943 #OK KEYS

SYNCSORT  3.5CRI TPF2A US PATENTS: 4210961,5117495, OTHER PAT. PEND.
(C
UNIVERSITY INFORMATION SYSTEMS (UIS)      ESA 4.
  PARMLIST :
  SORT FIELDS=(0001,0003,CH,A,0004,0002,CH,A)
  RECORD TYPE=F,LENGTH=00018
  WER177I  TURNAROUND SORT PERFORMED
  WER045C  END SORT PHASE
  WER055I  INSERT      14771, DELETE      14771
  WER246I  FILESIZE 265,878 BYTES
  WER054I  RCD IN          0, OUT          0
  WER169I  TPF LEVEL 2A
  WER052I  END SYNCSORT -
DSPNAT45,STEP020,,DIAG=AC02,2C03,42DE,EC3D,ADCE,

1180 SORT BY #COL #CLASS

SYNCSORT  3.5CRI TPF2A US PATENTS: 4210961,5117495, OTHER PAT. PEND.
(C
                           UNIVERSITY INFORMATION SYSTEMS (UIS)
ESA 4.
   PARMLIST :
   SORT FIELDS=(0001,0003,CH,A,0004,0002,CH,A)
   RECORD TYPE=F,LENGTH=00429
   WER124I  ESTIMATED PREALLOCATED/USED SORTWORK SPACE USAGE FACTOR
=5.1
   WER045C  END SORT PHASE
   WER055I  INSERT      14771, DELETE      14771
   WER418I  DATASPACE(S) AND/OR HIPERSPACE(S) USED
   WER246I  FILESIZE 6,336,759 BYTES
   WER054I  RCD IN          0, OUT          0
   WER169I  TPF LEVEL 2A
   WER052I  END SYNCSORT -
DSPNAT44,STEP020,,DIAG=8E00,8E50,60FE,AC8E,8FEE,
```

There is, however, one difference between a sort coded in the NATURAL program and an external sort (at least on MVS machines). The external sort step will utilize the sort program's 'block transfer' mode. An internal sort (within the NAT module) must use the sort input and output exit routines for that environment, which generally operate on a single record level. This could cause a performance difference due to increased I/O (physical and logical) between NATURAL and the sort routine.

D. Control

DECIDE FOR

The DECIDE FOR is different from the DECIDE ON. It is equivalent to a nested IF structure. It's useful for testing a number of logical conditions which need evaluating at the same point in the program. It is far more readable than a nested if structure. At first glance it can be confusing because of the various clauses. The only negative is that when compiled it requires more object space than an equivalent IF. This may be important if on compile you get NAT0888 error.

CLAUSE	EXPLANATION
EVERY	Indicates all conditions are tested. Each WHEN condition is tested and if true the statements following are executed. After all WHEN conditions are evaluated, ANY and ALL, if coded, are then evaluated.
FIRST	Discontinue testing as soon as one is true. ANY (see below), if coded, will be executed; ALL is not allowed with FIRST.
WHEN	Specifies the logical condition to be tested and in order of presentation. If true, following statements executed immediately.
WHEN ANY	Specifies the statement(s) to be executed m(after) if any of the evaluated conditions in previous WHEN clauses are true (OR logic).
WHEN ALL	Specifies the statement(s) to be executed (after) if all of the evaluated conditions in previous WHEN clauses are true (AND logic).
WHEN NONE	Is required. Specifies the statement(s) to be executed if all of the conditions are false.

Here is an example showing the equivalent IF structure in both report and structured mode:

```
0300 DECIDE FOR FIRST CONDITION
0310   WHEN #TASK NE #CURRENT-TASK
0320   TERMINATE
0330   WHEN #SONM042-HELP = #QUESTION
0340     CALLNAT #SONS003-HELP-IND #SONS003-TOPIC-PARM
0350     RESET #SONM042-HELP
0360   WHEN NOT (#SONM042-HELP = #QUESTION OR = #SPACES)
0370     MOVE #INVALID-HELP-MESSAGE          TO #MESSAGE
0380     MOVE #CURSOR-HELP                   TO #MARK
0390   WHEN NONE
0400     IF BREAK #SONM042-KEYWORD
0410       MOVE 1                       TO #P-IDX
0420       MOVE #SPACES                 TO #SONM042-NEXT-TASK
0430     END-IF
0440 END-DECIDE
```

Equivalent IF - Structured Mode

```
0200 IF #TASK NE #CURRENT-TASK
0210    TERMINATE
0220 ELSE
0230    IF #SONM042-HELP = #QUESTION
0240       CALLNAT #SONS003-HELP-IND #SONS003-TOPIC-PARM
0250       RESET #SONM042-HELP
0260    ELSE
0270       IF NOT (#SONM042-HELP = #QUESTION OR = #SPACES)
0280          MOVE #INVALID-HELP-MESSAGE TO #MESSAGE
0290          MOVE #CURSOR-HELP TO #MARK
0300       ELSE
0310          IF BREAK #SONM042-KEYWORD
0320             MOVE 1 TO #P-IDX
0330             RESET #SONM042-NEXT-TASK
0340          END-IF
0350       END-IF
0360     END-IF
0370 END-IF
```

Equivalent IF - Report Mode

```
0200 IF #TASK ¬= #CURRENT-TASK
0210    TERMINATE
0220 ELSE
0230    IF #SONM042-HELP = #QUESTION DO
0240     CALLNAT #SONS003-HELP-IND #SONS003-TOPIC-PARM
0250        RESET #SONM042-HELP
0260        DOEND
0270    ELSE
0280       IF NOT (#SONM042-HELP = #QUESTION OR = #SPACES) DO
0290         MOVE #INVALID-HELP-MESSAGE TO #MESSAGE
0300            MOVE #CURSOR-HELP TO #MARK
0310            DOEND
0320       ELSE
0330          IF BREAK #SONM042-KEYWORD DO
0340           MOVE 1 TO #P-IDX
0350           RESET #SONM042-NEXT-TASK
0360          DOEND
```

DECIDE ON

This statement is similar to the CASE structure in Pascal. It is used when there are multiple values possible for a single field. It allows for the testing of specific possible values. You are required to state the action taken if none of the listed values are found. A typical example of its usage is for testing codes entered on a menu screen.

CLAUSE	EXPLANATION
VALUE	Specifies the value to be evaluated, either constant or variable. Multiple values are listed separated with commas, or a range can be specified with ':'. Array references are not allowed.
ALL	Specifies the statement(s) to be executed if all values are encountered. These statements are executed IN ADDITION TO those in VALUE clause.
ANY	Specifies the statement(s) to be executed if any values are encountered. These statements are executed IN ADDITION TO those in VALUE clause.
NONE	This is required. Specifies the statement(s) to be executed if no value is encountered.
EVERY	This is indicates all conditions are tested.
FIRST	Discontinue testing as soon as a value is encountered.

Example:

```
0300 DECIDE ON FIRST VALUE OF #OPTION
0310  VALUE 'A'
0320     STACK TOP COMMAND 'NEWS'
0330   VALUE 'B'
0340     REINPUT ' FUNCTION NOT YET AVAILABLE.'
0350   VALUE 'C'
0360     STACK TOP COMMAND 'DRNPPL03'
```

```
0370   VALUE 'D'
0380     INPUT NO ERASE
0390        WITH TEXT
0400        ' Enter text member name, &/OR library or hit <ENTER>'
0410      15/33 'TEXT NAME:' #MEMBER-NAME
0420            'LIB NAME:' #LIB-NAME
0430     STACK TOP COMMAND 'MAILIT' #MEMBER-NAME #LIB-NAME
0440   VALUE #PERIOD
0450     STOP
0460   NONE VALUE
0470     REINPUT ' ERROR IN OPTION SELECTION'
0480 END-DECIDE
```

Equivalent IF - Report Mode

```
0200 IF #OPTION = 'A' THEN
0210   DO STACK TOP COMMAND 'NEWS'
0220   DOEND
0230 ELSE
0240 IF #OPTION = 'B' THEN
0250   DO REINPUT ' FUNCTION NOT YET AVAILABLE.'
0260   DOEND
0270 ELSE
0280IF #OPTION = 'C' THEN
0290   DO STACK TOP COMMAND 'DRNPPL03'
0300   DOEND
0310 ELSE
0320 IF #OPTION = 'D' THEN
0330    DO INPUT NO ERASE WITH TEXT
0340         ' Enter text member name, &/OR library or hit <ENTER>'
0350       15/33 'TEXT NAME:' #MEMBER-NAME ' LIB NAME:' #LIB-NAME
0360       STACK TOP COMMAND 'MAILIT' #MEMBER-NAME #LIB-NAME
0370    DOEND
0380 ELSE
0390 IF #OPTION = #PERIOD
0400    DO STOP
0410    DOEND
0420 ELSE
0430    DO REINPUT ' ERROR IN OPTION SELECTION'
0440    DOEND
```

Equivalent IF - Structured Mode

```
0200 IF #OPTION = 'A' THEN
0200    STACK TOP COMMAND 'NEWS'
0200 ELSE
0200    IF #OPTION = 'B' THEN
0200       REINPUT ' FUNCTION NOT YET AVAILABLE.'
0200    ELSE
0200       IF #OPTION = 'C' THEN
0200          STACK TOP COMMAND 'DRNPPL03'
0200       ELSE
0200          IF #OPTION = 'D' THEN
0300             INPUT NO ERASE WITH TEXT
0310               ' Enter text member name, &/OR library or hit <ENTER>'
0320               15/33 'TEXT NAME:' #MEMBER-NAME
0330                     ' LIB NAME:' #LIB-NAME
0340             STACK TOP COMMAND 'MAILIT' #MEMBER-NAME #LIB-NAME
```

```
0350          ELSE
0360             IF #OPTION = #PERIOD
0370                STOP
0380             ELSE
0390                REINPUT ' ERROR IN OPTION SELECTION'
0400             ENDIF
0410          ENDIF
0420       ENDIF
0430    ENDIF
0440 ENDIF
```

I have seen examples of DECIDE FOR which are better suited as DECIDE ON. The DECIDE ON is best suited when a potential value is checked for a simple data type. The proper action is listed for each one. Here is one such example:

```
0300 DECIDE FOR EVERY CONDITION
0310   WHEN #ITEM-CD = '11' OR = '12' OR = '13'
0320               OR = '14' OR = '15' OR = '16'
0330               OR = '17' OR = '18' OR = '19'
0340      MOVE 'TUITION AND FEES' TO #MSG
0350   WHEN #ITEM-CD = '21' OR = '22' OR = '23'
0360               OR = '24' OR = '25' OR = '26'
0370               OR = '27' OR = '28' OR = '29'
0380      MOVE 'HOUSING AND BOARD CHARGES' TO #MSG
0390   WHEN #ITEM-CD = '41' OR = '42' OR = '43'
0400               OR = '44' OR = '45' OR = '46'
0410               OR = '47' OR = '48' OR = '49'
0420      MOVE 'FINANCIAL AID CREDIT' TO #MSG
0430   WHEN #ITEM-CD = '61' OR = '62' OR = '63'
0440               OR = '64' OR = '65' OR = '66'
0450               OR = '67' OR = '68' OR = '69'
0460      MOVE 'PAYMENTS'  TO #MSG
0470   WHEN #ITEM-CD = '81' OR = '82' OR = '83'
0480               OR = '84' OR = '85' OR = '86'
0490               OR = '87' OR = '88' OR = '89'
0500      MOVE 'MISCELLANEOUS' TO #MSG
0510   WHEN NONE
0520      IGNORE
0530 END-DECIDE
```

More preferable is:

```
0300 DECIDE ON FIRST VALUE OF #ITEM-CD
0310   VALUE '11':'19'
0320      MOVE 'TUITION AND FEES' TO #MSG
0330   VALUE '21':'29'
0340      MOVE 'HOUSING AND BOARD CHARGES' TO #MSG
0350   VALUE '41':'49'
0360      MOVE 'FINANCIAL AID CREDIT' TO #MSG
0370   VALUE '61':'69'
0380      MOVE 'PAYMENTS'  TO #MSG
0390   VALUE '81':'89'
0400      MOVE 'MISCELLANEOUS' TO #MSG
0410   WHEN NONE
0420      IGNORE
0430 END-DECIDE
```

Be careful with range checking for an alphanumeric variable which has numeric values. In the above DECIDE ON statement, if #ITEM-CD is defined as (A3) and contained a value of '100', the value falls within the range of the first value statement. This may *not* be what is intended (this is also true for an IF statement). It is better to redefine #ITEM-CD as numeric and write:

```
0400 DECIDE ON FIRST VALUE OF #ITEM-CD-NUM
0410   VALUE 11:19
0420      MOVE 'TUITION AND FEES' TO #MSG
0430   VALUE 21:29
0440      MOVE 'HOUSING AND BOARD CHARGES' TO #MSG
0450   VALUE 41:49
0460      MOVE 'FINANCIAL AID CREDIT' TO #MSG
0470   VALUE 61:69
0480      MOVE 'PAYMENTS'  TO #MSG
0490   VALUE 81:89
0500      MOVE 'MISCELLANEOUS' TO #MSG
0510   WHEN NONE
0520      IGNORE
0530 END-DECIDE
```

A good example of a DECIDE FOR structure is one which replaces the following nested IF structure:

```
0200 IF *NUMBER = 0 THEN
0210    DO #MESSAGE = 'NO REQUEST EXISTS.'
0220       #ERROR-FLAG = ##YES
0230       #STOP = ##YES
0240    DOEND
0250 ELSE
0260 IF REQ-SIGNATURE-COUNT = 0 THEN
0270    DO #MESSAGE = 'ERROR. NO SIGNATURES EXIST.'
0280       #ERROR-FLAG = ##YES
0290       #STOP = ##YES
0300    DOEND
0310 ELSE
0320 IF REQ-CLOSEOUT-FLAG = 'R' THEN
0330   DO #MESSAGE  = 'NO ACTIVITY ALLOWED. REQUEST IS CLOSED.'
0340       #ERROR-FLAG = ##YES
0350       #STOP = ##YES
0360    DOEND
0370 ELSE
0380    DO #NEXT-CYCLE-NUM = REQ-CYCLE-SEQUENCE + 1
0390    DOEND
```

The previous example can be translated to a DECIDE FOR structure much easier to read, understand and maintain.

```
0200 DECIDE FOR FIRST CONDITION
0210   WHEN *NUMBER = 0
0220      #MESSAGE = 'NO REQUEST EXISTS.'
0230   WHEN REQ-SIGNATURE-COUNT = 0
0240      #MESSAGE = 'ERROR. NO SIGNATURES EXIST.'
0250   WHEN REQ-CLOSEOUT-FLAG = 'R'
0260      #MESSAGE  = 'NO ACTIVITY ALLOWED. REQUEST IS CLOSED.'
0270   WHEN ANY
0280      #ERROR-FLAG = ##YES
0290      #STOP = ##YES
```

```
0300   WHEN NONE
0310       #NEXT-CYCLE-NUM = REQ-CYCLE-SEQUENCE + 1
0320 END-DECIDE
```

IF

The IF statement is more generalized in its NATURAL capabilities. An empty IF or ELSE is no longer valid under Version 2.2 (if needed, an IGNORE can be inserted). First, a logical condition can include arithmetic expressions as operands. Array occurrences can also serve as operands. One must be careful in using range notation or the asterisk wildcard notation. The condition is true if any occurrence of the condition matches, not all the conditions. It is an OR condition, not an AND condition.

A complex IF involving both OR and THRU which does not repeat the variable name will get a NAT0001 at compile time. In other words,

```
IF #N = 10 OR = 20 OR = 100 THRU 200
```

is incorrect. The correct syntax is:

```
IF #N = 10 OR = 20 OR #N = 100 THRU 200
```

Array checking is similar to OR logic:
IF #ARRAY (*) = 10 asks if any position contains the number 10

SET CONTROL

This statement allows for terminal command to be set in a NATURAL program. The operand of the SET CONTROL can be provided in a variable. The is statement is also used in an IF structure.

SET	CONTROL	'MT'	Place message line on top of screen
SET	CONTROL	'WFLC11L28B06/49'	Creates a window with a frame positioned with upper left corner on line 6 column 49 and is 11 long and 28 wide.
SET	CONTROL	'C'	Copies current screen into source work area.
SET	CONTROL	'N'	Displays in non-conversational mode
SET	CONTROL	'Z'	Clear work area
SET	CONTROL	'K'	Set *PF-KEY value

Here is a bit of code which reflects a new approach in our user interface at Boston University. How often have we submitted work online, only to see 'X SYSTEM' on the bottom of the screen, not knowing what is happening. This code segment shows a file being read to extract data for a report sent to a local printer. While the report is being generated, the screen is sent an image in a window within time intervals to give the user the impression work is being done. This window is generated with SET CONTROL 'N' so that the user need not press the ENTER key.

```
2320 PRINT. READ FILE INV
2330  BY F015Z2-COURSE-INV-KEY STARTING FROM #CIF-KEY1
2340   MOVE F015AA-COL-COURSE               TO #READ-COL
2350   MOVE F015AB-DEPT-COURSE              TO #READ-DEPT
2360   IF #READ-COL-DEPT > #CIF-COL-DEPT2
2370     ESCAPE BOTTOM (PRINT.)
2380   END-IF
2390   PERFORM #100-EXAMINE-COURSE
2400   ADD 1                                TO #INTERVAL
2410   IF #INTERVAL GE 10
2420     MOVE 0                             TO #INTERVAL
2430     PERFORM #200-DISPLAY-PROGRESS
2440   END-IF
2450 END-READ
  ...
4100 DEFINE SUBROUTINE #200-DISPLAY-PROGRESS
4110 *
4120 COMPUTE ROUNDED #20THS
4130              = (#TOTAL-INVENTORY-READ / #TOTAL-INV-SELECTED) * 20
4140 FOR #IDX2 = 1 TO #20THS
4150   MOVE #ASTERISK                       TO #PCT-CELL (#IDX2)
4160 END-FOR
4170 SET CONTROL 'N'
4180 SET CONTROL 'WFC11L29B7/24'
4190 INPUT USING MAP 'SCNM002'
4200 SET CONTROL 'WB'
4210 END-SUBROUTINE
```

Another idea is to use SETTIME and *TIMD to point out slow response versus very quick response. Quick responses do not need messages.

There is some caution that one must exercise when using non-conversational writes. Non-conversational means screen I/O without an ENTER response. One, EXECUTE/FETCH REPEAT which involves a REINPUT or a tight loop with a logic error with SET CONTROL 'N' can cause your VTAM buffers to fill up too fast on output. This can tie up your entire environment; the whole machine comes to a standstill. The only way to correct this situation is to IPL.

The SET CONTROL affects the next buffer written while flushing the previous buffer. In the following example what seems obvious is not what happens. The affect is also in part affected by LS, PS and AVERIO.

```
0010 WRITE 'THIS IS MESSAGE LINE ONE.'
0020 SET CONTROL 'N'
0030 INPUT (AD=O) 'WISDOM SHOULD NOT APPEAR.'
0040 END
```

One user on SAG-L wanted to simulate a PF key to exit a map without user interaction, because REINPUT is need without perceiving an INPUT was already done. All of this was based on a code. It was thought that the following code might work, but doesn't:

```
IF #CODE NE 'B'                   /* If a certain code is received do the
   STACK TOP COMMAND '%K6'        /* input, but perform a PF6 simulation
END-IF                            /* at next.
INPUT USING MAP 'MAPNAME'
```

Instead, this works:

```
IF #CODE NE 'B'
  STACK TOP DATA ' '    /* Bypass next input statement
END-IF

INPUT USING MAP

IF #CODE NE 'B'
  SET CONTROL 'K6'   /* PF6
END-IF
```

If you have a map that's three physical screens wide and you want to provide a parameter to the program for which of the panels to start in - either the first, second, or third. How do you do this?

A tip I learned from the great Andreas Schutz, this can be done with a REINPUT statement with SET CONTROL 'Q' and SET CONTROL 'W+'. First, perform the SET CONTROL 'Q', then the INPUT to produce the wide map in the page buffer, then the SET CONTROL 'W+' to put the position for which panel you want, and then issue a REINPUT statement.

SET KEY

This statement establishes the functions of terminal keys, particularly PF keys at program level. The options are:

SET KEY OFF	turns off function key settings and returns control to terminal communications
SET KEY ON	reactivates usage assigned to PF/PA keys prior to OFF
SET KEY ALL	makes all keys program sensitive
SET KEY PAn	establishes program sensitivity for PA keys
SET KEY PFn	establishes program sensitivity for PF keys

Once a key is made program sensitive, pressing the PF key is equivalent to pressing the ENTER key. The NATURAL system variable *PF-KEY identifies which key has been hit. All of this allows for programs to interrogate key strokes and key assignments.

Terminal commands can also be assigned as values to PF keys. Any SET KEY assignment of a command is effective upon leaving execution and re-entering execution of any other program. When the defined PF-key is entered, the current executing program is interrupted and the assigned command is stacked for execution. You do not return to the program previously executing. The assignments are valid across levels until another SET KEY assignment is encountered. However, if the key is modified at a lower level, this assignment remains at all lower levels until return to a higher level which has the original SET KEY statement.

Example:

```
SET KEY PF10 = 'STARTUP'
SET KEY PF10 = 'STARTUP' NAMED 'MENU'
SET KEY PF1  = OFF turns off specific setting, making it equivalent to
hitting the e  key
SET KEY CLR
SET KEY PF1  = HELP
SET KEY PF12 = DATA 'string'
```

One fact about the set key statement may not be obvious to some. I know I misinterpreted how it works when I first encountered it several years ago. Here are two program listings. If you recognize the difference you can ignore the explanation afterwards.

```
0010 DEFINE DATA LOCAL
0020 1 #WORD (A10)
0030 END-DEFINE
0040 SET KEY ALL          /* MAKE ALL KEYS PGM. SENSITIVE
0050 *                     /* W/O SPECIFIC SETTING, KEY IS = <ENTER>
0060 SET KEY PF1 = 'TIME'
0070 INPUT 'TYPE>' #WORD
0080 *
0090 IF #WORD = 'GO' THEN
0100    DO FETCH RETURN 'TIME'
0110    DOEND
0120 *
0130 WRITE '=' #WORD
0140 END
```

```
0010 DEFINE DATA LOCAL
0020 1 #WORD (A10)
0030 END-DEFINE
0040 SET KEY ALL
0050 INPUT 'TYPE>' #WORD
0060 *
0070 IF *PF-KEY = 'PF1' THEN
0080    DO FETCH RETURN 'TIME'
0090    DOEND
0100 *
0110 WRITE 'RETURNED FROM TIME ROUTINE. WORD VALUE =' #WORD
0120 END
```

In program 1, PF key 1 is established via the SET statement. In program 2 the key is not set; it is otherwise made program sensitive and is equivalent to the the ENTER key. The two programs behave differently.

In program 1, hitting PF key 2 upon input causes the program 'TIME' to be invoked executed immediately. Return to this program is impossible, since it was invoked as if a FETCH statement was executed immediately after the INPUT.

However, in program 2, hitting PF key 1 on the INPUT sets the NATURAL value *PF-KEY. The check of the value for PF1 is true, allowing for the FETCH RETURN mechanism to take affect. The program 'TIME' is executed and control is returned to program 2.

```
0010 DEFINE DATA LOCAL
0020 1 #WORD (A10)
0030 END-DEFINE
0040 SET KEY ALL
0050 FORMAT KD=ON
0060 SET KEY PF2  = 'TIME' NAMED 'CLOCK'
0070 SET KEY PF10 = 'FIN'  NAMED 'QUIT'
0080 INPUT 'TYPE>' #WORD
0090 *
0100 IF #WORD = 'TIME' THEN
0110    DO FETCH RETURN 'TIME'
0120    DOEND
0130 WRITE '=' #WORD
0140 END
```

The NAMED option provides an identifier for PF keys when displaying function key specifications on the screen. To do this requires the FORMAT KD=ON statement. However, just including this option and FORMAT is NOT enough. If you display a map via the INPUT USING MAP statement, unless you set the STD KEYS option in the map default characteristics to the value 'Y', you will not get the display. This allows the map editor to keep the last two lines empty for the key display at execution time. Here is an example of the statements which are required to cause a key display.

```
SET KEY ALL    /* causes all keys to be program sensitive
FORMAT KD=ON   /* causes display of PF-key names
SET KEY PF1  = 'HELP'
SET KEY PF3  = 'FIN'      NAMED 'QUIT'
SET KEY PF4  = 'BUS'      NAMED 'BUS'
SET KEY PF5  = 'ADALOG'   NAMED 'LOG'
SET KEY PF9  = 'CALENDAR' NAMED 'CLNDR'
SET KEY PF10 = 'TIME'     NAMED 'CLOCK'
SET KEY PF12 = 'SYSMAIN'  NAMED 'GRAB'
SET KEY TREQ NAMED 'MORE'  /* NAMES THE <ENTER> KEY
```

E. Input / Output

DEFINE PRINTER

This statement allows for the definition of logical print queues. It is used to assign a symbolic name to a report number. The valid print numbers are 1 to 31 (I have heard of problems where n=0 under CICS). Multiple logical names can be assigned to the same printer number n. The equivalence is done at compile time.

The OUTPUT destination is determined at run time. This allows for any logical printer to be assigned to a destination as long as it is defined via parameters or JCL during the start-up of NATURAL. In SM05, you can name a logical printer 'HARDCOPY'. This defaults to the printer which is your current hard copy device. In COM-PLETE (SAG'S TP MONITOR), this designation can be done through the USCHC facility. For destination 'SOURCE', the output is generated into the NATURAL source area. Once in the Program Editor, you can use NATURAL commands to modify it.

The following shows the syntax of the DEFINE PRINTER statement. What it doesn't show is how the destination can be determined via INPUT or reading a table stored in ADABAS or on an OS data set.

```
DEFINE PRINTER (REP1 = 2) OUTPUT #DEST
DEFINE PRINTER (REP1 = 1) OUTPUT 'SOURCE'
```

The following two problems have plagued users in the past:
1). all line numbers are zero, so source can not be run >
2). */ inserted all over the place

Here's a sample program that deals with the 2 issues above:

```
0010*****************************************************************
*
0020  SET CONTROL 'Z'                       /* Clear source area so can
0030  DEFINE PRINTER(30) OUTPUT 'SOURCE'    /* create new program there
0040  FORMAT(30) LS=100 PS=0                /* PS=0 will generate only a
0050*                                       /* single page break on line
1
0060  RESET #I(N3)
0070  FOR #I = 2 TO 600
0080     WRITE (30) NOTITLE '** Now on line:' #I
0090  LOOP
0100  CLOSE PRINTER(30)
0110*
0120  STACK TOP COMMAND 'RENUM'              /* Renumber source when
done 0130*
0140  END
0150*****************************************************************
**
```

Darrell Davenport summed it up well:

> "One way is to write to the editor work space by using DEFINE PRINTER... OUTPUT 'SOURCE' then issue the STACK TOP COMMAND 'STOW' #PGM-NAME or callnat to USR0210N (as documented in SYSEXT library).
>
> Another is to use NATURAL ISPF if you have it. Build macros or write to the work space directly, then issue the ISPF command to SAVE/STOW it. Essentially anything you can do manually with the editor, you can now do programatically (e.g. find a string and insert lines after that line)."

However, do not go against the FUSER directly. You can really screw up the works. NATURAL has over time changed the rules on how it puts objects onto the system file. Plus some day you may not be able to write directly to NATURAL system files.

Another technique would be to do it in batch. Write the code to a work file, with the logon command, and an "EDIT" command in front of the source. After the source put a ".E" command, followed by a "SCRATCH xxxxxxxx" to guarantee that "SAVE pgmxxxx" command will take.

New DEFINE PRINTER options extend its processing capabilities. HARDCOPY routes output to the current hard copy device defined. INFOLINE routes output to the NATURAL Information Line. The INFOLINE is the same line which is used to display statistics about screen attributes and current I/O. This statistics line is executed by typing %X on the command line. %XI+ must be executed to switch to the INFOLINE mode and executing %XI- switches the information line into statistics mode (%XI toggles the mode).

Printer spooling system parameters are also supported when PROFILE is used to specify the name of a printer control character table. Other print control parameters available: FORMS, NAME, DISP, and COPIES (1-255). You need to check with your NATURAL Administrator on using these functions.

The DEFINE PRINTER statement in the following example assigns the output of printer 1 to the INFOLINE. The PRINT statement in the program places the data to printer 1, the INFOLINE in this example. Use the Terminal Command %X to control line placement and other information line attributes including switching between INFOLINE and statistics display modes.

```
DEFINE DATA LOCAL USING EMPLDA
END-DEFINE
DEFINE PRINTER OUTPUT 'INFOLINE'
READ (3) EMPLOYEE BY CITY = 'D' THRU 'DETROIT'
PRINT  'RECORDS READ:' *COUNTER (AD=L)
  'PRESS "ENTER" TO CONTINUE'
  INPUT (SG=OFF IP=OFF AD=MI'.')
  / '             NAME:' NAME
  / '       FIRST NAME:' FIRST-NAME
  / 'PERSONNEL NUMBER:' PERSONNEL-ID
  / '           GENDER:' SEX
END-READ
END
```

```
           NAME: SPEISER.............
     FIRST NAME: ANN.................
PERSONNEL NUMBER: 20020300
         GENDER: F







RECORDS READ: 1 PRESS 'ENTER' TO CONTINUE
```

What is more ideal is not to have to list different printer files in JCL, especially when only 31 can be listed. What if you want the batch job to spool to one of 100 printers?

First, the DEFINE PRINTER issues a close to the printer which may be important to get last report lines out without exiting the program or ending a NATURAL session. Also, COM-PLETE allows for dynamic LOGONs and allows the destination to correctly be assigned at runtime in NATURAL 2.2. This allows application/program to control routing if you disable %H and allow any sensitive reports to go where you want them to go. Also, this capability can replace the print spooling routines of COM-PLETE (SAG'S TP MONITOR), i.e., (PSPUT, PSWRITE, etc.).

For organizations that have COM-PLETE, you can link edit NATURAL with a module named NATCMPL which is delivered in the NATURAL link library. You must also have the COM-PLETE (SAG's TP MONITOR) parameter BATCH set to 'YES'. You can then modify the DEFINE PRINTER statement to include the destination using the OUTPUT clause. Finally, in your batch stream you must feed the PRINTER parameter as a dynamic parameter. Upon execution the job directs the output to the destination specified in the DESTINATION option of the DEFINE PRINTER statement. The JCL does NOT have to have a DD coded for the print data set.

Under normal circumstances, printers are closed when the program EOJs or another DEFINE PRINTER specifies the same printer again. CLOSE PRINTER closes the printer specified from within a NATURAL program. You may use a user-defined variable as established in a previous DEFINE PRINTER statement.

```
DEFINE PRINTER (SPOOL1=1)
     :
CLOSE PRINTER (SPOOL1)  or you may explicitly code the printer number:

CLOSE PRINTER (1)
```

DEFINE WINDOW

DEFINE WINDOW replaces the SET CONTROL 'W' window definition statement (SET CONTROL 'W' is still compatible). DEFINE WINDOW defines a window's size, location on the screen and attributes of the window. There may be multiple DEFINE WINDOW statements, each activated by a SET WINDOW statement or by using INPUT WINDOW.

The window name cannot be greater than 32 characters. These DEFINE WINDOW clauses are positional.

Window Dimensions

SIZE

This clause defines the width and depth of thewindow. This optional parameter is dynamic, that is, without SIZE coded thewindow will conform the the width and depth of the data displayed (seeAUTO).

AUTO

AUTO sets the window dimensions dynamically. Thewindow will be as wide and as deep as needed to present the fields andlabels defined to be placed in the window.

QUARTER

QUARTER creates a window 12 lines deep and 40 characters wide into which your data may be displayed.

Line * Column

You may specify the exact dimensions of your window by coding the number of characters per line as a non-decimal value and the number of lines as another non-decimal value. These two values are separated by an asterisk; spaces are allowed around the asterisk. These values may be specified as variables also.

The window's frame must be taken into consideration when setting thedimensions, that is, it is included in the values you defined. If framing is turned off, the minimum line*column setting is 2 * 10; with framing turned on, the minimum dimensions for a window is 4 * 13. No window may be defined larger than the monitor screen size.

Include the message line in your SIZE calculations unless the CONTROL SCREEN clause is specified.

Window Placement

The positioning of the window on the screen is determined by the BASE setting.

Cursor

If a BASE is not coded, the current cursor position is the default base parameter. If cursor position potentially forces the window off screen, NATURAL retrofits the window back onto the physical screen.

Line / Column

The upper left corner of the window will be positioned at the specified line and column on the screen. Attempts to position window off the physical screen will result in an error message. The line position must be supplied as a non-decimal number or variable. This is also true for the column setting. The values are separated by a slash and spaces may be placed to the left and right of the slash for readability if necessary.

Corners

TOP, BOTTOM, LEFT, and RIGHT position the window into one of the four corners of the screen. These four terms may not be abbreviated.

REVERSED

The window may also be displayed in reverse video on monitors supporting this function. The background color may also be set with the REVERSED clause by coding CD= followed by one of the supported color abbreviations within a set of parentheses.

TITLE

Windows may have a title defined by the TITLE clause. Titles are centered on the top frame line of window. The title will be truncated if it is wider than frame defined. TITLE is not valid unless FRAMED has been specified.

CONTROL

The message line, PF keys and command lines all take up space within the window. An alternative to cluttering the window and providing an extension to the user's productivity, CONTROL is used to determine if the PF key line, message line and statistics line are to be displayed in the window.

WINDOW

The default is WINDOW which places the displays(message line and PF key lines) inside the window. The PF key line will be truncated if the window is too narrow to allow it to display in its entirety.

SCREEN

SCREEN places the displays outside the window, that is, back on the main screen where they may be used in addition to the window display. Control is shifted to the PF keys displayed on the mainmap, the message line and the command/statistics line.

FRAMED

Windows are automatically framed (use the %F terminal command to change the frame characters). The frame may be displayed with a color supported by the monitors in use. OFF turns the frame display off.

Frame Sensitivity

The top and bottom of the frame is cursor sensitive. When there is more data than can fit into the window, NATURAL places information in the frame indicating a scrolling capability. If the data is wider than the window defined, the word "More" followed by the ">" symbol (the "greater than" symbol more commonly called the "right arrow") will appear in the upper, right portion of the frame. To scroll the remaining data into the window, position the cursor directly on top of the symbol and press the ENTER key. The symbol will eventually change to a"<" indicating you are at the far right of the data and may now scroll left if desired. Forward and backward scrolling operate in a similar fashion. The symbols appearing will be the familiar "+" and"-" sign following "More". The plus sign indicates page forward to see more data; the minus sign indicates page backward to return to previously viewed data. POSITION provides added control of the positions and instructions with this cursor sensitivity in the frame.

The window in the following example will be framed and positioned on the screen at the current cursor position. It will be sized to the data that will appear in the window which includes any prompting messages and data field values.

```
DEFINE DATA LOCAL
   1 #F1(A10) INIT <'CLASSLIB'>
   1 #F2(P3.2) INIT <49.95>
END-DEFINE
FORMAT AD=MI'.' SG=OFF
DEFINE WINDOW PORTAL
   SIZE AUTO
INPUT WINDOW='PORTAL' #F1 #F2
END
```

Changing SIZE AUTO to SIZE QUARTER modifies the SIZE to 4 * 20. Thc window has not been sized wide enough illustrating the cursor sensitive facility with the DEFINE WINDOW definition. "More" indicates that there is additional data to be displayed in this window.

Any titles specified with the window will be truncated if the POSITION message is requested. Be careful when setting titles when using POSITION messages/symbols.

BASE overrides the default window location options and places the window at your designated screen coordinates.

```
DEFINE DATA LOCAL
  1 #F1(A10) INIT <'CLASSLIB'>
  1 #F2(P3.2) INIT <49.95>
END-DEFINE
FORMAT AD=MI'.' SG=OFF
DEFINE WINDOW PORTAL
  SIZE 4*30
  BASE TOP LEFT
  FRAMED
INPUT WINDOW='PORTAL' #F1 #F2 /* '<'
END
```

```
DEFINE DATA LOCAL
   1 #VAR(A10/6) INIT <'ONE','TWO','THREE','FOUR','FIVE','SIX'>
END-DEFINE
DEFINE WINDOW PORTAL
  SIZE 4*20
  FRAMED POSITION SYMBOL TOP RIGHT
INPUT WINDOW='PORTAL' (AD=MI'.')
  #VAR(1) / #VAR(2) / #VAR(3) / #VAR(4) / #VAR(5) / #VAR(6)
END
```

If you execute the above program in the editor, scrolling is available by positioning the cursor at the first plus sign after the position message "More" and pressing the ENTER key.

Place the cursor under the minus sign and press the ENTER key, and it scrolls backwards.

SET WINDOW

SET WINDOW is used to activate and de-activatewindows. A SET WINDOW sets all subsequent windows until another SET WINDOW defines a different window or a SET WINDOW OFF is executed. Only one window can be active at a time.

```
DEFINE DATA LOCAL
1 #F1(A10) INIT <'CLASSLIB'>
1 #F2(P3.2) INIT <49.95>
END-DEFINE
FORMAT AD=MI'.' SG=OFF
DEFINE WINDOW CLASSX
  SIZE AUTO
SET WINDOW 'CLASSX'
INPUT #F1 / #F2 /* '<'
END
```

```
                                                              01
  +---------------+                                           02
  | #F1 CLASSLIB.. |                                          03
  | #F2 .49.95     |                                          04
  +---------------+                                           05
                                                              06
                                                              07
                                                              08
```

Here's another running example:

```
* SET WINDOW Example 1
DEFINE WINDOW W1
  SIZE QUARTER
  BASE UPPER RIGHT
INPUT 'Main' / 'Screen'
SET WINDOW W1
INPUT 'First' / 'Window'
INPUT 'Next' / 'Window'
SET WINDOW OFF
INPUT 'Next' / 'Main' / 'Screen'
```

```
                                                                        01
 Main                                                                   02
 Screen                                                                 03
                                                                        04
                                                                        05
                                                                        06
                                                                        07
                                                                        08
                                                                        09
                                                                        10
                                                                        11
                                                                        12
```

Press the ENTER key.

```
                                                                        01
 Main                             +------------------------------------+ 02
 Screen                           | First                              | 03
                                  | Window                             | 04
                                  |                                    | 05
                                  |                                    | 06
                                  |                                    | 07
                                  |                                    | 08
                                  |                                    | 09
                                  |                                    | 10
                                  |                                    | 11
                                  +------------------------------------+ 12
```

Press the ENTER key.

```
                                                                        01
 Main                             +------------------------------------+ 02
 Screen                           | Next                               | 03
                                  | Window                             | 04
                                  |                                    | 05
                                  |                                    | 06
                                  |                                    | 07
                                  |                                    | 08
                                  |                                    | 09
                                  |                                    | 10
                                  |                                    | 11
                                  +------------------------------------+ 12
```

Press the ENTER key.

```
                                                                        01
 Next                                                                   02
 Main                                                                   03
 Screen                                                                 04
                                                                        05
                                                                        06
                                                                        07
                                                                        08
                                                                        09
                                                                        10
                                                                        11
                                                                        12
```

```
* SET WINDOW Example 2
DEFINE WINDOW W1
  SIZE 5 * 20
  BASE UPPER RIGHT
DEFINE WINDOW W2
  SIZE 5 * 20
  BASE UPPER LEFT
INPUT 'Main' / 'Screen'
INPUT WINDOW='W1' 'First' / 'Window'
INPUT WINDOW='W2' 'Next' / 'Window'
```

```
                                                                 01
 Main                                                            02
 Screen                                                          03
                                                                 04
                                                                 05
                                                                 06
                                                                 07
                                                                 08
                                                                 09
                                                                 10
                                                                 11
                                                                 12
```

Press the ENTER key.

```
                                                                 01
 Main                                             +------------------+ 02
 Screen                                           | First            | 03
                                                  | Window           | 04
                                                  |                  | 05
                                                  +------------------+ 06
                                                                 07
                                                                 08
                                                                 09
                                                                 10
                                                                 11
                                                                 12
```

Press the ENTER key.

```
                                                                 01
 +------------------+                             +------------------+ 02
 | Next             |                             | First            | 03
 | Window           |                             | Window           | 04
 |                  |                             |                  | 05
 +------------------+                             +------------------+ 06
                                                                 07
                                                                 08
                                                                 09
                                                                 10
                                                                 11
                                                                 12
```

Press the ENTER key.

```
                                                                 01
 Next                                                            02
 Main                                                            03
 Screen                                                          04
                                                                 05
                                                                 06
                                                                 07
                                                                 08
                                                                 09
                                                                 10
                                                                 11
                                                                 12
```

In example 2, each INPUT is evaluated separately. The dimensions of each window used only applies to the INPUT statement to which it is attached. Without WINDOW= or a preceding SET WINDOW, a full screen is generated.

INPUT

The INPUT statement may now use windows setup by the new DEFINE WINDOW statement in NATURAL 2.2 with "WINDOW='window-name' " clause.

MARK POSITION

The POSITION clause now permits the MARK option to position the cursor in a selected field position. The default is the first position of the field.

NO PARAMETER

NO PARAMETER provides a facility with which one may STOW a program when a map object does not exist. NATURAL requires the map to be STOWed before the program itself can STOW successfully. When using NO PARAMETER, the map may not have any input fields.

REINPUT

Any modifications made to the variables of an INPUT statement after it was executed are applied immediately as a result of **REINPUT FULL**. POSITION permits the MARK option to position the cursor in a selected field position. The default is the first position of the field.

This statement is used to provide an interactive mode for data entry. It is also used to remove data elements off the NATURAL stack. The formatted screen creation is of interest here. Screen capabilities is discussed in Chapter 5, but here I will briefly describe some options of the INPUT statement.

Of course, the most widely used form of the INPUT statement is the INPUT USING MAP. This allows for the invocation of an external map created in the map editor, which is discussed in more detail in chapter 5. The mapname is provided either in a string or as a value stored in a user variable. If the name is provided as a string, the map must be in the current library or steplib upon compile of the unit which invokes it. Otherwise, the program will stow but you can get a NAT1117 at runtime if the map is not available.

The **MARK** positions the cursor either by numeric reference or by position relative to field on screen by using *field-name. If you use position by numeric value, this references modifiable fields only.

The NO ERASE option allows for screen overlay without erasing the contents of old screen. Any field on the old screen is flagged as protected, or you can define new fields to overlay old fields.

The **WITH TEXT** option allows for the display of text on the message line. The position of the message line is determined by the value of '%M' (top/bottom/line number). You can suffix up to seven elements with the text by using ':n:' as part of the text. The order of the data fields in the text depends on the notation ':n:'. In the following example, #NAME is suffixed as 1, #TITLE as 2, etc.

Example:

```
0010 DEFINE DATA LOCAL
0020 1 #NAME   (A10) INIT <'JIM WISDOM'>
0030 1 #TITLE  (A14) INIT <'NATURAL EXPERT'>
0040 1 #SUFFIX (A14) INIT <'NONPAREIL'>
0050 1 #AGE    (N3)
0060 END-DEFINE
0070 INPUT WITH TEXT 'HISTORY FOR:1::2::3:',#NAME,#TITLE,#SUFFIX
0080  'AGE>' #AGE (SG=OFF)
0090 END
```

Help routines are defined at the field level by defining the HE= 'parameter'. Help is invoked via PF keys, '?' in the field and <ENTER> or REINPUT USING HELP.

By using control variables, selected field attributes can be changed dynamically. Field modes such as modifiable or input/output only, display characteristics and color are definable. Attribute definitions are discussed in chapter III.

A Special Display Scenario

A user wanted to display blank for a numeric value that was used as a date field. The way to accomplish is to:

```
RESET YEAR (N4)
INPUT (AD=MIZT'_' ZP=OFF) YEAR
END
```

Another user asked the following question:

Is there a way to change the input length of a field on a map, ie. To allow the user in one instance to consider the length to be 7 characters, and in another input of the same map only 3 characters? The answer lies in understanding the dynamic attribute as specified by DY=

```
DEFINE DATA LOCAL
1 FIELD(A20)
1 LEN (P2)
END-DEFINE
INPUT 'Actual string length>' LEN
MOVE '<' TO SUBSTRING(FIELD,LEN,1)
INPUT (AD=MI'_' DY=<NP>) 'Input:' FIELD
END
```

As my friend Dennis Hamilton suggests, if a 12 was typed into the parameter on the first INPUT statement, without the "N" portion of the DY parameter the remainder of the field would still display the filler characters albeit in a non-input color.

REINPUT

Much like the INPUT statement, REINPUT is used to re-execute an INPUT statement. It has all of the options of the INPUT statement. It cannot be executed with any intervening WRITE/DISPLAY statements. It does not show changed values in the map fields. REINPUT is invalid in batch.

The ability for REINPUT to re-execute an INPUT statement crosses subroutine boundaries. It does not, however, allow for a REINPUT to an INPUT which has a status of terminated. There are some special new and useful options with REINPUT.

REINPUT *nnnn

This statement allows a NATURAL program to issue an error message. The number nnnn represents the error message number created in the SYSERR facility.

REINPUT USING HELP MARK

This option invokes the help routine established for the marked field. The MARK option allows for positioning of the cursor on the map at the specified field. The value can be supplied with a constant, a user variable (scalar or array), or the notation **field-name*. If the statement parameter of 'AD=P' is supplied, then fields which are marked 'AD=A' or 'AD=M' can be protected except those listed in the MARK option.

REINPUT and Control Variables

There is an existing situation that NATURAL resets control variables as a result of a REINPUT or REINPUT FULL statement. One suggestion from our friends in Australia is to stick the INPUT into a loop and simply ESCAPE TOP instead of using REINPUT.

Steve Robinson describes it adaquately this way:

> "There are two aspects to this problem.
>
> Suppose you have a field with contents ABC. You change the field to DEF. Now, following a REINPUT you change back to ABC. A control variable will test as MODIFIED, when in reality it should show as NOT MODIFIED.

The converse is also a problem. You start with ABC and change to DEF. After a REINPUT you do not change the field. It now shows as NOT MODIFIED, when in reality it is MODIFIED. Define a variable such as #OLD-FIELD and move the #FIELD to it before the screen goes up. Then compare the two after all REINPUT logic. Not pretty, especially if you have a lot of fields."

PRINT

This statement functions similarly to the WRITE statement. PRINT truncates the trailing blanks from fields / text; splits a field across lines if it does not fit; leading blanks / zeroes are included (unless AD=L is included).

```
0010 DEFINE DATA LOCAL
0020 1 #NAME (A20) INIT <'JAMES THOMAS WISDOM'>
0030 1 #AGE  (N5)  INIT <40>
0040 END-DEFINE
0050 WRITE 'THE MASTER GAMESMAN      ' #NAME 'IS' #AGE
0060       'YEARS OLD AND PLANS TO STAY ON TOP'
0070 PRINT 'THE MASTER GAMESMAN' #NAME 'IS' #AGE
0080       'YEARS OLD AND PLANS TO STAY ON TOP'
0090 END
```

```
THE MASTER GAMESMAN      JAMES THOMAS WISDOM  IS     40
YEARS OLD AND PLANS TO STAY ON TOP
THE MASTER GAMESMAN JAMES THOMAS WISDOM IS
40 YEARS OLD AND PLANS TO STAY ON TOP
```

WRITE FORM

The Map Editor provides full functionality to build a report layout. To accomplish this requires a WRITE statement to be generated and not an INPUT statement. This is done in the map setting definitions screen. You can use the editor PF keys to shift to utilize the entire 133 columns for editing. PF11 shifts to the right; PF10 shifts to the left. The fields required for the report can be brought in from an external data source - a program or an external local / global data area.

The negative aspect of WRITE FORM USING is that it does not recognize carriage control. This means that forms with a great deal of white space generates more I/O. Also, two logical forms on the same physical page can have a page break between them. However, NATURAL can be tricked to prevent this from occurring.

II. NATURAL System Functions

The Software AG NATURAL Reference Manual excellently describes the system variables and the system functions. You may also try WH&O's *NATURAL Developers Handbook*.

Function	Fld Length	Description
COUNT(field)	P7	incremented by 1 for each pass thru loop; reset after control break
NCOUNT(field)	P7	same as COUNT except nulls excluded; reset after control break
MAX(field)	field f.l.	maximum value found in field; reset after control break
MIN(field)	field f.l.	minimum value found in field; reset after control break
NMIN(field)	field f.l.	same as MIN except nulls excluded; reset after control break
AVER(field)	field f.l.	average of all values for field; reset after control break
NAVER(field)	field f.l.	same as AVER except nulls excluded; reset after control break
SUM(field)	field f.l.	sum of all values for field; reset after control break
TOTAL(field)	field f.l.	total of all values for field;NOT reset after control break
OLD(field)	field f.l.	value of field from previous record

```
0010 DEFINE DATA LOCAL
0020 1 HISTOGRAM-BY-MAKE VIEW OF VEHICLES
0030  2 MAKE
0040 END-DEFINE
0050 HISTOGRAM HISTOGRAM-BY-MAKE MAKE
0060 DISPLAY MAKE *NUMBER
0070 AT BREAK OF MAKE /1/
0080   WRITE COUNT(MAKE) 22T SUM(*NUMBER)
0090 * COUNT THE NUMBER OF MANUFACTURERS IN EACH GROUP
0100 * TALLY NUMBER OF VEHICLES IN EACH GROUP
0110 END
```

The SUM of a field has the length internally defined for the operand field, which has the potential to overflow. You can correct this problem by simply typing (NL=x) following the System function specification:

```
WRITE 10T OLD (CITY) SUM(LEAVE-DUE) (NL=6)
```

This syntax changes the internal size of the SUM(LEAVE-DUE) allowing calculations to exceed the N2 default.

III. Terminal Commands

In addition to what I write here you might want to read the section on terminal commands in the SAG NATURAL Reference Manual or the appendix in WH&O's *NATURAL Developers Handbook*.

All terminal commands are issued at the command level prefixing the command with the the control character defined for the session. Most default to the default control character '%'. You can execute terminal commands programmatically by issuing the SET CONTROL. When the SET CONTROL is used, you omit the control character. For example, %M allows you to set the message line. The program equivalent is SET CONTROL 'M'.

Command	Description
C	Copy current screen into NATURAL work space.
D	Activate keyword/delimiter mode
E	Redisplay stored screens (NATPAGE Facility)
F	Activate forms/screen display
H	Activate send to destination (COM-PLETE, SAG's TP Monitor)
I	Activate storing current screen (NATPAGE Facility
Knn	Permits PF keys 13-24 to be assigned to PF 1-12
KPn	Simulate program attention keys
KN, KO,KS	Siemens terminal function key logic
L	No lower to upper case translation
M,MT,MB	Message line control The message line is unprotected; to protect, add 'P' as suffix to command
N	Activate non-conversational mode The next output screen will be sent without waiting for a response from the user. Much likeEXECUTE-REPEAT
O	Terminate the storing of screens (NATPAGE Facility)
P	Activate storing subsequent screens (NATPAGE Facility) All previous screens are cleared.
Q	Suppress screen printing in batch
R	Resend last screen
S	Resume saving screens (NATPAGE Facility) All previous screens are kept.
T	Set cursor position to top of screen
Trr/cc	Set cursor position to row/column position
T=	Activate converter routine for specific device
U	Activate lower to upper case translation
V	Switch print directional mode
W	Activate window handling (see Chapter 5)
X	Statistics line control
XI+	Switch to infoline mode
XI	Switches infoline into stats mode
Y	Control function key line (New options in SM05)
Z	Used with SET CONTROL only, clear source area
*	In batch, suppress printing of next read input record
%	Interrupt NATURAL processing Interrupt processing while preserving stack
/	Forces EOF condition while using INPUT in batch
+,-	Enable/disable NATURAL CONNECTION
=	Assign colors to field attributes
?	Invoke help for terminal commands
Z	Terminate current processing loop (COM-PLETE, SAG's TP Monitor only)

Here are some additional terminal commands introduced to enhance terminal I/O under CICS.

%P=V Prevents thread block (awaiting user input) when a NATURAL program calls a non-NATURAL program which issues a conversational terminal I/O. Parameterdata is copied out of the thread and the thread is rolled out making it available to another user. When the called program releases control to the NATURAL program, the thread is rolled back in, the parameter data (modified and unmodified) is copied back and the NATURAL program continues processing.

%P=S Call a non-NATURAL program using standard linkage convention and standard register usage instead of the normal"EXEC CICS LINK"request.

%P=C Instead of passing the CALL parameter address list into the CICS COMMAREA, executing this terminal command before the call permits passing the parameter values themselves into the COMMAREA.

%.P An enhancement to stack processing. This allows a facility to read the STACK without erasing the top entry.

%.S An enhancement to stack processing. This allows a facility to pop the STACK entry without execution.

Rules for using terminal commands:

Rule 1: You can enter any terminal command into an unprotected field.

Rule 2: The command cannot span fields, except the control character which can be entered into the preceding field.

Rule 3: The control character which precedes any terminal command, but is omitted when using the SET CONTROL statement, can be redefined using the session parameter CF, or through GLOBALS, NATURAL Security, or SET GLOBALS; % is the default.

IV. Report Mode/Structured Data Mode/Structured Mode

NATURAL application development should use the appropriate programming techniques and constructs which follow the tenets of structured programming. Part of the design of the NATURAL 2 is to provide a system which supports the development of programs which are well-designed, readable, maintainable and follows standards. However, it would be remiss of Software AG to allow NATURAL 2 to not support any development done in NATURAL 1.

Therefore, the need to support both old and new development is handled by the programming modes of NATURAL 2 - report and structured. Any NATURAL version 1.2 object executes under NATURAL 2. You also have the option to compile under NATURAL 2 and recatalog. If the program requires a map, then be sure to compile and recatalog the maps as well

In essence, the difference between Report Mode and Structured Mode is a compile time issue only. The resulting stowed objects execute the same. Objects created under either mode can be executed in an application. However, this is said with caution. Any application which relies on dynamic global variables, if it is not converted with a global data area, may cause headaches for maintenance. Also, if you decide to have all new development in structured mode, you must reconcile NATURAL's future support of dynamic global variables in structured mode.

Structured Data Mode is really reporting mode with one added feature - you can build data areas using the DEFINE DATA statement. This is very useful for properly constructing the data requirements for your programs. By adopting this approach, dynamic definition of user-defined variables is disallowed. Also, any ADABAS file is accessed by a view name defined in the DEFINE DATA block instead of referencing the DDM in the traditional NATURAL 1.2 way.

Structured Mode provides a disciplined approach to applying systems design to applications coding and implementation. It forces the planning of relationships between program modules and data beforehand. By doing this, large, complex systems can be implemented and maintained much more easily.

There is a list of Structured Mode restrictions in NATURAL.

1. DO...DOEND is not allowed.
2. MOVE INDEXED is not available.
   Real array processing takes its place.
3. OBTAIN is not allowed.
   All occurrences of the fields are determined by the views.
4. A subroutine must have the header DEFINE SUBROUTINE.
5. The FIND FIRST, SET GLOBALS, and LOOP statements are not allowed.
6. The ESCAPE mechanism requires TOP, BOTTOM, ROUTINE or ROUTINE IMMEDIATE.
7. ADABAS 2 character names are prohibited.
8. All variables must be defined in the DEFINE DATA blocks.
9. Format parameters PC and MC are disallowed.
10. SORT statement requires the USING clause.

Loop Constructs

HISTOGRAM............................END-HISTOGRAM
READ ....END-READ
FIND ....END-FIND
SORT ....END-SORT
READ WORK FILE.................END-WORK
CALL FILE ....END-FILE
CALL LOOP.............................END-LOOP
REPEAT...................................END-REPEAT
FOR.................. ....END-FOR
END-ALL ....Used with SORT to close all loops opened prior to SORT.

Conditional Blocks

DECIDE............... ....END-DECIDE
IF.................. ....END-IF
IF NO RECORDS FOUND ......END-NOREC
AT BREAK.............. ....END-BREAK
AT START OF DATA ....END-START
AT END OF DATA ....END-ENDDATA
AT TOP OF PAGE...................END-TOPPAGE
AT END OF PAGE ....END-ENDPAGE
ON ERROR ....END-ERROR

Functional blocks

DEFINE DATA.........................END-DEFINE
DEFINE SUBROUTINE..........END-SUBROUTINE

V. Global Data Areas - GDA

Considerations of data requirements and use of GDAs needs to be done early in the implementation stage of a system using NATURAL. Specific person(s) ought to be responsible for evaluating and making changes to the GDA since changes have the potential for creating havoc throughout an application.

Here are several characteristics of the GDA:

1. A GDA must be separately definable and compilable. Only one can be referenced per program and must be external.
2. Recompilation requires recompilation of all objects (programs, subprograms, etc.) which use it.
3. A GDA can be divided into “blocks” to create overlay data areas
4. The maximum size of a GDA is 64k source & 32k object (However, practice shows object has a maximum of 28K)
5. When used by a subprogram, a GDA is created at one “GDA level” below the lowest active “GDA level". This means that any stored values in the lower level GDA are not related to those in the higher level GDA. The maximum size of all GDAs is 60K.(Not 64K since 4K is used by NATURAL.)
6. If a group contains a constant ("CONST") definition then the group must contain only constants.

Here are some guidelines for using GDAs:

1. Use one common GDA at "GDA level 1" for all programs in an on-line application. This GDA should contain only data that is global to all programs in the application such as user command line, current/previous/next program/function/application, record keys, function key definitions, etc. In batch, it provides an alternate area for work files to store report counts, import totals, passed parameters, etc.
2. Group variables together logically under level 1 groups with meaningful names. Avoid level 1 elementary variables (wastes spaces due to compiler-generated "fillers" for word alignment which increases memory requirements at execution time (ESIZE). Make the group names small in length since it serves as a qualifier index level in the symbol table. There are people who say "Do not use groups"
3. Use the application startup program to initialize session/user-dependent variables. Use INIT and CONST for all other definitions. This also enables the use of the RESET INITIAL statement. Avoid naming the GDA Common or Acommon.
4. Consider prefixing GDA elements to ensure that future enhancements to NATURAL will not create a new reserved word which may conflict and to make it easier to determine where global variables and constants are referenced. Use meaningful but succinct names.
5. If GDA contents are required in a sub program, it can be passed as a parameter to the sub program through the PDA.

VI. Local Data Areas - LDA

LDAs provide the means to define the data requirements particular to one or more programs. Multiple LDAs - defined externally - can be linked to a single program along with an internally defined LDA. Here are several characteristics of the LDA:

1. They can be separately defined and compiled or can be coded directly within programs.
2. Recompilation does not require recompilation of all objects using it (but is highly suggested).
3. The maximum size per LDA: 32K object (USIZE) , is in reality 28K.
4. When a program invokes another module, the program's LDA remains active as well as that of the invoked module.
5. If a group contains a constant ("CONST") definition then the group must contain only constants.
6. Views contain current security access authorization. If the security access changes, the LDA must be recompiled.

Here are some good guidelines for using LDAs:

1. Prefix all local variables and constants with "#" to ensure future releases of NATURAL will not create a new reserved word which conflicts. Use meaningful but succinct names.
2. Group variables together logically under a level 1 group with a short but meaningful name. Avoid level 1 elementary variables (wastes space due to compiler-generated "fillers" for word alignment).
3. Use one common external LDA for all variables common throughout an application (e.g., flags, constants). Do not go overboard; logically related fields belong in an external LDA.

Here is an example of a common LDA used in many applications:

```
1 CONSTANT-CHAR
2 #BLANK                          A    1
2 #ASTERISK                       A    1 INIT<'*'>
2 #QUESTION-MARK                  A    1 INIT<'?'>
2 #DASH                           A    1 INIT<'-'>
2 #COLON                          A    1 INIT<':'>
2 #SEMI-COLON                     A    1 INIT<';'>
2 #DOLLAR                         A    1 INIT<'$'>
2 #PERCENT                        A    1 INIT<'%'>
2 #EXCLAMATION                    A    1 INIT<'!'>
2 #PLUS                           A    1 INIT<'+'>
2 #MINUS                          A    1 INIT<'-'>
2 #HYPHEN                         A    1 INIT<'-'>
2 #AMPERSAND                      A    1 INIT<'&'>
2 #EQUAL                          A    1 INIT<'='>
2 #CENT                           A    1 INIT<']'>
2 #QUOTE                          A    1 INIT<''''>
2 #SLASH                          A    1 INIT<'\'>
2 #COMMA                          A    1 INIT<','>
2 #PERIOD                         A    1 INIT<'.'>
2 #DECIMAL                        A    1 INIT<'.'>
2 #YES                            L      INIT<TRUE>
2 #NO                             L      INIT<FALSE>
```

4. Put views used into one or more LDAs. There is a controversy whether or not views belong in external LDAs or GDAs.
5. One advantage for using the data editor to define an LDA is that an external LDA can always be ".i"d into a program. If you code data internally to a program and then decide that you want an external LDA, you have to manually type the definitions into an LDA using the data editor.
6. Views of DDMs are required. Be conscious of the tendency to "pull in" more fields than are necessary. This is often overdone.
7. Structured array definitions are possible.
8. MOVE BY NAME is available because of definable group data structures.

VII. Parameter Data Areas - PDA

The parameter data area provides a mechanism for passing data to subprograms and helproutines. They may be defined within the subprogram or in a separately defined and compiled PDA. Although a PDA can be defined and edited using the Data Editor, there is no separate "data area" created at execution time. The NATURAL compiler sets up and uses the addresses of the passed parameters.

Here are several characteristics of the PDA:

1. Parameters in the subprogram define the order, format and length of the parameters specified in the "CALLNAT" statement in the invoking program.
2. A passed parameter can be an elementary item or a group item or an array.
3. You cannot redefine in a PDA as you can in an LDA/GDA.

Here are some guidelines for using PDAs:

1. For maintenance and readability, code "AD=O" or "AD=M" for each parameter to document whether its contents is changeable or not. This is written as part of the CALLNAT statement. (optional of course)
2. Only use an external PDA when it is used by more than one subprogram or when an externally defined PDA will significantly reduce the size of the source of the subprogram. This way it can double as an LDA in other modules.
3. Make sure that whatever data is passed is needed by the subprogram. Do not use groups arbitrarily.
4. If a subprogram is to be widely used, it is easier and faster to create a COPYCODE object for its "CALLNAT" statement, which you can ".I" or INCLUDE.

This also guarantees that the sequence of the parameters is correct. A sample CALLNAT statement might be:

```
CALLNAT 'LKUPF5S' /* LOOKUP SINGLE OCCURRENCE ON TABLE FILE 5
  #TABLE-NAME     /* (AD=M) (A6)    - PART I  TABLE INDEX
  #TABLE-EXT      /* (AD=M) (A6)    - PART II TABLE INDEX EXTENSION
  #RET-VAL1       /* (AD=O) (A30)   - RETURN VALUE 1
  #RET-VAL2       /* (AD=O) (A30)   - RETURN VALUE 2
```

Chapter 5

Maps and Windows

I. The Map Editor

Without a doubt, the newest and “biggest” of the editors is the Map Editor. This material complements Software AG’s NATURAL Reference Manual as does WH&O’s NATURAL Developers Handbook. The Map Editor allows you to create map displays which are invoked by the INPUT USING MAP option. In the editor you can define your screen form and its fields, any processing rules and their levels, help text, the specification of helproutines, and various other map characteristics.

There will be no complete descriptions since Software AG's NATURAL Utilities Manual and WH&O's NATURAL Developers Handbook do so in their notes. First, a Map Editor overview is provided which you can keep pinned by your bed post.

Map Editor Overview

All field commands are prefixed with '.', where as all line commands are prefixed with '..' (or whatever control character is currently defined in the edit session).

Editor's Note: This chapter extends some of the topics mentioned in Chapter 4 with more detail on how to use some of the features of the NATURAL 2.2 Map Editor and windowing techniques that illustrate many of the new features introduced with the release of NATURAL 2.2.

Field Command	Line Command	Description
.A	..A	Invokes array definition menu
.C	..C	Center current line(s) or field(s)
.D	..D	Delete current line(s) or field(s)
.E	..E	External (extended) field editing of field or fields on current line
-	..Fc	Fill blanks with character 'c'
-	..I	Insert single line
.J	..J	Join current line/field with following line/field
.M	..M	Move current line/field to current screen position of the cursor
.P	..P	Processing rule editor
.R	..R	Repeat text at current cursor position
-	..S	Split line at cursor position
.T	-	Truncate from current field to end of line

Figure 5.1. Map Editor Command Summary

Line Commands permit an * suffix which indicates that the operation is to include the current line and all lines that follow. They also permit a number suffix which indicates that the operation is to include the current line and all the n -1 lines that follow.

Two simultaneous Field Commands indicate a range of fields for execution of the operation, that is, two .C field commands indicate centering of all the fields between the two .C fields. Some Fields Commands are cursor sensitive.

The Map Profile

The Map Editor allows you to define default map settings of your own or use a default set named 'SYSPROF'. These map settings are clearly described in the Software AG NATURAL reference manual section on the Map Editor. Let me highlight some of the features.

The editor profile is divided into four sections: delimiters, format, context and filler characters. The delimiter definitions define the attributes for fields as well as the color attributes for display.

A. Do not use the '%' as any attribute character in the Map Editor. The use of the symbol which also serves as the terminal control character can cause confusion.

B. Because of window processing, the page size and line size can be 250 and 250 respectively; however, the size of the logical screen is limited by the page buffer. I tend to stay away from defining screens larger than 23 by 79 (the default). Without the terminals which allow oversize display you have to engineer windowing around the physical screen. This can be done with PF keys, but some shops do not allow them. Therefore, I avoid the whole issue by relying on NATURAL's reduced data transmission and design screen overlays to give the impression of scrolling right and left.

C. If a layout is defined, a standard form is brought into the work area. However, under normal circumstances you cannot remove anything which is part of the predefined layout.

  Static Layouts - predefined map elements which are incorporated into the current map definition
  Dynamic Layouts - predefined map elements which are stored externally and invoked during execution of the map

D. Under COM-PLETE, upper and lower case for maps does not take affect until you toggle the TP session. This can be done through calling the MODIFY routine. Then, if you default the map case setting to LC, an individual field can be upper case by attaching the field attribute 'T' for translate.

E. Manual skip 'Y' can be important for data entry screens. the only drawback to this parameter is that is applies to all screen fields, that is, it cannot be selectively applied.

F. The Std Key option establishes the availability of the last two lines on the physical screen for function key display. The combination of events one must do to get screen display may not be obvious. First, the map must have the Std Key option set to yes. The page size must include the actual screen plus two for the function key specifications. Last, of course, are the PF key definitions in the program.

G. You can assign special filler characters to display in fields for which mandatory or complete filling is required. The choice of characters is done on the filler characters portion of the profile screen. Assigning the character is not enough. You must also attach the appropriate attribute to the field - AD=E for partial filled but required, AD=G for completely filled but optional to type, and AD=EG means completely filled and mandatory.

H. Control variables may be defined and attached to multiple fields through a special user-defined delimiter character. Alternatively, control variables may also be set for fields individually.

I. Edit masks are easily applied to fields although for input they come up short in their approach for handling data. Each edit mask may be coded manually or selected from a menu invoked when help is requested.

J. Help may be attached to all fields on a map or attached to a specific field. A separate menu is available if the number of additional parameters required by the help member are greater than the normal HE parameter size.

K. Application of dynamic parameters to a specific field is available as either a set of escape sequences set manually or as selected from a menu invoked when help is requested.

Additional Notes About The Map Editor

A. You can attach a NATURAL system variable to a map. Remember to prefix the variable with the output delimiter defined for the map.

B. You do not have to define fields in the Map Editor as the source. If they are predefined in a data area or in a program, you can select the object and choose the fields from this source. Since you have to define the type of the object, type '?' on the 'Ob:' line and a window with this list appears:

D	(Delimiters)	P	Program
V	View	N	Subprogram
G	Global	S	Subroutine
L	Local	M	Map
A	Parameter	H	Helproutine

C. There are two ways to change the length of a field. By using extended field editing, a redefinition of the length results in a redefinition in the parameter data area. However, you can also alter the length on the screen by erasing characters from the screen. This results only in changing the display by attaching the AL parameter as a field attribute. The length of the passed parameter is not changed.

Processing Rules

Processing rules fall into three categories - inline rules, free rules defined in the Map Editor loaded into PREDICT, and automatic rules which are created in PREDICT. Any change to a PREDICT rule cannot take affect without recompiling the map(s) to which the field is attached. Inline rules only pertain to the map which you are editing. Free rules are created and named in the Processing Rule Editor and subsequently stored in PREDICT. Inline rules are not named and are not stored in PREDICT, however, inline rules can be saved as copycode.

A processing rule can be created within the Map Editor. By typing '.P' on the field or by invoking Extended Field Editing ('..E') on the line. Figure 5.2 illustrates a typical Processing Rule Editor screen. If the rule is defined to PREDICT as an automatic rule, then after selecting it from a view, all of the automatic rules linked to the field are linked to the map definition. They cannot be modified or unlinked, but you can change the rank. The rank determines the order in which a processing rule is evaluated. The character '&' in the source of the processing rule is dynamically replaced with the field to which it is assigned at compile time. In another variation, '&.' used as a prefix will fully qualify a database field to the current view permitting broader sharing of inline rules.

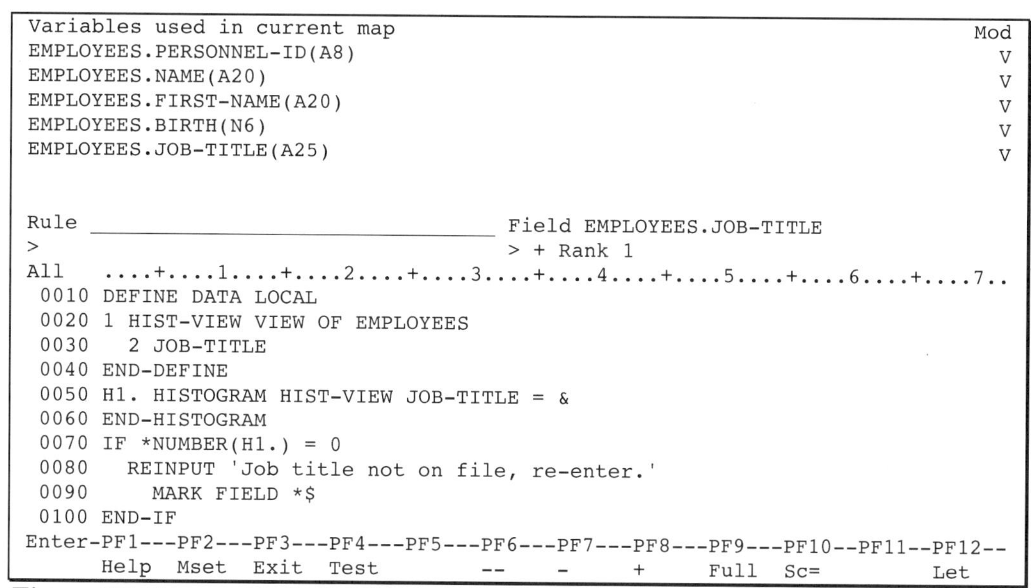

```
Variables used in current map                                         Mod
EMPLOYEES.PERSONNEL-ID(A8)                                              V
EMPLOYEES.NAME(A20)                                                     V
EMPLOYEES.FIRST-NAME(A20)                                               V
EMPLOYEES.BIRTH(N6)                                                     V
EMPLOYEES.JOB-TITLE(A25)                                                V


Rule ________________________________ Field EMPLOYEES.JOB-TITLE
>                                     > + Rank 1
All   ....+....1....+....2....+....3....+....4....+....5....+....6....+....7..
 0010 DEFINE DATA LOCAL
 0020 1 HIST-VIEW VIEW OF EMPLOYEES
 0030   2 JOB-TITLE
 0040 END-DEFINE
 0050 H1. HISTOGRAM HIST-VIEW JOB-TITLE = &
 0060 END-HISTOGRAM
 0070 IF *NUMBER(H1.) = 0
 0080   REINPUT 'Job title not on file, re-enter.'
 0090     MARK FIELD *$
 0100 END-IF
Enter-PF1---PF2---PF3---PF4---PF5---PF6---PF7---PF8---PF9---PF10--PF11--PF12--
      Help  Mset  Exit  Test        --    -     +     Full  Sc=         Let
```

Figure 5.2. Processing Rule Editor

A special processing rule is available specific to Program Function keys giving the programmer, and the user, an added level of control in the application.

```
 Rule ________________________________ Field *PF-KEY
 >                                     > + Rank 0
 All
....+....1....+....2....+....3....+....4....+....5....+....6....+....7..
```

Figure 5.3. PF Key Processing Rule Editor

Map Variables

The ability to share data across processing rules is provided through a Local Data area defined for the map. Variables may also be defined to pass data from the map to the invoking module through a Parameter Data Area defined for the map. Both these new editors are invoked from the Field and Variable Definitions Summary Menu (type 'D' while at the Edit Map Menu).

Parameter Data

By default, all data base fields and program variables incorporated into the map potentially pass data to and from the map. Additional variables may be defined in the map to pass more data than originally defined. The invoking program will have to define these fields to execute successfully. Both scalar and array definitions are permitted.

```
12:34:56          Field and Variable Definitions - Summary          97-04-01

Cmd Field Name (Truncated)                                Mod  Typ    Ru Lin Col
___ EMPLOYEES.PERSONNEL-ID                                D    A08        1   20
___ EMPLOYEES.NAME                                        D    A20        2   20
___ EMPLOYEES.FIRST-NAME                                  D    A20        3   20
___ EMPLOYEES.BIRTH                                       D    N06        4   20
___ EMPLOYEES.JOB-TITLE                                   D    A25        5   20
+--------------------------- PARAMETER DEFINITIONS ------------------------+
| Cmd Name of Parameter (Truncated)                            Format Ar |
| ___ ___________________________________________________     ______    |
| ___ ___________________________________________________     ______    |
| ___ ___________________________________________________     ______    |
| ___ ___________________________________________________     ______    |
| ___ ___________________________________________________     ______    |
| ___ ___________________________________________________     ______    |
| ___ ___________________________________________________     ______    |
| ___ ___________________________________________________     ______    |
| ___ ___________________________________________________     ______    |
|                                                                       |
+-----------------------------------------------------------------------+
Enter-PF1---PF2---PF3---PF4---PF5---PF6---PF7---PF8---PF9---PF10--PF11--PF12--
      Help  Mset  Exit              --                Parm  Local       Let
```

Figure 5.4. Map Variables Editor - Press PF9 or PF10 to Invoke the Appropriate Editor

Local Data

Variables defined specifically for sharing data between processing rules. Both scalar and array definitions are permitted.

Other Map Objects

Help text

Using the same set of facilities to build a map, help text are mini-maps invoked as help. The same steps are generally followed in creating a help text member as maps. About the only exception is that help text, since they are commonly defined as windows, needs to be positioned on the screen when called upon. By default, the base of the help text window is positioned just under the first position of the field to which it is attached. As an alternative, explicit line and column values may be coded thus fixing the help text at the same screen location each time it is invoked. See Figure 5.5 for the new helptext delimiter setting menu.

Forms

Report layouts are easily defined using the Map Editor. These are executed with the WRITE USING FORM statement thus reducing the amount of source code required in the program. Since they are WRITE oriented, only text and output data may be defined in the layout (see Figue 5.6 for the typical delilmiter settings menu associated with forms definition). Line size and page size parameters define the depth of the data and the report width at the same time. The full complement of facilities used to define maps may be used to define a report layout. There is one advantage to this - WRITE statements cannot display a three dimensioned array but since these arrays can be laid out in the Map Editor, a FORM has the capability of displaying three dimensions of an array. Since they are external, FORMS may be used in more than one program or more than once in the same program.

```
12:34:56              Define Map Settings for HELPMAPS                97-04-01

 Delimiters              Format                         Context
 -----------------       ---------------------------    -------------------------
 Cls Att CD  Del         Page Size ...... 23_           Device Check .... _______
  T   D      BLANK       Line Size ...... 79_           WRITE Statement   _
  T   I      ?           Column Shift ... 0 (0/1)       INPUT Statement   X
  A   D      _           Layout ......... ________
  A   I      )            dynamic ....... N (Y/N)
  A   N      ^           Zero Print ..... N (Y/N)       Position Line ___ Col ___
  M   D      &           Case Default ... UC (UC/LC)
  M   I      :           Manual Skip .... N (Y/N)       Automatic Rule Rank  1
  O   D      +           Decimal Char ... .             Profile Name .... SYSPROF
  O   I      (           Standard Keys .. N (Y/N)
                         Justification .. L (L/R)       Filler Characters
                         Print Mode ..... __            ------------------------
                                                        Optional, Partial .... _
                         Control Var .... ________      Required, Partial .... _
                                                        Optional, Complete ... _
                                                        Required, Complete ... _

Enter-PF1---PF2---PF3---PF4---PF5---PF6---PF7---PF8---PF9---PF10--PF11--PF12--
      Help        Exit                                                 Let
```

Figure 5.5. Help text Delimiter Settings Menu. Notice the absence of the Help Parameter and the Addition of the Position Parameter.

```
12:34:56              Define Map Settings for MAP                     97-04-01

 Delimiters              Format                         Context
 -----------------       ---------------------------    -------------------------
 Cls Att CD  Del         Page Size ...... 10_           Device Check .... ______
  T   D      BLANK       Line Size ...... 133           WRITE Statement   x
  T   I      ?           Column Shift ... 0 (0/1)       INPUT Statement   _
                         Layout ......... ________      Help ___________________
                          dynamic ....... N (Y/N)        as field default N (Y/N)
                         Zero Print ..... N (Y/N)
                         Case Default ... UC (UC/LC)
                         Manual Skip .... N (Y/N)       Automatic Rule Rank  1
  O   D      +           Decimal Char ... .             Profile Name .... SYSPROF
  O   I      (           Standard Keys .. N (Y/N)
                         Justification .. L (L/R)       Filler Characters
                         Print Mode ..... __            ------------------------
                                                        Optional, Partial .... _
                         Control Var .... ________      Required, Partial .... _
                                                        Optional, Complete ... _
                                                        Required, Complete ... _

Enter-PF1---PF2---PF3---PF4---PF5---PF6---PF7---PF8---PF9---PF10--PF11--PF12--
      Help        Exit                                                 Let
```

Figure 5.6. When the WRITE Statement Parameter is Selected, the Delimiters Section Loses the A and M Classes- WRITE Only Displays Output Only Values

II. Case Studies

Case Study #1

Creating A Map With 4 Single Dimension Lists

There are several different ways to accomplish this. The technique presented here is one of the easiest for me. I have a tendency to forget how to define arrays on maps. Too much time seems to pass before I need the capability again, and I seem to sit at my terminal, scratching my head trying to remember. So I let *NATURAL* remind me!

I want a screen to have four columns, three columns of display values and one selection column to 'X' any record for which the view is to be exploded. The columns are to be 15 elements in length.

Assume ')' is the character for AD=A
'+' is the character for AD=O

Step 1. Create one line reflecting how the columns should look.

```
 )X +XXXXXXXXXXXXXXXXXXXX   +XXXXXXXXXXXXXXXXXXXXXXXXXXXXXX   +XXXXXXXX
```

Figure 5.7. Base Line for Array Definition Example

Make sure you create the first line accurately, positioning the fields in the columns you want them. I suggest drawing the display before you start. Repositioning the columns can be tricky after you are done.

Step 2. Type .A in the first position of the first field in the first column.

```
Name #001                                               Upper Bnds 1____  1____  1____

Dimensions                         Occurrences    Starting from       Spacing
0 . Index vertical                 ___            ________            0   Lines
0 . Index horizontal               ___            ________            0   Columns
0 . Index (h/v) V                  ___            ________            0   Cls/Ls

001    --010---+----+----+---030---+----+----+---050---+----+----+---070---+---


  .A +XXXXXXXXXXXXXXXXXXXX   +XXXXXXXXXXXXXXXXXXXXXXXXXXXXXX   +XXXXXXXX








Enter-PF1---PF2---PF3---PF4---PF5---PF6---PF7---PF8--PF9---PF10--PF11--PF12--
      HELP  MSET  QUIT              TOP    -     +          <     >      LET
```

Figure 5.8. Invoking Array Definition Menu

Step 3. Type 15 in first position for "Upper Bnds:", 1 in the "Dimensions" column adjacent to "Index vertical", and type 15 in the "Occurrences" column. Your screen should now look like:

```
Name #001                                              Upper Bnds 15___ 1____ 1____

Dimensions                           Occurrences    Starting from      Spacing
1 . Index vertical                   15_            ________           0    Lines
0 . Index horizontal                 ___            ________           0    Columns
0 . Index (h/v) V                    ___            ________           0    Cls/Ls

001    --010---+----+----+---030---+----+----+---050---+----+----+---070---+---


 .A +XXXXXXXXXXXXXXXXXXXXX   +XXXXXXXXXXXXXXXXXXXXXXXXXXXXXXX   +XXXXXXXX
 )X
 )X
 )X
 )X
 )X
 )X
 )X
 )X
 )X
Enter-PF1---PF2---PF3---PF4---PF5---PF6---PF7---PF8---PF9---PF10--PF11--PF12--
      Help  Mset  Exit              --     -     +                     Let
```

Figure 5.9. Layout of Array As Defined in Array Definition Menu

Step 4. Press the ENTER key. Type '..A' on the top row of this display and the following screen will appear.

```
12:34:56                  ***** NATURAL MAP EDITOR *****                97-04-01
                            - Array Table Definition -

Main   Index:  Vert. Occur.   15   Starting from ________   Spacing 0   Lines
Second Index:  Direction(H/V) V                  ________           0   Cls/Ls
Third  Index:  Direction(H/V) V                  ________           0   Cls/Ls
------------------------------------------------------------------------------
Name of Variable                  Col   Dimension Size     Order   2.Ind  3.Ind
(truncated)                       Pos   Ind1  Ind2  Ind3   M S T   Occ.   Occ.
------------------------------------------------------------------------------
 #001                              11     15     1     1   1
 #002                              14      1     1     1   1
 #003                              31      1     1     1   1
 #004                              54      1     1     1   1








Enter-PF1---PF2---PF3---PF4---PF5---PF6---PF7---PF8---PF9---PF10--PF11--PF12--
      Help  Mset  Exit              --     -     +                     Let
```

Figure 5.10a. Table Definition Menu

```
12:34:56                 ***** NATURAL MAP EDITOR *****                97-04-01
                          - Array Table Definition -

Main    Index:  Vert. Occur.    15    Starting from ________    Spacing 0    Lines
Second  Index:  Direction(H/V) V                    ________            0    Cls/Ls
Third   Index:  Direction(H/V) V                    ________            0    Cls/Ls
-------------------------------------------------------------------------------
Name of Variable                  Col   Dimension Size       Order   2.Ind  3.Ind
(truncated)                       Pos   Ind1   Ind2   Ind3   M S T   Occ.   Occ.
-------------------------------------------------------------------------------
 #001                              11     15     1      1   1
 #002                              14     15     1      1   1
 #003                              31     15     1      1   1
 #004                              54     15     1      1   1




Enter-PF1---PF2---PF3---PF4---PF5---PF6---PF7---PF8---PF9---PF10--PF11--PF12--
      Help  Mset  Exit                --    -     +                      Let
```

Figure 5.10b. Table Definition Menu

Step 5. Now that I see what NATURAL wanted for the original display, I can now idiotically repeat it for the other fields. The result is the following:

```
Ob: _                                  Ob: D CLS ATR DEL       CLS ATR DEL
.                                      .      T   D   BLNK      T   I   ?
.                                      .      A   D   _         A   I   )
.                                      .      A   N   ^         M   D   &
.                                      .      M   I   :         O   D   +
.                                      .      O   I   (
001   --010---+----+----+---030---+----+----+---050---+----+----+---070---+----


  )X +XXXXXXXXXXXXXXXXXXXX    +XXXXXXXXXXXXXXXXXXXXXXXXXXXXXX    +XXXXXXXX
  )X +XXXXXXXXXXXXXXXXXXXX    +XXXXXXXXXXXXXXXXXXXXXXXXXXXXXX    +XXXXXXXX
  )X +XXXXXXXXXXXXXXXXXXXX    +XXXXXXXXXXXXXXXXXXXXXXXXXXXXXX    +XXXXXXXX
  )X +XXXXXXXXXXXXXXXXXXXX    +XXXXXXXXXXXXXXXXXXXXXXXXXXXXXX    +XXXXXXXX
  )X +XXXXXXXXXXXXXXXXXXXX    +XXXXXXXXXXXXXXXXXXXXXXXXXXXXXX    +XXXXXXXX
  )X +XXXXXXXXXXXXXXXXXXXX    +XXXXXXXXXXXXXXXXXXXXXXXXXXXXXX    +XXXXXXXX
  )X +XXXXXXXXXXXXXXXXXXXX    +XXXXXXXXXXXXXXXXXXXXXXXXXXXXXX    +XXXXXXXX
  )X +XXXXXXXXXXXXXXXXXXXX    +XXXXXXXXXXXXXXXXXXXXXXXXXXXXXX    +XXXXXXXX
  )X +XXXXXXXXXXXXXXXXXXXX    +XXXXXXXXXXXXXXXXXXXXXXXXXXXXXX    +XXXXXXXX
  )X +XXXXXXXXXXXXXXXXXXXX    +XXXXXXXXXXXXXXXXXXXXXXXXXXXXXX    +XXXXXXXX
  )X +XXXXXXXXXXXXXXXXXXXX    +XXXXXXXXXXXXXXXXXXXXXXXXXXXXXX    +XXXXXXXX
  )X +XXXXXXXXXXXXXXXXXXXX    +XXXXXXXXXXXXXXXXXXXXXXXXXXXXXX    +XXXXXXXX
Enter-PF1---PF2---PF3---PF4---PF5---PF6---PF7---PF8--PF9---PF10--PF11--PF12--
      Help  Mset  Quit  Test  Edit  Top   -     +          <     >     Let
```

Figure 5.11. Tables (Arrays) Laid Out From Table Definition Selections

This could have all been accomplished by going directly into the '..A' command, but I am so forgetful sometimes on using the various powerful NATURAL facilities, that I prefer it does the work for me!

Case Study #2

Creating Map With Two One-Dimensional Lists Wrapping Around the Screen

This is similar to the above case study 1. The only difference is that the list is defined for a length enough to guarantee the need for multiple columns. Be careful not to cause overlapping of positions by not calculating the spacing correctly. A Map Editor error message of 4622 causes consternation, for it seems you can not get out of it.

Example: Define map to have 2 lists of 48 elements each. One is for selection (A1), the other displays a field of format A8.

```
_ ________   _ ________   _ ________   _ ________
_ ________   _ ________   _ ________   _ ________
_ ________   _ ________   _ ________   _ ________
_ ________   _ ________   _ ________   _ ________
_ ________   _ ________   _ ________   _ ________
_ ________   _ ________   _ ________   _ ________
_ ________   _ ________   _ ________   _ ________
_ ________   _ ________   _ ________   _ ________
_ ________   _ ________   _ ________   _ ________
_ ________   _ ________   _ ________   _ ________
_ ________   _ ________   _ ________   _ ________
_ ________   _ ________   _ ________   _ ________
```

Assume '&' is the character for AD=A
'(' is the character for AD=O

Step 1. Place `&X (XXXXXXXX` positioned as illustrated on the screen. Again, position carefully and know the spacing you want between columns. This will minimize the need to redo or to move anything.

Step 2. Type ..A on this line. The screen shown in Figure 5.12 will appear .

```
12:34:56              ***** NATURAL MAP EDITOR *****                97-04-01
                        - Array Table Definition -

Main   Index: Vert. Occur.   1    Starting from ________   Spacing 0   Lines
Second Index: Direction(H/V) V                  ________           0   Cls/Ls
Third  Index: Direction(H/V) V                  ________           0   Cls/Ls
------------------------------------------------------------------------------
Name of Variable                   Col  Dimension Size      Order  2.Ind 3.Ind
(truncated)                        Pos  Ind1  Ind2  Ind3    M S T  Occ.  Occ.
------------------------------------------------------------------------------
 #001                              10     1     1     1    1
 #002                              14     1     1     1    1









Enter-PF1---PF2---PF3---PF4---PF5---PF6---PF7---PF8---PF9---PF10--PF11--PF12--
      Help  Mset  Exit                --    -     +                       Let
```

Figure 5.12. Overlapping Arrays Defined in Table Definition Menu

Step 3. By entering the following, you will create two lists which wrap around the screen.

```
12:34:56              ***** NATURAL MAP EDITOR *****                97-04-01
                        - Array Table Definition -

Main   Index: Vert. Occur.   12   Starting from ________   Spacing 0   Lines
Second Index: Direction(H/V) h                  ________          12   Cls/Ls
Third  Index: Direction(H/V) V                  ________           0   Cls/Ls
------------------------------------------------------------------------------
Name of Variable                   Col  Dimension Size      Order  2.Ind 3.Ind
(truncated)                        Pos  Ind1  Ind2  Ind3    M S T  Occ.  Occ.
------------------------------------------------------------------------------
 #001                              10    48     1     1            4
 #002                              14    48     1     1            4









Enter-PF1---PF2---PF3---PF4---PF5---PF6---PF7---PF8---PF9---PF10--PF11--PF12--
      Help  Mset  Exit                --    -     +                       Let
```

Figure 5.13. Multiple Array Layout

Step 4. Press the ENTER key. The following screen appears:

```
Ob: _                                          Ob: D CLS ATR DEL      CLS ATR DEL
.                                              .      T   D  BLNK      T   I   ?
.                                              .      A   D   _        A   I   )
.                                              .      A   N   ^        M   D   &
.                                              .      M   I   :        O   D   +
.                                              .      O   I   (
001  --010---+----+----+---030---+----+----+---050---+----+----+---070---+----

        &X  (XXXXXXXX  &X  (XXXXXXXX  &X  (XXXXXXXX  &X  (XXXXXXXX
        &X  (XXXXXXXX  &X  (XXXXXXXX  &X  (XXXXXXXX  &X  (XXXXXXXX
        &X  (XXXXXXXX  &X  (XXXXXXXX  &X  (XXXXXXXX  &X  (XXXXXXXX
        &X  (XXXXXXXX  &X  (XXXXXXXX  &X  (XXXXXXXX  &X  (XXXXXXXX
        &X  (XXXXXXXX  &X  (XXXXXXXX  &X  (XXXXXXXX  &X  (XXXXXXXX
        &X  (XXXXXXXX  &X  (XXXXXXXX  &X  (XXXXXXXX  &X  (XXXXXXXX
        &X  (XXXXXXXX  &X  (XXXXXXXX  &X  (XXXXXXXX  &X  (XXXXXXXX
        &X  (XXXXXXXX  &X  (XXXXXXXX  &X  (XXXXXXXX  &X  (XXXXXXXX
        &X  (XXXXXXXX  &X  (XXXXXXXX  &X  (XXXXXXXX  &X  (XXXXXXXX
        &X  (XXXXXXXX  &X  (XXXXXXXX  &X  (XXXXXXXX  &X  (XXXXXXXX
        &X  (XXXXXXXX  &X  (XXXXXXXX  &X  (XXXXXXXX  &X  (XXXXXXXX
        &X  (XXXXXXXX  &X  (XXXXXXXX  &X  (XXXXXXXX  &X  (XXXXXXXX
Enter-PF1---PF2---PF3---PF4---PF5---PF6---PF7---PF8--PF9---PF10--PF11--PF12--
      Help  Mset  Quit  Test  Edit  Top    -     +          <     >     Let
```

Figure 5.14. Multiple Arrays as Defined to the Map Editor

Important items:

1. The vertical occurrence count is the length in each column.

2. The direction of the secondary index reflects the columns going across the screen.

3. The spacing count is the distance between each LIKE column.

4. The dimension of Ind1 is the length of the list.

5. The "2.Ind Occ" represents the number of columns.

Case Study #3

Creating Map with Multiple Two-Dimensional Arrays in Standard Row-Column Order

Once you master multiple two-dimensional arrays, one is no problem. For this example, I wanted to make a clock (foolishness, but I had to learn how to handle this situation somehow and I figured I might as well have fun doing it!). Here is a "still" image of the clock, since it executed with the "REPEAT" option.

```
97-04-01                     Boston University                        DRMJTW
                                                                    CLOCKPGM




     1 1      8 8 8 8 8          1 1      3 3 3          2 2 2     0 0 0 0 0
 1 1 1 1      8 8 8 8 8      1 1 1 1    3 3   3 3      2 2 2 2 2   0 0 0 0 0
     1 1      8 8   8 8          1 1    3 3   3 3      2 2   2 2   0 0   0 0
     1 1      8 8   8 8          1 1          3 3      2     2 2   0 0   0 0
     1 1      8 8 8 8 8          1 1      3 3 3 3            2 2   0 0   0 0
     1 1      8 8 8 8 8          1 1      3 3 3 3          2 2     0 0   0 0
     1 1      8 8   8 8          1 1          3 3        2 2       0 0   0 0
     1 1      8 8   8 8          1 1    3 3   3 3      2 2 2       0 0   0 0
 1 1 1 1 1    8 8 8 8 8      1 1 1 1 1  3 3   3 3      2 2 2 2 2   0 0 0 0 0
 1 1 1 1 1    8 8 8 8 8      1 1 1 1 1    3 3 3        2 2 2 2 2   0 0 0 0 0



NEXT FUNCTION  _
```

Figure 5.15. The B.U. Clock, An Example of Array Usage

To define this screen, I needed a map which had 6 two-dimensional arrays to hold the hour, minute and second digits. Each array was defined as (A1/10,5), that is, 10 rows, 5 columns, each element having format A1.

Again, I used the NATURAL Map Facility to help me do it correctly. I defined the leftmost array first using the '.A' command, positioned the (1,1) occurrence of other five, and entered '..A' on the first line.

At this point, the screen looked like this:

```
 001    --010---+----+----+---030---+----+-----+---050---+----+----+---070---+--
--
 (XXXXXXXX                        ?Boston?University                         (XXXXXXXX
                                                                             (XXXXXXXX




 ..a(X(X(X(X  (X              (X           (X              (X           (X
  (X(X(X(X(X
  (X(X(X(X(X
  (X(X(X(X(X
  (X(X(X(X(X
  (X(X(X(X(X
  (X(X(X(X(X
  (X(X(X(X(X
  (X(X(X(X(X
  (X(X(X(X(X



?NEXT?FUNCTION&XXXXXXXXXXXXXXXXXXXXXXXXXXXXXXXXXXXXXXXXXXXXXXXXXXXXXXXXXXXXXXXX
```

Figure 5.16. The Preliminary Clock Screen.

After typing ..A and pressing the ENTER key, the following screen appears:

```
12:34:56                  ***** NATURAL MAP EDITOR *****                  97-04-01
                           - Array Table Definition -

Main   Index:  Vert. Occur.   10   Starting from ________    Spacing 0   Lines
Second Index:  Direction(H/V) h                  ________            0  Cls/Ls
Third  Index:  Direction(H/V) V                  ________            0   Cls/Ls
------------------------------------------------------------------------------
Name of Variable                  Col  Dimension Size     Order   2.Ind  3.Ind
(truncated)                       Pos  Ind1  Ind2  Ind3   M S T   Occ.   Occ.
------------------------------------------------------------------------------
 #CLOCK-ARRAY1                      2   10     5     1   1 2     5
 #006                              14   10     1     1
 #002                              30   10     1     1
 #007                              42   10     1     1
 #008                              58   10     1     1
 #009                              70   10     1     1




Enter-PF1---PF2---PF3---PF4---PF5---PF6---PF7---PF8---PF9---PF10--PF11--PF12--
      Help  Mset  Exit              --     -     +                      Let
```

Figure 5.17. Defining the Clock Elements in the Table Definition Menu

Once seeing the definition of the first array, I repeated it for the remaining five.

```
12:34:56                 ***** NATURAL MAP EDITOR *****                97-04-01
                           - Array Table Definition -

Main   Index:  Vert. Occur.   10   Starting from ________    Spacing 0   Lines
Second Index:  Direction(H/V) H                  ________            0  Cls/Ls
Third  Index:  Direction(H/V) V                  ________            0  Cls/Ls
-------------------------------------------------------------------------------
Name of Variable                  Col  Dimension Size     Order   2.Ind  3.Ind
(truncated)                       Pos  Ind1  Ind2  Ind3   M S T   Occ.   Occ.
-------------------------------------------------------------------------------
 #CLOCL-ARRAY1                      2    10     5     1   1 2     5
 #006                              14    10 5         1   1 2     5
 #002                              30    10 5         1   1 2     5
 #007                              42    10 5         1   1 2     5
 #008                              58    10 5         1   1 2     5
 #009                              70    10 5         1   1 2     5_



Enter-PF1---PF2---PF3---PF4---PF5---PF6---PF7---PF8---PF9---PF10--PF11--PF12--
      Help  Mset  Exit              --    -     +                       Let
```

Figure 5.18. Completing the Clock Elements.

When completed, the screen should look like:

```
001    --010---+----+----+---030---+----+-----+---050---+----+----+---070---+--
(XXXXXXXX                     ?Boston?University                      (XXXXXXXX
                                                                      (XXXXXXXX




  (X(X(X(X(X  (X(X(X(X(X   (X(X(X(X(X  (X(X(X(X(X   (X(X(X(X(X  (X(X(X(X(X
  (X(X(X(X(X  (X(X(X(X(X   (X(X(X(X(X  (X(X(X(X(X   (X(X(X(X(X  (X(X(X(X(X
  (X(X(X(X(X  (X(X(X(X(X   (X(X(X(X(X  (X(X(X(X(X   (X(X(X(X(X  (X(X(X(X(X
  (X(X(X(X(X  (X(X(X(X(X   (X(X(X(X(X  (X(X(X(X(X   (X(X(X(X(X  (X(X(X(X(X
  (X(X(X(X(X  (X(X(X(X(X   (X(X(X(X(X  (X(X(X(X(X   (X(X(X(X(X  (X(X(X(X(X
  (X(X(X(X(X  (X(X(X(X(X   (X(X(X(X(X  (X(X(X(X(X   (X(X(X(X(X  (X(X(X(X(X
  (X(X(X(X(X  (X(X(X(X(X   (X(X(X(X(X  (X(X(X(X(X   (X(X(X(X(X  (X(X(X(X(X
  (X(X(X(X(X  (X(X(X(X(X   (X(X(X(X(X  (X(X(X(X(X   (X(X(X(X(X  (X(X(X(X(X
  (X(X(X(X(X  (X(X(X(X(X   (X(X(X(X(X  (X(X(X(X(X   (X(X(X(X(X  (X(X(X(X(X
  (X(X(X(X(X  (X(X(X(X(X   (X(X(X(X(X  (X(X(X(X(X   (X(X(X(X(X  (X(X(X(X(X



?NEXT?FUNCTION&XXXXXXXXXXXXXXXXXXXXXXXXXXXXXXXXXXXXXXXXXXXXXXXXXXXXXXXXXXXXXXXX
```

Figure 5.19. B.U. Clock Elements As Defined to the Map Editor

Again, be careful not to cause overlapping! This can easily happen, especially if the screen width in the map settings isn't large enough to handle the definitions. Also, beware of 'INVALID RANK SETTING', or 4630. Press PF3 to quit does not seem to cure the problem. Either use PF12 to reset, or figure out your mistake and correct it (good luck).

Case Study #4

Creating Map With 2 Two-Dimensional Arrays in Different Arrangements.

There are four basic screen displays presented here and examples of how each was defined. Here are all of the screen displays and the array definitions so you can compare the differences.

```
                                TEST                              MAP: MAPTST01

                S E M E S T E R   I             S E M E S T E R  II
               GRADES      SUBJECT             GRADES      SUBJECT
                 A         CALCULUS              A         CALCULUS

                 A         ENGLISH               A         ENGLISH

                 B         HISTORY               A         HISTORY

                 A         GEOGRAPHY             A         GEOGRAPHY

                 C         SCIENCE               A         SCIENCE
```

Figure 5.20. Two Two-Dimensional Arrays Defined For Use To The Map Editor Arranged Side-By-Side Vertically.

```
12:34:56                    - NATURAL MAP EDITOR -                    97-04-01
                           - Array Table Definition -
------------------------------------------------------------------------------
Main Index: Vert occur.  : 5        Secondary Index: Starting from : 1
            Starting from: 1                         Direction(H/V): H
            Line Spacing : 1                         Spacing       : 25
------------------------------------------------------------------------------
Name of Variable                   Col   Dimension Size    Order   2.Ind
(truncated)                        Pos   Ind1  Ind2  Ind3  M S F   Occ
------------------------------------------------------------------------------
#GRADE                              18      5     2       1 1 2     2
#SUBJECT                            27      5     2       1 1 2     2





Enter-PF1---PF2---PF3---PF4---PF5---PF6---PF7---PF8---PF9---PF10--PF11--PF12--
      Help  Mset  Exit              --    -     +                       Let
```

Figure 5.21. Two Two-Dimensional Arrays Defined By The Table Definition Menu

```
                              TEST                           MAP: MAPTST01

          SEMESTER I      GRADE     A     SUBJECT     CALCULUS
          SEMESTER I      GRADE     A     SUBJECT     CALCULUS

          SEMESTER I      GRADE     A     SUBJECT     ENGLISH
          SEMESTER II     GRADE     A     SUBJECT     ENGLISH

          SEMESTER I      GRADE     B     SUBJECT     HISTORY
          SEMESTER II     GRADE     A     SUBJECT     HISTORY

          SEMESTER I      GRADE     A     SUBJECT     GEOGRAPHY
          SEMESTER II     GRADE     A     SUBJECT     GEOGRAPHY

          SEMESTER I      GRADE     C     SUBJECT     SCIENCE
          SEMESTER II     GRADE     A     SUBJECT     SCIENCE
```

Figure 5.22. Two Two-Dimensional Arrays Defined For Use To The Map Editor Rearranged Side-By-Side Horizontally.

```
12:34:56                   - NATURAL MAP EDITOR -                    97-04-01
                          - Array Table Definition -
-----------------------------------------------------------------------------
Main Index: Vert occur.   : 5       Secondary Index: Starting from : 1
            Starting from: 1                         Direction(H/V): V
            Line Spacing : 1                         Spacing       : 0
-----------------------------------------------------------------------------
Name of Variable                  Col   Dimension Size     Order    2.Ind
(truncated)                       Pos   Ind1  Ind2  Ind3   M S F    Occ
-----------------------------------------------------------------------------
#GRADE                             40      5     2         1 1 2     2
#SUBJECT                           58      5     2         1 1 2     2







Enter-PF1---PF2---PF3---PF4---PF5---PF6---PF7---PF8---PF9---PF10--PF11--PF12--
      Help  Mset  Exit              --     -     +                      Let
```

Figure 5.23. Two Two-Dimensional Arrays Defined By The Table Definition Menu.

```
                         TEST                                  MAP: MAPTST01

      SEMESTER I       GRADE     A     SUBJECT     CALCULUS
      SEMESTER I       GRADE     A     SUBJECT     CALCULUS

      SEMESTER I       GRADE     A     SUBJECT     ENGLISH
      SEMESTER II      GRADE     A     SUBJECT     ENGLISH

      SEMESTER I       GRADE     B     SUBJECT     HISTORY
      SEMESTER II      GRADE     A     SUBJECT     HISTORY

      SEMESTER I       GRADE     A     SUBJECT     GEOGRAPHY
      SEMESTER II      GRADE     A     SUBJECT     GEOGRAPHY

      SEMESTER I       GRADE     C     SUBJECT     SCIENCE
      SEMESTER II      GRADE     A     SUBJECT     SCIENCE
```

Figure 5.24. Two Two-Dimensional Arrays Defined For Use To The Map Editor Rearranged One On Top Of Another.

```
12:34:56                    - NATURAL MAP EDITOR -                   97-04-01
                          - Array Table Definition -
------------------------------------------------------------------------------
Main Index: Vert occur.  : 2         Secondary Index: Starting from : 1
            Starting from: 1                          Direction(H/V): V
            Line Spacing : 3                          Spacing       : 0
------------------------------------------------------------------------------
Name of Variable                 Col   Dimension Size     Order    2.Ind
(truncated)                      Pos   Ind1   Ind2  Ind3  M S F    Occ
------------------------------------------------------------------------------
#GRADE                            32      2      5       1 1 2      5
#SUBJECT                          42      2      5       1 1 2      5






Enter-PF1---PF2---PF3---PF4---PF5---PF6---PF7---PF8---PF9---PF10--PF11--PF12--
      Help  Mset  Exit                --    -     +                      Let
```

Figure 5.25. Two Two-Dimensional Arrays Defined By The Table Definition Menu.

```
                              TEST                         MAP: MAPTST01

                          S E M E S T E R  I
SUBJECT:    CALCULUS   ENGLISH     HISTORY     GEOGRAPHY    SCIENCE
 GRADES:        A          A           B            C            B




                          S E M E S T E R  II
SUBJECT:    CALCULUS   ENGLISH     HISTORY     GEOGRAPHY    SCIENCE
 GRADES:        A          A           A            A            A
```

Figure 5.26. Two Two-Dimensional Arrays Defined For Use To The Map Editor Rearranged Stacked & Side-By-Side.

```
12:34:56                 - NATURAL MAP EDITOR -                   97-04-01
                       - Array Table Definition -
-----------------------------------------------------------------------------
Main Index: Vert occur.  : 2       Secondary Index: Starting from :
            Starting from:                          Direction(H/V): H
            Line Spacing : 7                        Spacing       : 11
-----------------------------------------------------------------------------
Name of Variable                 Col   Dimension Size     Order    2.Ind
(truncated)                      Pos   Ind1  Ind2  Ind3   M S F    Occ
-----------------------------------------------------------------------------
#SUBJECT                          16      2     5         1 1 2     5
#GRADE                            21      2     5         1 1 2     5








Enter-PF1---PF2---PF3---PF4---PF5---PF6---PF7---PF8---PF9---PF10--PF11--PF12--
      Help  Mset  Exit                --     -     +                     Let
```

Figure 5.27. Two Two-Dimensional Arrays Defined By The Table Definition Menu.

Likewise, #SUBJECT is defined the same.

Now each display is created slightly differently in the Map Editor. After mastering the technique to do these four, you will be able to develop any screen with any number of two-dimensional arrays.

Case Study #5

Creating Map With 2 Two Dimensional Arrays Not In Row-Order On The Screen.

Ninety percent of the time the physical layout of an array is displayed in exactly that order. However, this is certainly not mandatory. This need is not easily handled through the Map Editor because it wants to do exactly that - match the physical definition with the layout requirements. Suppose you want a screen with the following definition:

```
Row 1 | A | B | C | D |
Row 2 | E | F | G | H |
Row 3 | I | J | K | L |
Row 4 | M | N | O | P |
Row 5 | Q | R | S | T |
Row 6 | U | V | W | X |
```

to actually look like this:

```
Row 1 | A | B | C | D |          Row 2 | E | F | G | H |

Row 3 | I | J | K | L |          Row 4 | M | N | O | P |

Row 5 | Q | R | S | T |          Row 6 | U | V | W | X |
```

How to do this is not intuitively obvious. Fundamentally, one can define a two-dimensional array which is programmatically manipulated, and redefine it into sub-lists which are useful for display purposes. The order of events is to set up the appropriate definitions in the program first, then enter the Map Editor and use the 'P' option (P followed by the program name typed in the "Ob:" window) and "pull" (or incorporate) the definitions from the program into the screen. Here is a program example of these definitions to accomplish the above display.

```
0010 DEFINE DATA LOCAL
0020 1 #ARR (A2/1:6,1:4)
0030 1 REDEFINE #ARR          /* REDEFINITION FOR MAP DISPLAY
0040   2 #ROW1 (A2/1:8)
0050   2 #ROW2 (A2/1:8)
0060   2 #ROW3 (A2/1:8)
0070 END-DEFINE
0080 *
0090 #ARR (1,*) = 'AB'
0100 #ARR (2,*) = 'CD'
0110 #ARR (3,*) = 'EF'
0120 #ARR (4,*) = 'GH'
0130 #ARR (5,*) = 'IJ'
0140 #ARR (6,*) = 'KL'
0150 *
0160 INPUT USING MAP 'TESTMAP '
0170 END
```

After that, invoke the '.A' field option to properly define the list to the Map Editor. The definition should be done like the following to guarantee the direction of the list is horizontal, NOT vertical.

Once completed, the last step is to move the portion of the list (in this case, elements 6 through 10) to the appropriate position on the screen. It is this requirement that can cause a problem.

An attempt to issue the '.A' command on any array causes the Map Editor to attempt to reorganize the array according to its original definition, regardless of any change is made or not. If there is any text on the screen that "gets in the way" of the reorganization, errors occur. Basically, if you decide to do this, make sure you know what you want the display to be and do everything to reduce the need for modification.

Here is the display when running the program listed on the previous page:

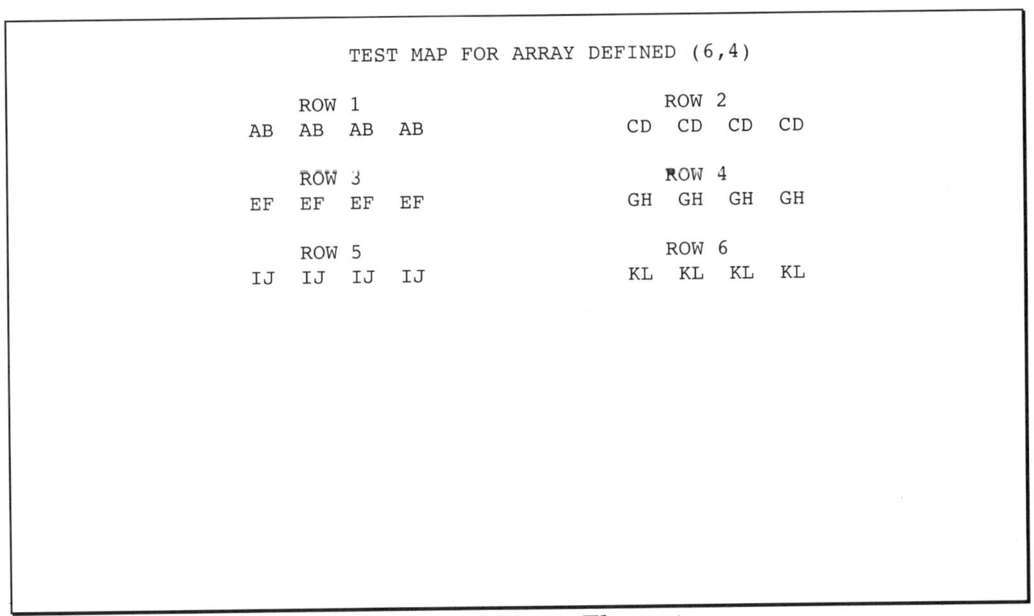

TEST MAP FOR ARRAY DEFINED (6,4)
ROW 1
AB AB AB AB
ROW 2
CD CD CD CD
ROW 3
EF EF EF EF
ROW 4
GH GH GH GH
ROW 5
IJ IJ IJ IJ
ROW 6
KL KL KL KL

Figure 5.28. Map of Arrays Ordered In Column Elements

III. Windows

A window gives you the ability to view a segment of a logical screen. The processing is controlled by 'W' Terminal Commands. Let's assume for the rest of this discussion that the Terminal Control Character is '%'.

Definition of Terms

1. Logical screen size is the original LS, PS settings for doing the current DISPLAY, WRITE, INPUT, PRINT.
2. Window width is the number of lines (Horizontal extension).
3. Window length is the number of columns(Vertical extension).

Parameters need to be provided to tell the window facility:

1. the length and width of the window;
2. an orientation as to where on the physical terminal screen the window will display.
3. and whether or not the window is displayed with a frame.

There are some rules which relate to sizing and positioning that could affect your expectations.

Rule 1: The minimum size for a window is 3 lines by 13 columns.

Rule 2: The maximum size for a window is the logical screen size.

Rule 3: If you want framing, you need to add 2 to the actual width and length to account for framing. Also, add one for the message line.

Example: You want to define a window which is 8 by 15. To frame it, you must declare the number of lines as 10 and the number of columns as 17.

Rule 4: NATURAL will not shift the window to extend "off" the physical screen. It will retrofit the window to display fully on the screen.

Example: Your physical screen is the normal 23 by 80. You try to shift a 10 by 20 window (this includes the frame) so that its upper left hand corner is matched to row 1 column 70. This is impossible; your window is 20 columns wide. If you tried to place it at column position 70 it would want to extend itself to position 90 which is "to the right" of the physical border.

Window Definition Commands

Dimension and Location

Command	Description
`%WB`	Window size set to physical screen size
`%WBll/cc`	Window top left corner positioned to line/column on physical screen
`%WBnnn`	Exact line and column screen location to position the window. Both `%WB6/30` and `%WB510` set a map in the same screen position.
`%WLnn`	Window line count is nn (horizontal)
`%WCnn`	Window column count is nn (vertical)
`%WF`	Window framing on
`%WM`	Window framing off Frames on 3270 terminals are plus signs at the corners, dashes across the top and bottom and vertical bars on the left and right sides; the Terminal command %F sets the frame characters. Frames on 3278 and 3279 type terminals is a white framed box. %F can be used to fill in the white box with frame characters.
`%WO`	Suppresses PF key line, message line or statistics line (introduced at SM05)

Window Movement Commands

Shifting the Contents of the Window

Command	Description
`%W<`	Window shift left one window length
`%W<<`	Window shift far left of physical screen
`%W<n`	Window shift left n columns
`%W>`	Window shift right one window length
`%W>>`	Window shift far right of physical screen
`%W>n`	Window shift right n columns
`%W+`	Window shift down one window length
`%W++`	Window shift down to bottom of physical screen
`%W+n`	Window shift down n lines
`%W-`	Window shift up one window length
`%W--`	Window shift up to bottom of physical screen
`%W-n`	Window shift up n lines
`%W*`	Window contents shift to match cursor position. Wherever the cursor is pointing in the window, that position becomes the new upper left of the window. Window movement commands - shifting the physical window on the screen
`%WLL/CC`	Window shift upper left corner to line LL column CC.
`%WB0`	Window shift to upper left corner of physical screen

Window Movement Commands

Shifting the Contents of the Window

`%W#`	Window shift upper left corner to cursor position The window does not generally change its size but may have to reduce to fit. Any current windows do not clear so it can get confusing.
`%W?`	This is a counterpart to '%W#'. The command allows you to "stretch" or expand the window. You can treat the frame like a rubber band. The window shifts the lower right to the cursor position. The top left remains unchanged and the size is adjusted if needed.

Other Window Commands

`%WN`	Set stay off, pressing the ENTER key will return control to program
`%WS`	Set stay on, pressing the ENTER key causes continued paging vertically until you reach the end of the logical page. The next time the ENTER key is pressed control is returned program.
`%WA`	Saves all the active screen data that gets overlaid when a window is opened. If that window is shifted or another window is opened on the screen, reconstruction of the overlaid screen is possible using this command.

Observations on Their Use

1. Always leave at least two positions in a window for entering '%%'. It is awfully embarrassing to open a window and you cannot get out. Window processing is invoked with terminal commands and you can create some bizarre situations while working in a NATURAL edit session.

2. I love the almost PC look you get with windows. But always remember the interface must meet the users' satisfaction as well as adhere to standards. Too many windows at any one time can be distracting.

3. Any shift within the window is reset when you start at the top of a new logical page.

4. You can combine multiple window commands in one string.

5. You can both construct and position windows dynamically.

Examples

```
SET CONTROL 'WFC10L20B2/45'
```

Establishes a framed window 10 columns deep with 20 characters on each line and the base of the window begins on line 2 at position 45.

```
SET CONTROL 'W+5>20'
```

Shifts the window 20 columns to the right and down 5 lines.

```
SET CONTROL 'W<<--'
```

Shifts the window to the extreme left margin and to the uppermost position, that is, the upper left-hand corner of the screen.

Case Study #6

Providing a window for selection to fill in the main screen

A question mark in the array field with the prompt 'OBJECT TYPE>' on the main screen initiates a window for selection. You can see this in Figures 5.29 and 5.30.

The short segment below demonstrates the CALLNAT to #OBJHELP.

```
1270 REPEAT
1280    #NO-HELP-REQUESTED = #YES
1290    INPUT MARK #MARK-FIELD USING MAP 'DRMSCT00'
1300          #START-LIB  #PGM-PREFIX   #DBID
1310          #NAT2-FLAG  #OBJ-TYPE (*) #COMMENTS #ABS-FLAG
1320          #SCAN-TEXT (*)
1330          #AND-OR-FLAG #FUNCTION-REQUEST
1340    IF #OBJ-TYPE (*) = #QUESTION-MARK THEN DO
1350       #NO-HELP-REQUESTED = #NO
1360       CALLNAT #OBJHELP #OBJ-TYPE (*)
1370       #MARK-FIELD = 5
1380       DOEND  /* (1340)
      :
1620 UNTIL #NO-HELP-REQUESTED
1630 LOOP
```

Here is a listing of the CALLNAT routine which displays the selection menu and allows for the selection to pass back to the main screen.

```
0010 DEFINE DATA PARAMETER
0020 1 #OBJ-TYPE    (A1/1:7)
0030 LOCAL
0040 1 #I           (P1)
0050 1 #K           (P1)
0060 1 #SEL-OBJ     (A1/1:8)
```

```
0070 1 #OBJ          (A1/2:8)
0080     INIT (2) <'P'>
0090          (3) <'M'>
0100          (4) <'S'>
0110          (5) <'C'>
0120          (6) <'H'>
0130          (7) <'G'>
0140          (8) <'L'>
0150 END-DEFINE
0160 *
0170 SET CONTROL 'WFL30C14B06/29'
0180 *
0190 INPUT USING MAP 'DRMSCT03'
0200 *
0210 SET CONTROL 'WB'
0220 IF #SEL-OBJ (1) = 'A'
0230   #OBJ-TYPE (1) := 'A'
0240 ELSE
0250   FOR #I = 2 TO 8
0260     IF #SEL-OBJ (#I) = 'X' THEN DO
0270       ADD 1 TO #K
0280       #OBJ-TYPE (#K) := #OBJ (#I)
0290     END-IF
0300   END-FOR
0310 END-IF
0320 ESCAPE ROUTINE
0330 END
```

Because help is requested, the main loop has not met the exit condition and the screen is reset. Any data that is changed is picked up. A REINPUT after the CALLNAT would not have the same results. Here is an image of the screens created and the result of the selection.

```
97-04-01                       Boston University                          DRMJTW
12:34:56                       ScanText Facility                          DRMSCT00

          *****************************************************
          *                                                   *
          *       LIBRARY ID> OISPROD_    START PT> SAN_____  *
          *             DBID> 215_                            *
          *       NAT 2(Y/N)> Y                               *
          *      OBJECT TYPE> P / M / _ / _ / _ / _ / _       *
          *  INCLUDE COMMENTS(Y/N)> _                         *
          *  ABSOLUTE (Y/N) > _                               *
          *  SEARCH STRING 1> ________________________________*
          *  SEARCH STRING 2> ________________________________*
          *  SEARCH STRING 3> ________________________________*
          *  SEARCH STRING 4> ________________________________*
          *  SEARCH STRING 5> ________________________________*
          *  SEARCH STRING 6> ________________________________*
          *  AND OR LOGIC<&,|)> |                             *
          *                                                   *
          *****************************************************

Command: _____________________________________________________________________
```

Figure 5.29. The Selection Menu As Initially Presented

```
97-04-01                        Boston University                         DRMJTW
12:34:56                        ScanText Facility                         DRMSCT00

           ******************************************************
           *                                                    *
           *       LIBRARY ID> OISPROD_     START PT> SAN_____  *
           *             DBID> 215_                             *
           *       NAT 2(Y/N)> Y                                *
           *      OBJECT TYPE> P / M / _ / _ / _ / _ / _        *
           *  INCLUDE COMMENT +---------------------------+     *
           *  ABSOLUTE (Y/N)  |                           |     *
           *  SEARCH STRING 1 |  SEL      TYPE            |____ *
           *  SEARCH STRING 2 |   _       A - ANYTHING    |____ *
           *  SEARCH STRING 3 |     OR                    |____ *
           *  SEARCH STRING 4 |   x       P - PROGRAM     |____ *
           *  SEARCH STRING 5 |   x       M - MAP         |____ *
           *  SEARCH STRING 6 |   _       S - SUBROUTINE  |____ *
           *  AND OR LOGIC<&, |   _       C - CALLNAT SUBPGM|   *
           *                  |   _       H - HELPROUTIE  |     *
           ****************** |   _       G - GDA         |*****
                              |   _       L - LDA         |
Command: ____________________ |                           |______________
                              +---------------------------+
```

Figure 5.30. Display of Selection Menu Invoked by Typing a Question Mark in the Object Type Field.

DEFINE WINDOW

You can set aside all the %W work in lieu of DEFINE WINDOW which permits all the %W functions and then adds titles, reverse video, simultaneous main screen functions and cursor sensitive scrolling.

DEFINE WINDOW is discussed in detail in chapter III. The DEFINE WINDOW statement can be used in place of the %W command in the previous example producing an enhancement to the window displayed in Figure 5.30.

```
DEFINE WINDOW
  SIZE 14 * 30
  BASE 6 / 29
  TITLE
  CONTROL SCREEN
  FRAMED ON
*
INPUT WITH TEXT 'Mark the objects types for the scan text operation.'
  USING MAP 'DRMSCT03'
*
```

Produces the following window as shown in Figure 5.31.

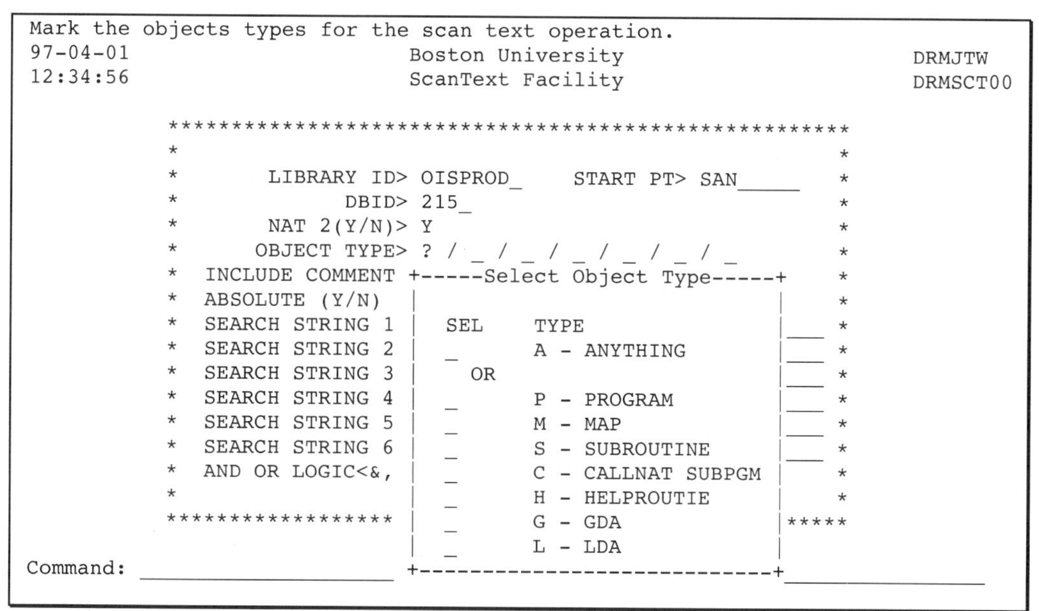

```
Mark the objects types for the scan text operation.
97-04-01                     Boston University                     DRMJTW
12:34:56                     ScanText Facility                     DRMSCT00

        ******************************************************
        *                                                    *
        *      LIBRARY ID> OISPROD_    START PT> SAN_____    *
        *            DBID> 215_                              *
        *      NAT 2(Y/N)> Y                                 *
        *     OBJECT TYPE> ? / _ / _ / _ / _ / _ / _ / _     *
        *  INCLUDE COMMENT +-----Select Object Type-----+    *
        *  ABSOLUTE (Y/N)  |                            |    *
        *  SEARCH STRING 1 |  SEL     TYPE              |___ *
        *  SEARCH STRING 2 |  _       A - ANYTHING      |___ *
        *  SEARCH STRING 3 |    OR                      |___ *
        *  SEARCH STRING 4 |  _       P - PROGRAM       |___ *
        *  SEARCH STRING 5 |  _       M - MAP           |___ *
        *  SEARCH STRING 6 |  _       S - SUBROUTINE    |___ *
        *  AND OR LOGIC<&, |  _       C - CALLNAT SUBPGM |   *
        *                  |  _       H - HELPROUTIE    |    *
        ****************** |  _       G - GDA           |*****
                           |  _       L - LDA           |
Command: ____________________ +--------------------------+______________
```

Figure 5.31. Revision of Selection Menu Invoked by Typing a Question Mark in the Object Type Field.

The title is always centered at the top of the frame. The message line and program function keys utilized by the main screen remain active when the window is opened up with the CONTROL SCREEN clause (notice the blank line at the bottom of the window in Figure 5.30, the message line, has disappeared in the window in Figure 5.31.) WITH TEXT permits the message line to be utilized by the INPUT statement even when a window is used instead of the full screen. The cursor is positioned in the first parameter under the SEL column in the window.

Scrolling Windows

Several techniques are available to generate and maintain scrolling of the data displayed in a window. Which technique depends on the data begin displayed and the needs of the user. The following is a brief overview of the more common techniques. These techniques are detailed in WH&O's NATURAL Developers Handbook.

Scrolling Report Output

A normally coded DISPLAY, WRITE, or PRINT can be output in a window and scrolled forward with the ENTER key. Either a %W command or a combination of DEFINE WINDOW and SET WINDOW determine the window attributes.

```
* Scroll Program One
DEFINE DATA LOCAL
1 STAFF VIEW OF EMPLOYEES-FILE
  2 PERSONNEL-ID
  2 NAME
  2 FIRST-NAME
  2 JOB-TITLE
  2 SALARY(1:6)
  2 BONUS(1:6,1:3)
END-DEFINE
DEFINE WINDOW SCROLL
  SIZE 12 * 65
  BASE 8 / 8
  TITLE 'Current Employee Job Listing'
  CONTROL SCREEN
  FRAMED ON POSITION OFF
SET WINDOW 'SCROLL'
READ STAFF BY NAME STARTING FROM 'A'
  PRINT NOTITLE PERSONNEL-ID NAME FIRST-NAME 35T JOB-TITLE
END-READ
END
```

```
+----------------Current Employee Job Listing-----------------+
|_60008339 ABELLAN KEPA              MAQUINISTA                |
| 30000231 ACHIESON ROBERT           DATA BASE ADMINISTRATOR   |
| 20005700 ADKINSON TIMMIE           SALES PERSON              |
| 20008600 ADKINSON MARTHA           MANAGER                   |
| 20008800 ADKINSON JEFF             PROGRAMMER                |
| 20009800 ADKINSON PHYLLIS          DBA                       |
| 20011000 ADKINSON BOB              SALES PERSON              |
| 20012700 ADKINSON HAZEL            MANAGER                   |
| 20013800 ADKINSON DAVID            SECRETARY                 |
| 20019600 ADKINSON CHARLIE          PROGRAMMER                |
+--------------------------------------------------------------+
```

Figure 5.32. Simple Scrolling Window.

Scrolling Arrays

A separate map is constructed with the array(s) laid out in a vertical perspective and then displayed in a window. Program function keys should be implemented to effect forward and backward scrolling. Its best to use DEFINE WINDOW in this scenario which permits the PF keys of the main screen to perform for the window as well.

The following example helproutine illustrates a typical scrolling array. A map is created beforehand the stacks several elements of an array. The code within the helproutine defines a window to display a portion of the array, populate the array, set scrolling functions, and some form of selection (optional).

```
DEFINE DATA LOCAL
1 RVIEW VIEW OF EMPLOYEES
  2 NAME
  2 FIRST-NAME
1 #NAME(A25/100)
END-DEFINE
DEFINE WINDOW ROLLDATA
SIZE 12*32
BASE 8/35
TITLE '- E-Mail Addresses -'
CONTROL SCREEN
FRAMED ON POSITION OFF
READ (100) RVIEW BY NAME STARTING FROM 'A'
  #I := *COUNTER
  COMPRESS RVIEW.NAME ',' INTO RVIEW.NAME LEAVING NO
  COMPRESS RVIEW.NAME RVIEW.FIRST-NAME INTO #NAME (#I)
END-READ
INPUT WINDOW='ROLLDATA' USING MAP 'ROLLMAP'
IF *PF-KEY = 'PF3'
  ESCAPE ROUTINE
END-IF
END
```

```
                                              +------- E-Mail Addresses -------+
                                              |     Adams, Stephanie           |
                                              |     Adams, Trent               |
                                              |     Adkinson, George           |
                                              |     Atkinson, Ralph            |
                                              |     Attor, Jesse               |
                                              |     Aubrey, Jeremy             |
                                              |     Augustine, Felipe          |
                                              |     Ayers, Steven              |
                                              |     Black, Robert              |
                                              |     Blake, James               |
                                              |     Blake, Lorraine            |
                                              +--------------------------------+

Enter-PF1---PF2---PF3---PF4---PF5---PF6---PF7---PF8---PF9---PF10--PF11--PF12--
      Help        Exit                        -     +
```

Figure 5.33. Simple Scrolling Window. CONTROL SCREEN Coded as a DEFINE WINDOW Parameter Utilizes the PF Key Assignments Established for the Main Program's Map.

Selecting Data from a Scrolling Window

Type a Value

This is a traditional technique that simply prompts the user to make a selection by typing a value into an input field. In the program "Scroll Program One", insert FORMAT PS=9 after the END-DEFINE statement following the PRINT statement insert:

```
AT END OF PAGE
  INPUT 'Type Personnel ID to Select:' PERSONNEL-ID
  IF PERSONNEL-ID NE ' '
    ESCAPE ROUTINE
  END-IF
END-ENDPAGE
```

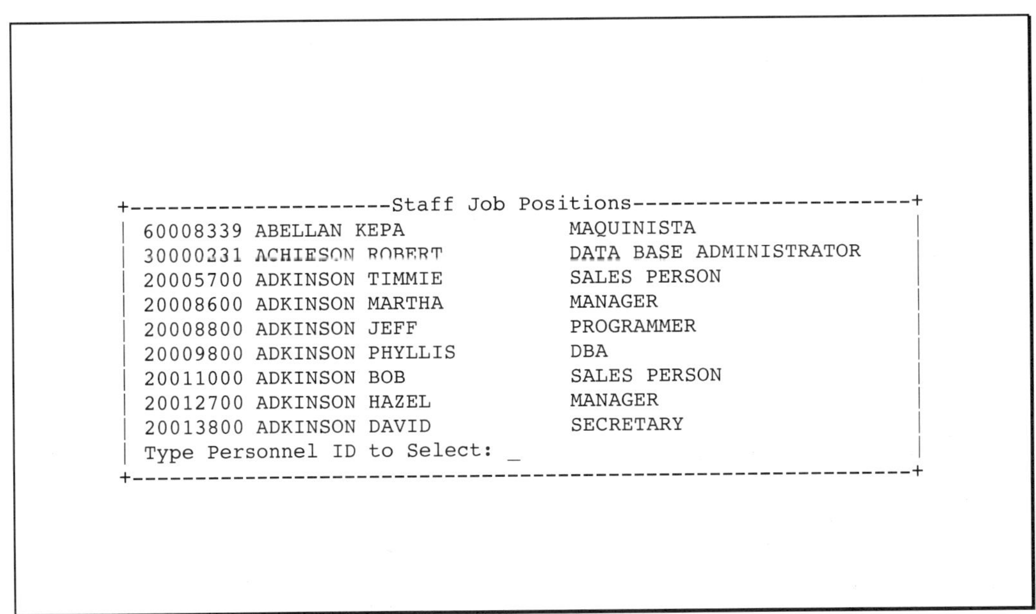

Figure 5.34. - Scoll Window with Data Input/Selection Field.

Mark a Field

A few added lines of code will convert the previous program example (and its window) into more useful dialog. By adding a selection field (an array with as many elements as the name array) in the map and program, the user can now select an appropriate record by simply marking the field.

```
DEFINE DATA
PARAMETER
1 #ADDRESSEE(A25)
LOCAL
1 RVIEW VIEW OF EMPLOYEES
  2 NAME
  2 FIRST-NAME
1 #I (I2)
```

```
1 #SELECT(A1/100)
1 #NAME(A25/100)
END-DEFINE
DEFINE WINDOW ROLLDATA
SIZE 12*32
BASE 8/35
TITLE '- E-Mail Addresses -'
CONTROL SCREEN
FRAMED ON POSITION OFF
READ (100) RVIEW BY NAME STARTING FROM 'A'
  #I := *COUNTER
  COMPRESS RVIEW.NAME ',' INTO RVIEW.NAME LEAVING NO
  COMPRESS RVIEW.NAME RVIEW.FIRST-NAME INTO #NAME (#I)
END-READ
INPUT WINDOW='ROLLDATA'
  WITH TEXT 'Select addressee with an "X" and the press the ENTER key.'
  USING MAP 'ROLLMAP'
EXAMINE #SELECT(*) FOR 'X' GIVING INDEX #I
IF #I = 0
  #ADDRESSEE := #NAME(#I)
END-IF
END
```

```
Select addressee with an 'X' and then press the ENTER key.




                                   +------- E-Mail Addresses -------+
                                   |  _    Adams, Stephanie         |
                                   |  _    Adams, Trent             |
                                   |  _    Adkinson, George         |
                                   |  _    Atkinson, Ralph          |
                                   |  _    Attor, Jesse             |
                                   |  _    Aubrey, Jeremy           |
                                   |  _    Augustine, Felipe        |
                                   |  _    Ayers, Steven            |
                                   |  _    Black, Robert            |
                                   |  _    Blake, James             |
                                   |  _    Blake, Lorraine          |
                                   +--------------------------------+



Enter-PF1---PF2---PF3---PF4---PF5---PF6---PF7---PF8---PF9---PF10--PF11--PF12--
      Help        Exit                          -     +
```
Figure 5.35. Scrolling Window with Selection Fields

Cursor Sensitive

The previous helproutine can be converted to a cursor sensitive selection process by modifying the code that scanned for a marked field with the three statements in the helproutine below starting with SET CONTROL 'CCC1'. This Terminal Command captures the value of the field in which the cursor is positioned into the NATURAL System Variable *COM. All that remains is to copy that value into a user-defined variable for processing.

```
DEFINE DATA
PARAMETER
1 #ADDRESSEE(A25)
LOCAL
1 RVIEW VIEW OF EMPLOYEES
  2 NAME
  2 FIRST-NAME
1 #I (I2)
1 #SELECT(A1/100)
1 #NAME(A25/100)
END-DEFINE
DEFINE WINDOW ROLLDATA
SIZE 12*32
BASE 8/35
TITLE '- E-Mail Addresses -'
CONTROL SCREEN
FRAMED ON POSITION OFF
READ (100) RVIEW BY NAME STARTING FROM 'A'
  #I := *COUNTER
  COMPRESS RVIEW.NAME ',' INTO RVIEW.NAME LEAVING NO
  COMPRESS RVIEW.NAME RVIEW.FIRST-NAME INTO #NAME (#I)
END-READ
INPUT WINDOW='ROLLDATA'
  TEXT 'Place cursor on selection and press [ENTER].'
  USING MAP 'ROLLMAP'
SET CONTROL 'CCC1'
#ADDRESSEE := *COM
RESET *COM
ESCAPE ROUTINE
END
```

```
Place cursor on selection and press [ENTER].




                             +----- E-Mail Addresses -----+
                             |_ Adams, Stephanie          |
                             |  Adams, Trent              |
                             |  Adkinson, George          |
                             |  Atkinson, Ralph           |
                             |  Attor, Jesse              |
                             |  Aubrey, Jeremy            |
                             |  Augustine, Felipe         |
                             |  Ayers, Steven             |
                             |  Black, Robert             |
                             |  Blake, James              |
                             |  Blake, Lorraine           |
                             +----------------------------+


Enter-PF1---PF2---PF3---PF4---PF5---PF6---PF7---PF8---PF9---PF10--PF11--PF12--
      Help        Exit                     -     +
```

Figure 5.36. Scrolling Window with Cursor Sensitive Selection

Chapter 6

NATURAL / ADABAS Communication

Efficiency issues fall into various categories when considering all of the ADABAS world. This chapter considers efficiency from one main perspective - NATURAL coding in support of minimal ADABAS processing.

> **The *GOLDEN GUIDELINE***
> **Live in the nucleus as little as possible**

NATURAL programs should be developed keeping the number of calls to ADABAS to a minimum. The number of calls to ADABAS adds significantly to the CPU usage and physical input / output. Each request to ADABAS can be identified by a two character command id. Below is the list of commands issued by NATURAL and a short description of what they do. The commands which are issued by NATURAL can be viewed using DBLOG, described in a later section in this chapter. Through DBLOG the programmer can see the commands issued and the buffers necessary to communicate to ADABAS.

I. ADABAS Commands

ADABAS Command Code	Purpose
A1	This command allows for the update of one or more fields on a record. The NATURAL statement that generates this ADABAS call is UPDATE. In NATURAL, a record cannot be be updated without being accessed first. ADABAS holds the record ISN until an ET/BT is issued.

ADABAS Command Code	Purpose
E1	This command allows for deletion of a record from a particular file. The NATURAL statement that generates this ADABAS call is DELETE. In NATURAL, a record cannot be deleted without being accessed first. ADABAS holds the record ISN until an ET/BT is issued.
L1/L4	The L1 command allows for a record to be read by ISN. The NATURAL statements which generate this ADABAS call are READ filename BY ISN, GET, GET SAME, and the GET NEXT logic due to FIND loop processing. If a record is to be read by ISN with hold (and NATURAL figures that out) an L4 is issued.
L2/L5	The L2 command allows for a record to be read in physical sequence. The NATURAL statement that generates this ADABAS call is READ filename. If a record is to be read physically with hold (and NATURAL figures that out) an L5 is issued.
L3/L6	The L3 command allows for a record to be read in a logical sequential order. The NATURAL statement that generates this ADABAS call is 'READ filename BY descriptor'. If you wish to read the record logically with hold, an L6 is issued.
L9	This command allows for the read of the Associator. The NATURAL statement that generates this ADABAS call is HISTOGRAM.
N1	This command requests a new record to be added to the file an creates a unique Internal Sequence Number for the new record. The NATURAL statement that generates this ADABAS call is STORE.
N2	This command requests a new record to be added to the file with a user generated ISN. The NATURAL statement that generates this ADABAS call is STORE...GIVING ISN.
RC	This command allows for a command ID to be released. NATURAL generates this at the completion of an ADABAS loop and when a program ends.

ADABAS Command Code	Purpose
RE	This command allows user data to be read which has been previously stored in an ADABAS system file by issuing a close, a C3 command (see ADABAS Command Reference Manual for this one), or ET command. This is NATURAL's way of supporting restartability after an abend. In batch, the user data can be read only if the same USERID with an open (OP) is provided, i.e., the job name must be the same. The NATURAL statement that generates this command is GET TRANSACTION DATA.
RI	This command releases a record from hold status. No NATURAL statement generates this command. However, if a READ / FIND loop is initiated in a NATURAL program with hold status on the records due a file modification (e.g., UPDATE), a record NOT updated has an RI issued before the reading of the next record. NATURAL provides the NATPARAM RI. When 'RI=ON', an RI is issued for held records not modified or deleted due to a WHERE clause, or an ACCEPT/REJECT statement
S1/S4	The S1 command allows for the selection of records which satisfy search criteria. The NATURAL statement that generates this call is FIND. If a record needs to be selected with hold, the ADABAS call S4 is generated.
S2	This command allows for a set of records to be selected which satisfy search criteria and which are to be returned in a user-specified order. The NATURAL statement that generates this call is FIND ... SORTED BY.
S8	This command allows for the resolution of ISN lists which have been previously created with FIND commands. The resulting list is in ISN sequence. A NATURAL statement example: FIND ... WITH ISNSET1 OR ISNSET2.
S9	Sorts an ISN list which is stored either in the ISN buffer or on ADABAS Work Part III. The Sort value is provided in the Additions 1 field. Command option 2 defines the sort order.

ADABAS Command Code	Purpose
ET	Indicates the end of a logical transaction. The NATURAL statement that generates this command is END TRANSACTION. This command commits pending deletes, adds, and updates. The End Transaction statement also releases any ISNs in hold status and alternatively write to the Protection Log.
BT	Indicates a backout of a logical transaction. The NATURAL statement that generates this command is BACKOUT TRANSACTION.
HI	Places a record with a specific ISN on hold.

II. Guidelines on READ, FIND, and HISTOGRAM

PAGE

Introduction

Many beginning NATURAL programmers have stumbled about in writing NATURAL programs to do record selection. The only basis upon which a programming decision could be made was to rely on past programming experiences, e.g.,. COBOL, possibly against IMS or some other data base system. First, one must get a feel for what the various access paths into the data base are via the NATURAL language. This is fundamental in writing good NATURAL programs.

ADABAS classifies all commands as simple, complex, or update. Instead of explaining how ADABAS does this, I will simply classify the NATURAL commands into Category I (least amount of work), Category II (more work), and Category III (file modification - potentially lots of work).

Category I		**Category II**		**Category III**	
READ PHYSICAL	(L2)	FIND - Alg. 4	(S1)	STORE	(N1)
READ BY ISN	(L1)	FIND - Alg. 1/2*	(S1)	UPDATE	(A1)
HISTOGRAM	(L9)			DELETE	(E1)
GET	(L1)			FIND...SORTED BY	(S2)
READ LOGICAL*	(L3)			END TRANSACTION	(ET)
				BACKOUT TRANSACTION	(BT)

* The category level between FIND using algorithm 1 or 2 and a READ LOGICAL is changeable depending on file conditions and search criteria.

The NATURAL verb FIND presents the most varied possibilities of levels of work demanded to complete the execution of all the verbs affecting the access path to the data base. FIND commands built by NATURAL are classified as complex operations to ADABAS. READ and FIND have their own merit for usage in programs. To decide on which to use requires some knowledge of each.

To complete a FIND loop, NATURAL generates an S1 command to build the ISN (internal sequence number) list for lookup. An ISN is either an ADABAS or user-assigned record number. Every record on an ADABAS file has such a unique assignment. An S1 command also returns the first record of the list automatically as part of the processing of an S1 command.

The FIND may now take advantage of the soft coupling facility of ADABAS 5. The COUPLED VIA clause permits a maximum of 42 search criterion specifications under strict conditions:

- no file may be specified more than once,
- only data from first view is available,
- cannot be used with RETAIN and SORTED BY clauses.

COUPLED uses the physical coupling facility. Only data from first view is available and a maximum of 4 files may be specified. COUPLED cannot be used with RETAIN and SORTED BY clauses.

```
FIND EMPLOYEE WITH NAME = 'JONES'
    AND COUPLED TO VEHICLE
    VIA PERSONNEL-ID = PERSONNEL-ID
    WITH MAKE = 'GENERAL MOTORS'
```

Non-Descriptor Searches

NATURAL 2.2 supports specification of a non-descriptor field as part of a FIND...WITH. The field must be defined with Index Type "N". Use of this function may use fewer calls to ADABAS than performing selection within program

```
FIND EMPLOYEE WITH PERSONNEL-ID = '20000000' THRU '29999999'
    AND POST-CODE = '87000' THRU '88999'
```

Hyperdescriptors now appear in the DDM with an"H" entered in the Index Type column. Hyperdescriptor searches are now fully supported under NATURAL 2.2/ADABAS 5. Hyperdescriptors are keys to the application but involve user code (Assembler, etc.) which is invoked by ADABAS upon STORE / UPDATE of an ADABAS record.

To retrieve the remaining records in ISN order, L1 ADABAS commands are issued by NATURAL to get the next record. An L1 command is generally less demanding on ADABAS than the READ LOGICAL (L3). Although both must be converted from ISN to physical address, the L1 does not involve inverted lists (descriptors) as does the L3 command.

I/O to other than Data Storage is required to successfully execute each command. ALL variations of the FIND write to the Work Data Set except FIND UNIQUE which does no I/O to Work. The FIND FIRST and FIND NUMBER options do no reads from Work, although they both write. FIND (1) also does no read from Work.

The READ LOGICAL verb reads the Associator as well as Data Storage. If you require records in a keyed sequence then READ can do it (assuming the inverted list for the key is defined). Programmers discover quickly that READ ... BY *descriptor* = *value* **does not** act like FIND but is a starting point for sequential reads. READ ... THRU also requires one additional command since NATURAL decides the ending point (unless you're doing a READ (1) ... THRU).

A question was once asked since ASCII and EBCDIC collating sequences are different, how can you get the display to be the same on an Unix machine as on an MVS machine? The suggested solution is to try:

```
READ automobiles by make starting from 'A' ending at '9'
```

you must replace with :

```
READ automobiles by make starting from H'00' ending at H'FF'
```

The programmer has several methods of retrieval. Here is a chart listing the paths of access to ADABAS.

NATURAL Statement	ADABAS Access Path
READ PHYSICAL	Data Storage only
READ BY ISN	Associator and Data Storage
READ LOGICAL	Associator and Data Storage
HISTOGRAM (minus WHERE)	Associator only
FIND NUMBER (minus WHERE)	Associator only
FIND	Associator and Data storage

VARIOUS LEVELS (ALGORITHMS) FOR FIND

I. Level 1 - single criteria (one descriptor)

Example:

```
FIND EMPLOYEES WITH NAME = 'SMITH'
```

II. Level 2 - values in a range (one descriptor) minor OR logic

Example:

```
FIND EMPLOYEES WITH NAME = 'SMITH' OR = 'WILSON'
FIND EMPLOYEES WITH BIRTH = 48000 THRU 529999
```

III. Level 3 - AND logic (2 descriptors, 1 value each)

Example:

```
FIND EMPLOYEES WITH NAME = 'SMITH'
AND LANG = 'ENG'
```

IV. Level 4 - Compound AND/OR logic (everything else)

Example:

```
FIND EMPLOYEES WITH NAME = 'SMITH' OR = 'JONES'
AND LANG = 'ENG' OR = 'FRE'
```

GUIDELINE #01: Fundamental Access Theorem

The 80/20 rule states that if you need to process more than 80% of your file do NOT do so by invoking S1's (FIND) or L3's (READ in logical sequence) but do L2's (READ PHYSICAL) and then apply ACCEPT / REJECT logic or a WHERE clause. If less than 20% of the file is to be processed, L3's (READ LOGICAL) or S1's (FIND) should be less intensive for the ADABAS nucleus. However, no rule is sacred. A time test may help you to decide the best access path.

GUIDELINE #02: Single Criterion search

Use a READ rather than a FIND when selecting a range of values for a single descriptor. This decision is however affected by; (1) file size; (2) volatility due to update; (3) number of records; (4) distribution of ISNs in physical storage; (5) if sequence matters. Any one of these might push the decision to use FIND instead of READ.

The timing in many situations seems better for the READ than for the FIND. However, under some conditions it was noticed that FIND performed faster. When the count of records selected is small, algorithm 1/2 FIND is very quick.

When doing a READ for a range of values for a superdescriptor or subdescriptor, the THRU option produces an ADABAS error. You must code your own ESCAPE logic. Consider below the case for which the field of choice is a super/subdescriptor.

Example:

```
READ EMPLOYEES BY BIRTH
    STARTING FROM 480000 THRU 529999
```

is preferred to

```
FIND EMPLOYEES WITH BIRTH = 480000 THRU 529999
```

The problem with the FIND is that it appeals to the understanding of the word due to common usage. The word is often used in other query languages for other data base systems. However, its meaning is not consistent across systems. In NATURAL, it can be very CPU intensive and often I/O intensive, especially if algorithms 3/4 are invoked and/or the number of resulting ISNs is large.

A college Biographical file (ALUMNI) uses a descriptor ALUMNI-CD, a 2 character (byte) field. The number of records on the file is 400,000 plus. The value of 'AL' for the field ALUMNI-CD represents an alumni record.

Example:

```
READ ALUMNI BY ALUMNI-CD = 'AL' THRU 'AL'
```

is preferred to

```
FIND IN ALUMNI WITH ALUMNI-CD = 'AL'
```

However, since more than 80% of this file is alumni, even better might be:

```
READ ALUMNI
    ACCEPT IF ALUMNI-CD = 'AL'
```

This points out a basic credo - KNOW THY DATA! If a file's physical sequence also reflects the logical order of the file, then READ PHYSICAL would suffice. There is a danger if you must rely on your database maintenance when unloading and reloading this order is preserved.

If a decision is made to code a FIND, for instance in the event complex criteria returns a small percentage of the file, the next three guidelines are useful.

GUIDELINE #03: FIND - Range Checking

When doing inclusive range checking, always use the THRU range feature instead of AND logic for a FIND.

Example:

```
FIND EMPLOYEES WITH BIRTH 480000 THRU 499999
```

and not

```
FIND EMPLOYEES WITH BIRTH >= 480000 AND
                    BIRTH <= 499999
```

The preferred choice uses an algorithm 2 FIND, requiring only one descriptor list. The second method results in an algorithm 3 FIND. The AND logic must then be performed on the two lists to get the ISNs in common. This is excessive work that should be avoided.

GUIDELINE #04: READ(1) vs FIND FIRST or FIND UNIQUE

When doing a search for a single record, use FIND FIRST (report mode only-no loop), FIND .... ESCAPE END-FIND (Structured mode-loop), or if the ISN is known, use the GET statement (no loop).

If the search criteria is complex or the program requires a count (*NUMBER), again a choice is not easy. This concern is more important for repeated searches within a loop. Also, the descriptor needed for the search makes a difference.

Example:

```
FIND FIRST EMPLOYEES WITH PERSONNEL-ID = 123456789
** Not allowed in structured mode
```

is preferred to

```
READ (1) EMPLOYEES BY PERSONNEL-ID
         STARTING FROM 123456789 THRU  123456789
```

However, consider this example - the descriptor LANG could return a large ISN list making the FIND FIRST rather expensive.

Then,

```
READ (1) EMPLOYEES BY LANG
      STARTING FROM 'ENG'
```

is preferred to

```
FIND FIRST EMPLOYEES WITH LANG = 'ENG'
```

An alternative is to set up a normal READ loop and do an escape at the bottom of the loop (see example below). This allows for the programmer not to be concerned about the LE option. If LE = ON, then an error message is issued by NATURAL.

Example: An alternative to READ (1) is

```
READ EMPLOYEES BY LANG STARTING FROM 'ENG'
    ESCAPE BOTTOM
END-READ
```

GUIDELINE #05: Minor OR vs Major OR

Example:

```
FIND EMPLOYEES WITH NAME = 'SMITH' OR = 'DANIELS'
** Algorithm 2
```

is preferred to

```
FIND EMPLOYEES WITH NAME = 'SMITH' OR NAME = 'DANIELS'
** Multiple calls to ADABAS
```

GUIDELINE #06: Using the WHERE Clause

Avoid FIND selections which return large ISN lists if possible. A program using an algorithm 4 FIND can be rewritten to an algorithm 1/2 FIND plus a WHERE clause. Although this may cause unwanted records to be read your chances are good that the resources needed to resolve an ISN list are diminished. This can lower CPU utilization and reduce I/O.

In the example below, the file contains 100,000 automobile records. The major key is the registration number (REGNUM). NATURAL is responsible for resolving the WHERE clause, not ADABAS.

Many batch jobs have had long run times when the second choice of code was employed.

Example:

```
FIND AUTOFILE WITH REGNUM = '339AQ99P'
                      OR  = '26QBM775'
          ACCEPT IF MAKE  = 'FORD' AND COLOR = 'RED'
```

is preferred to

```
FIND AUTOFILE WITH REGNUM = '339AQ99P'
                       OR = '26QBM775' /* SMALL LIST
                 AND MAKE = 'FORD'     /* LARGE LIST
                AND COLOR = 'RED'      /* LARGE LIST
```

GUIDELINE #07: Use of Parentheses in Search Criteria

Use parentheses ONLY when necessary in delimiting search criteria. They are necessary when you wish to change the order for logical operations or for NOT logic.

This guideline has caused much furor with some of my colleagues. It has been RIGHTFULLY pointed out to me that the parentheses cause the FIND algorithm 4 to be reinterpreted as a series of algorithm 1 FINDs with intermittent S8 calls to resolve the various ISN lists. Hence, the simpler calls may outperform the more complex algorithm 4 call. However, every command needing ADABAS' attention requires thread availability, and this could be more difficult in an environment with much resource contention.

They may look pretty and make your programs readable, but they can hurt in processing time! NATURAL generates extra ADABAS calls to resolve the selection. In the preferred example below, 1 S1 call is generated before a series of L1 calls are done. However, in the version with the parentheses, 4 additional calls are generated before an L1 (get record with ISN). Imagine if this FIND statement was being executed within a READ/FIND loop !!

Example:

```
            FIND EMPLOYEES WITH NAME = 'ANDERSEN' OR = 'ADKINSON'
                                  OR = 'JONES'
                            AND CITY = 'BEVERLEY HILLS'
```

is preferred to

```
            FIND EMPLOYEES WITH NAME = 'ANDERSEN' OR = 'ADKINSON'
                                  OR = 'JONES'
                           AND (CITY = 'BEVERLEY HILLS')
```

GUIDELINE #08: Solution to Negating Minor OR

For the FIND, avoid not equal (NOT EQUAL or NE) relationships for single values in the search criteria.

NATURAL does not allow for the negation of a minor OR. Use a FIND NUMBER with the RETAIN option and use the retained ISN set in a single AND NOT 'setname' clause.

Rather than using a series of "AND ... NE ..." consider doing a FIND NUMBER on the minor OR selection with the RETAIN option, and do a second FIND ANDing the NOT RETAINed ISN set with the other basic search criteria.

> ******* WARNING ********
>
> The outcome is unpredictable if the field being minor OR'ed is also a multiple valued field or periodic group occurrence.

Example:

```
0010 DEFINE DATA LOCAL USING ST-REG-FI
0020 END-DEFINE
0030 FIND NUMBER STUDENT-REG-INFO WITH
0040 YRSEM-CD = '001853'
0050       OR = '002853'
0060       OR = '007853'
0070       OR = '011853'
0080       OR = '012853'
0090 RETAIN AS 'YRSEM-SET'
0100 WRITE *NUMBER (0030)
0110 FIND STUDENT-REG-INFO WITH DISP-CODE = '10'
0120     AND FS-CODE = '10'
0120     AND NOT 'YRSEM-SET'
0130   MOVE C*ACCOUNT-DETAIL TO #ACCT-DETAIL (N4)
0140   DISPLAY (ES=ON) IDNO L-NAME ITEM-CODE (1:5)
0150     YR-SEM #ACCT-DETAIL
0160 END-FIND
0170 END
```

is preferred to

```
0010 DEFINE DATA LOCAL USING ST-REG-FI
0020 END-DEFINE
0030 FIND STUDENT-REG-INFO WITH DISP-CODE = '10'
0040     AND FS-CODE = '10'
0050     AND YRSEM-CD NE '001853'
0060     AND YRSEM-CD NE '002853'
0070     AND YRSEM-CD NE '007853'
0080     AND YRSEM-CD NE '011853'
0090     AND YRSEM-CD NE '012853'
0080   MOVE C*ACCOUNT-DETAIL TO #ACCT-DETAIL (N4)
0090   DISPLAY (ES=ON) IDNO L-NAME ITEM-CODE (1:5)
0100     YR-SEM  #ACCT-DETAIL
0110   AT END OF DATA
0120     WRITE *NUMBER (0030)
0130   END-ENDDATA
0140 END-FIND
0150 END
```

GUIDELINE #09: Use of FIND NUMBER vs HISTOGRAM

Use FIND NUMBER for counts for descriptors for unique values or for a list of counts per descriptor value. Since 2.2, NATURAL constructs the correct format buffer to not write to the WORK dataset.

Example:

```
FIND NUMBER EMPLOYEES WITH AGE = 30 THRU 50
```

is preferred to

```
HISTOGRAM EMPLOYEES FOR AGE STARTING FROM 30 THRU 50
** By taking TOTAL(*NUMBER) or adding up *NUMBER in the loop
```

Use HISTOGRAM and FIND NUMBER for counts as often as possible. These two methods do not require accessing data storage. Inverted lists are accessed to get the counts. The FIND NUMBER may generate more I/O, whereas the HISTOGRAM as written results in two L9 ADABAS commands and one RC ADABAS command to be issued. For an interesting scenario see Case Study #1 at end of this chapter.

To test if a keyed record is on file or not is best done by one of two methods. The first and foremost method is to use a FIND NUMBER statement. This statement allows for the reading of the associator for a descriptor value which points to an existing record on file. However, one does not have to read data storage to decide this. An example of an application employing this method is the purging of records from a master file coordinated with subordinate files.

Assume you want to read the PERSONNEL file and display only those who do not have a Finance record. Here is one solution.

```
0010 DEFINE DATA LOCAL
0020 1 EMP VIEW OF EMPLOYEES
0030   2 PERSONNEL-ID
0040   2 NAME
0050 1 CAR VIEW OF VEHICLES
0060   2 PERSONNEL-ID
0070 END-DEFINE
0080 FORMAT PS=21
0090 * THE PURPOSE OF THIS EXAMPLE PROGRAM IS TO EXPLAIN THE PROBLEMS
0100 * WITH ASKING THE QUESTION 'IS SOMEONE ON FILE?'. THIS PROGRAM
0110 * COORDINATES LOOKUP ON THE VEHICLES FILE AND PRINT OUT ANYONE WHO
0120 * DOES NOT HAVE AN AUTOMOBILE
0130 *
0140 EMP. READ EMP    /* BY NAME , IF YOU WANT NAME ORDER
0150 HIS.
0160   HISTOGRAM CAR FOR PERSONNEL-ID
0170       STARTING FROM EMP.PERSONNEL-ID THRU EMP.PERSONNEL-ID
0180   END-HISTOGRAM
0190   IF *NUMBER (HIS.) = 0 THEN
0200      WRITE 'LAST NAME:' NAME 'ID:'
0210            PERSONNEL-ID 'HAS NO AUTO RECORD.'
0220   END-IF
0230 END-READ
0240 END
```

```
PAGE      1                                                  97-04-01  12:34:56
LAST NAME: TURNER                    ID: 30000007 HAS NO AUTO RECORD.
LAST NAME: HAIBACH                   ID: 11100302 HAS NO AUTO RECORD.
LAST NAME: CRIST                     ID: 9307801  HAS NO AUTO RECORD.
LAST NAME: SMITH                     ID: 99999999 HAS NO AUTO RECORD.
LAST NAME: CRIST                     ID: 9307803  HAS NO AUTO RECORD.
LAST NAME: CRIST                     ID: 9307804  HAS NO AUTO RECORD.
LAST NAME: BROWN                     ID: 10000300 HAS NO AUTO RECORD.
```

The fundamental problem with this approach is the number of ADABAS calls required to do the job. If the record in question is present on the vehicles file, three ADABAS calls are generated; if not, two are generated.

You can reduce the calls somewhat by altering the code to:

```
0010 DEFINE DATA LOCAL
0020 1 #ID (A8)
     ....
0180 READ EMP BY PERSONNEL-ID
0190  STARTING FROM EMP.PERSONNEL-ID
0200      #ID = PERSONNEL-ID
0210      ESCAPE BOTTOM
0220   END-HISTOGRAM
0230   IF #ID NE EMP.PERSONNEL-ID THEN
0240      WRITE 'LAST NAME:' NAME 'ID:'
0250           PERSONNEL-ID 'HAS NO AUTO RECORD.'
0260   END-IF
     ....
```

The code is slightly more complicated in the second example. As the programmer you must know that the HISTOGRAM loop may be beyond your starting point and the file key may not match your search key. This is not a concern in the previous coded example, but you guarantee that you never issue more than one L9 and one RC per HISTOGRAM call.

Simplest and best is:

```
0170 EMP. READ EMP   /* BY NAME , IF YOU WANT NAME ORDER
0180 CAR. FIND NUMBER CAR WITH PERSONNEL-ID = EMP.PERSONNEL-ID
0190   IF *NUMBER (CAR.) = 0
0200     WRITE 'LAST NAME:' NAME 'ID:'
0210          PERSONNEL-ID 'HAS NO AUTO RECORD.'
0220   END-IF
0230 END-READ
```

GUIDELINE #10: FIND NUMBER with RETAIN

If a selected group of records are required for several processes in the same session, use a FIND NUMBER WITH ... RETAIN AS 'setname'. Then the FIND WITH 'setname' can be used as needed. This results in not repeating work on the inverted lists.

Example:

```
FIND NUMBER EMPLOYEES WITH SALARY > 40000
                           AND SEX = 'M'
      RETAIN AS 'SETAA'
*****************************************************
FIND EMPLOYEES WITH 'SETAA'    /* HAPPENS LATER */
                               /* IN PGM/APPL   */
```

GUIDELINE #11: FIND NUMBER Instead of HISTOGRAM for Ranges

Use FIND NUMBER to get a count for a range of values for a single descriptor or any complex Basic Search Criteria. If you want the individual counts per value in the range, the HISTOGRAM statement is what you want.

Example:

```
FIND NUMBER FINANCE WITH
     NETWORTH = 10000 THRU 10999
```

is preferred to

```
HISTOGRAM EMPLOYEES FOR NETWORTH
    STARTING FROM 10000 THRU 10999
AT END OF DATA
*  PROCESS TOTAL(*NUMBER)
```

GUIDELINE #12: SORTED BY vs SORT BY

Use the SORT BY statement rather than the SORTED BY clause of the FIND statement.

Avoid the use of FIND ... SORTED BY (S2 ADABAS call) and utilize the SORT verb or add a SORT step within your job stream to order your data to your specification. The SORTED BY is accomplished by ADABAS ONLY on descriptor fields. For non-descriptor fields, SORT is the only way to go. The sort method is dependent upon whether the program runs in batch or on-line. It has its own peculiarities, but the system or NATURAL does the work, not ADABAS !

Example:

```
FIND EMPLOYEES WITH BIRTH > 35000
SORT BY DEPT NAME
```

is preferred to

```
FIND EMPLOYEES WITH BIRTH > 35000
     SORTED BY DEPT NAME
```

III. File Modification

For most of the following discussion of file modification, I will use this simple table which exist on both development and production data bases.

```
0010 DEFINE DATA PARAMETER
            ....
0250 ****************************************
0260 **      V  I  E  W  S
0270 ****************************************
0280 1 TB5-P VIEW OF UISTBL005AP
0290   2 TABLE-KEY
0300   2 TABLE-RESULT-1
0310   2 TABLE-RESULT-2
0320 1 TB5-T VIEW OF UISTBL005AT
0330   2 TABLE-KEY
0340   2 TABLE-RESULT-1
0350   2 TABLE-RESULT-2
0360 END-DEFINE
```

STORE / STORE SAME

```
0800 READ TB5-T BY TABLE-KEY STARTING FROM #KEY THRU #END-KEY
0810   MOVE TABLE-KEY       TO #AA
0820   MOVE TABLE-RESULT-1 TO #BA
0830   MOVE TABLE-RESULT-2 TO #BB
0840 *
0850   IF #ACTUAL-RUN-FLAG = 'A' THEN
0860     MOVE BY NAME TB5-T TO TB5-P
0870     STORE TB5-P
0880     END TRANSACTION
0890   END-IF
0900   WRITE 'NEW TABLE KEY:' #AA /
0910         '      RESULT 1:' #BA /
0920         '      RESULT 2:' #BB
0930   SKIP 1
0940 *
0950   IF *LINE-COUNT > 20 THEN NEWPAGE
0960   END-IF
0970 *
0980   ADD 1 TO #NEW-COUNT
0990 END-READ
```

You may use the same view area defined in a DEFINE DATA GLOBAL/LOCAL for both READ and STORE purposes. Just remember to null out any fields you do not intentionally plan to put into a record. This way you do not accidently put field values into a new record which uses the same format / record buffers.

STORE SAME uses format / record buffers like those defined by an ADABAS access (READ, FIND). It provides a programmer convenience in that a list of variables need not be provided. The same field values as are read, unless modified, are stored as a result of the STORE SAME process. This capability is often used for a "rollover" concept. I define "rollover" as creating a duplicate record which is originally different by a date / time consideration. A student's housing record by semester may be such a consideration.

The STORE SAME syntax is not permitted in either Structured Mode or Structured Data Mode, i.e., Report Mode with a DEFINE DATA statement. However, the functionality is incorporated by using views in a DEFINE DATA structure. The ISN of the record stored is available using the NATURAL system variable *ISN.

DELETE

The delete scenario is the easiest to understand. A record cannot be deleted unless it is initially accessed through READ, FIND or GET. Syntactically, you may include a reference back in the form of a line number or line label or via statement qualification. Here is a short segment of code illustrating a few points.

```
0520 RD-5.
0530 READ TB5-P BY TABLE-KEY
0540  = #KEY THRU #END-KEY
0550   ADD 1 TO #REPLACE-COUNT
0560   IF #ACTUAL-RUN-FLAG = 'A' THEN
0570     DELETE (RD-5.)
0580     END TRANSACTION
0590     WRITE 'DELETED PRODUCTION KEY:'
0600        TABLE-KEY (RD-5.) /
0610           'DELETED PRODUCTION TABLE RESULT-1:'
0620        TABLE-RESULT-1 (RD-5.) /
0630           'DELETED PRODUCTION TABLE RESULT-2:'
0640        TABLE-RESULT-2 (RD-5.) /
0650   ELSE
0660     WRITE 'CURRENT PRODUCTION KEY:' TABLE-KEY /
0670         'CURRENT TABLE RESULT 1:' TABLE-RESULT-1 /
0680         'CURRENT TABLE RESULT 2:' TABLE-RESULT-2 /
0690     SKIP 1
0700   END-IF
0710 *
0720   IF *LINE-COUNT > 20
0730     THEN NEWPAGE
0740   END-IF
0750 END-READ
```

First, any record read with a DELETE statement pointing to it causes NATURAL to issue a READ / FIND with hold. In this example, a READ LOGICAL with hold results in an L6 command issued to ADABAS. Also, the DELETE statement (E1 ADABAS command) refers back to the READ loop. Since the DELETE is being done based on an input parameter to the subprogram, there is a better way to code this segment. Before reading on, see if you detect a "better" method.

```
0520 IF #ACTUAL-RUN-FLAG = 'A' THEN
0530 RD-5.
0540   READ TB5-P BY TABLE-KEY = #KEY THRU #END-KEY
0560     ADD 1 TO #REPLACE-COUNT
0570     DELETE (RD-5.)
0580     END TRANSACTION
0590     WRITE 'DELETED PRODUCTION KEY:'
0600       TABLE-KEY (RD-5.) /
0610          'DELETED PRODUCTION TABLE RESULT-1:'
0620       TABLE-RESULT-1 (RD-5.) /
0630          'DELETED PRODUCTION TABLE RESULT-2:'
0640       TABLE-RESULT-2 (RD-5.) /
0650     IF *LINE-COUNT > #DEVICE-LINE-CNT THEN NEWPAGE
0660     END-IF
0670   END-READ
```

```
0680 ELSE
0690   RD-5B.
0700   READ TB5-P BY TABLE-KEY = #KEY THRU #END-KEY
0710     ADD 1 TO #REPLACE-COUNT
0720     WRITE 'CURRENT PRODUCTION KEY:' TABLE-KEY /
0730           'CURRENT TABLE RESULT 1:' TABLE-RESULT-1 /
0740           'CURRENT TABLE RESULT 2:' TABLE-RESULT-2 /
0750     SKIP 1
0760     IF *LINE-COUNT > #DEVICE-LINE-CNT THEN NEWPAGE
0770     END-IF
0780   END-READ
0790 END-IF
```

The END TRANSACTION can also provide restart capability. ET data can be stored as part of the ET operation. Up to 40 fields can be stored as part of the ET statement. These can be reread via GET TRANSACTION DATA statement (ADABAS RE command). This data is identified by the *ETID value. This value can be set in NATURAL Security to default to the LOGON user id (*INIT-USER) or can be set to a value either user-defined or generated by NATURAL Security. Many shops generate ETID values and provide them as a dynamic parameter. This avoids the problem of two LOGON with the same user id getting a blank ETID value for a subsequent LOGON. Needless to say, the data read by GET TRANSACTION must be in same order, format and length as that which is written via END TRANSACTION.

Even though this style of writing is available in report mode, it isn't desirable. The two READ loops causes NATURAL to construct two different sets of format buffers for the program if you do not use a DEFINE DATA construct or you do not use structured mode. Using this style, the correct READ loop is entered depending on the conditional check and the appropriate command is issued to the situation - either a normal READ LOGICAL (L3) or a READ LOGICAL with the hold option (L6).

A delete can not be issued outside of the data base loop; otherwise a NATURAL error 45 occurs during the compile.

UPDATE / UPDATE SAME

It is important, in either structured mode or report mode with a DEFINE DATA statement, to set up the fields for an update before the UPDATE statement is issued. The WITH and USING SAME options are not allowed in these situations. Remember that if an UPDATE is done using a view set up for both READ and UPDATE, then whatever values are stored in the buffer at the time of the STORE are applied to the data base. Sometimes it is wise to define different areas for READ/FIND and for UPDATE/STORE, especially if key values are *not* updated.

Beware of trying to update in physical order. It is possible that records which are updated may migrate to other file data blocks. If the update process re-encounters records because they have migrated higher in the file they could be processed more than once. This would cause inconsistent results. If you need to update an entire file, then the safest route is to READ the file by ISN and UPDATE.

Issuing ETs

The ET command generated for an END TRANSACTION is a very expensive command to complete. It requires a lot of system resources and therefore should not be overused. Too often ET/BTs are issued to release held records, or to attempt to lower ET time. This should not be the case.

First, it is not good to hold records which are not intended to be updated! If the application can afford to not keep a record on hold (I have seen such applications in other shops which require this), then it is better to reread the record than to fill up the hold queue or cause unnecessary ADABAS overhead.

Second, there is a better approach to handling the transaction timer in ADABAS. Programmatically, if you are maintaining an ET counter, you might find that the transaction time limit expires before an ET is issued. The usual practice is to lower the ET counter. This is NOT necessary. Instead, using an approach I learned from Cornell University, it is better to programmatically keep track of the time passed since the last ET. If the time is about to expire, then issue an ET and reset your program's timer. See SETTIME and *TIME-OUT.

IV. Concurrency vs. response code 145

There is always the conflict arising from online transaction requirements and batch transaction requirements. The length of a transaction in batch can cause concurrency problems for online users. From NATURAL, this is avoidable if wall time is not an issue for batch update you want to run concurrently with online access. To avoid an inevitable response code 145, the batch NATURAL session can be initialize with the parameter WH=ON.

If you run a lot on non-NATURAL code against ADABAS, you might consider a modification to the ADABAS link routine on order to interrogate control blocks and for the appropriate commands blank out the command option 1 field. This is equivalent to WH=ON.

The parameter WH set to ON forces the application to issue any modification command with command option 1 field in the ADABAS control block to be blank, instead of 'R' which implies to return a response code 145 immediately. The blank value signals to ADABAS to place the user in a wait state until the record that is simultaneously requested for update is released for the batch application.

A user on SAG-L asked how to do the following:

> "We are a little stumped with this one:
> -doing a search to a file based on two descriptors,
> -occasionally getting 145 (record held)
> -would like to skip over the record and continue on
> -ISN, unique values of the record not available
> How do we get to the next record? The ON ERROR block could bring us back into the program, however, how do we start it at the proper place?"

A respondent from the development team in Darmstadt said:

> The solution is quite simple provided you have NATURAL 2.2. You code a main program

```
READ file BY desc
 move *ISN to #ISN
 CALLNAT 'callnat' #ISN #ERROR
END-READ
```

and a subprogram

```
DEFINE DATA PARAMETER
1 #ISN (a10)
1 #ERROR (L)
LOCAL
1 file ...
 ...
END-DEFINE
*
G1. GET file #ISN
 ... do updates ...
UPDATE (G1.)
*
ON ERROR IF *ERROR-NR = 3145 ESCAPE ROUTINE END-IF
END-ERROR
END
```

That's it! NATURAL 2.2 allows you to 'escape routine' an ON-ERROR-block, i.e. you can test anything which might lead to an error in a subprogram without any danger.

Steve Robinson, consultant and author of the NATURAL Newsletter dreams about what could be:

> "While your technique works, you are issuing two READ commands per record (one READ or FIND and one GET). This is potentially very expensive. In Direct Call Mode this would be simple since you have access to the ISN list. You would issue just one ADABAS call per record, not two. That is what is really needed here; a similiar capability to what exists in Direct Calls, but in NATURAL. Perhaps something like a variation on the RETAIN AS clause which allows you to specify an array for the ISN's to be stored in."

Here is a suggestion which improves the above:

> "Make that program a level 2 program (ie. FETCH RETURN/CALLNAT/PERFORM) from a controlling program). In the level 2 program, instead of EXIT from the ON ERROR block, use the ESCAPE ROUTINE statement. This will return the flow of processing to the controlling program, where you can check for 3145 and repeat the call to the level 2 program. No need to break out of your program structure via FETCH.
>
> Feel free to use the ESCAPE ROUTINE from within the ON ERROR block as a mean of turning your NATURAL applications a little bit more fault tolerant."

Sometimes, a program can get into trouble by issuing holds against too many records that may cause 145 return code for other applications. This happens often with this scenario.

Here is some sample code that best describes how to issue updates to a record that is not yet on the file. Notice the use of GET which issues a re-read of the record.

```
RESET #HOLD-ISN
FD1. FIND (1) /* on some key value
   IF NO RECORDS FOUND
      ST1. STORE RECORD
      MOVE *ISN(ST1.) TO #HOLD-ISN
   END-NOREC
   IF #HOLD-ISN LE 0
      MOVE *ISN(FD1.) TO #HOLD-ISN
   END-IF
G1. GET FILE #HOLD-ISN

        make changes...

   UPDATE FILE (G1.)
END-FIND
```

Tip: How to put a record on hold in NATURAL without modifying or updating it.

There is no mystery here. The following code will place a record on hold without issuing an actual update. However, all considerations about hold logic and being at ET status must be considered.

```
RD. READ FILE …
   …
   IF TRUE = FALSE
         UPDATE (RD.)
   END-IF
```

V. ADABAS Interface Buffers

There are five buffers and a control block used by application programs to communicate to ADABAS. They are

Control Block	Contains the Command ID, DB and File IDs, return code, etc. The layout of the Control Block is listed later at the end of Section V.
Format Buffer	Contains the field names, format information for any data base read / modify operation.
Record Buffer	Contains the decompressed values of the fields (in order) specified in the format buffer.

Search Buffer	Descriptor field names and Boolean relationships needed for retrieval. This buffer works in conjunction with the value buffer to define the search criteria to be used for selecting a group of records via a FIND command. This and the value buffer are also used in READ LOGICAL and HISTOGRAM operations.
Value Buffer	The values which correspond to the search expressions in the search buffer are found here.
ISN Buffer	The current ISN resulting from a successful FIND with GET NEXT processing. The ISNs in the buffer are 4 byte binary.

Being knowledgeable about the contents and uses of the control block and these buffers is critical in understanding how NATURAL communicates to ADABAS. To understand this communication mechanism is to make the debugging process less time consuming and more productive. The contents of the control block will be shown in Section V on TEST DBLOG. Let's see some examples of the buffers before discussing DBLOG.

```
0010 DEFINE DATA LOCAL
0020 1 #KEY (A8) INIT <'ACH'>
0030 1 EMP VIEW OF EMPLOYEES
0040   2 NAME
0050   2 SEX
0060   2 LANG (1:2)
0070   2 COUNTRY
0080 END-DEFINE
0090 *
0100 READ EMP BY NAME
0110   STARTING FROM #KEY
0120   DISPLAY NAME SEX LANG (1:2) COUNTRY
0130 END-READ
0140 *
0150 END
```

At compile time, NATURAL builds a Format Buffer containing all the data base fields required by this program. The field names included are in the order as they appear in the program! Here is a snapshot of the Format Buffer and Record Buffer for this program. The Record Buffer contains the information of the first record retrieved by the READ request.

Format Buffer:

```
(AE,020,A,AG,001,A,AZ0001-0002,0003,A,AL,0003,A.)
```

The period marks the end of the byte string in the buffer. The parentheses are mine.

Record Buffer:

```
(ACHIESON            MENGFREUK)
```

Search Buffer:

```
(AE ,020,A.)
```

If the READ statement is changed to:

```
READ EMP BY NAME STARTING FROM 'SMITH'
```

then the contents of the Value Buffer contains:

```
(SMITH)
```

Changing the search criteria to include 'ENDING AT' does not change the contents of the Value Buffer, since NATURAL is responsible for checking the 'THRU' option.

Change the program slightly to report mode without a data definition:

```
0010 * WRITTEN IN REPORT MODE, 1.2 STYLE
0020 READ EMPLOYEES BY NAME
0030     STARTING FROM 'A'
0040   IF LANG (2) NE ' ' THEN DO
0050     DISPLAY NAME SEX LANG (1:2) COUNTRY
0060   DOEND
0070 LOOP
0080 *
0090 END
```

For this program, the Format Buffer contains:

```
(AE,020,A,AZ0002,0003,A,AE,020,A,AG,001,A,AZ0001-0002,003,A,AL,003,A.)
```

Example:

```
0010 DEFINE DATA LOCAL
0020 1 CAR VIEW OF VEHICLES
0030   2 MAKE
0040   2 MODEL
0050   2 REG-NUM
0060   2 COLOR
0070 END-DEFINE
0080 *
0090 FIND CAR WITH MAKE = 'JAGUAR' OR = 'LANCIA'
0100   DISPLAY REG-NUM MAKE MODEL COLOR
0110 END-FIND
0120 *
0130 END
```

For this example the ISN Buffer contains the ISN of the returned record.

The ISN Buffer:

`(0000003C)` which is the hex display for 60.

The Search buffer:

`(AD,020,A,O,AD,020,A.)` where , 'O' is the character which signifies OR logic.

The Value Buffer:

`(JAGUAR LANCIA)`

An interesting exercise is to discover the various ways in which NATURAL builds the various buffers for the other logical relations (e.g., >, <=, etc.). Better yet, see how NATURAL constructs buffers for algorithm 3 & 4 FIND operations. DBLOG provides the means to do this.

To avoid the infamous NAT902, simply record a selection of elements ranges within a PE group at the elementary field level. It is my understanding that NATURAL is totally re-constructing the format buffers to communicate to ADABAS more correctly in 2.3.

VI. The TEST & TEST DBLOG Facilities

The DBLOG facility is invoked by entering 'TEST DBLOG?' as a command at the 'NEXT' prompt. The DBLOG facility can be invoked at NEXT/Command level using keywords.

DBLOG ON - starts up an DBLOG session
DBLOG OFF - discontinue logging
DBLOG SHOW - stop logging and display results.

It is possible to request a logging of :

1. specific command
2. specific commands to a specific file
3. specific commands generating a response code within a certain range
4. specific commands to a specific file generating a response code within a certain range.

It is also possible to request a 'snapshot dump' of the ADABAS Control Blocks and buffers for a specific ADABAS call. If you need to log against the system files, you must initiate the session with LOG=ON. All of these options are specified using the '?' option of the TEST DBLOG command. The entry: 'TEST DBLOG ?' produces the menu in Fig. 6.1.

```
12:34:56              ***** NATURAL Test Utilities *****               97-04-01
User APDDBS                     - DBLOG Menu -                   Library APDDBS


                  Code  Function
                  ----  -----------------------------------
                   B    Begin Logging of ADABAS Commands
                   E    End and Display Log Records
                   S    Snapshot of Specific ADABAS Commands
                   .    Exit
                  ----  -----------------------------------
          Code .. _

      Command .. __       Skip ....... _____    Program .... ________
      DBID ..... _____    FNR ........ _____    Line from .. 0000
      Low Resp . _____    High Resp .. 9999_    Line to .... 0000

      Optional Buffers for Code B
        FB .. _ RB .. _ SB .. _ VB .. _ IB .. _


Command ===>
Enter-PF1---PF2---PF3---PF4---PF5---PF6---PF7---PF8---PF9---PF10--PF11--PF12---
      Help  Print Exit  Begin End   Snap                                  Canc
```

Figure 6.1. DBLOG Menu

Code

IF 'S' is specified, a snapshot of all ADABAS buffers is taken for the specified command. With SNAP=YES, the execution of the program will immediately be interrupted after the specified command has been executed by ADABAS and the snapshot of the ADABAS Control Block will be displayed. After a snapshot has been taken, the DBLOG facility is turned off and may be reactivated by reentering the DBLOG command.

A Command ID conflict *may* occur with the next program invoked after a snapshot has been taken. For example, a list of ISNs created by a FIND command will not be released when a snapshot is taken and will still exist: 1) when the same program is executed again after the SNAP, or 2) when a different program is invoked. To avoid unpredictable results, it is recommended that the NATURAL session be terminated and re-invoked after a SNAP.

Command

Required two-character alphanumeric literal (e.g., L3). Allows selective logging of commands with the specified command code.

Skip

This is an optional three-digit numeric literal. It allows the specification of the occurrence of a command to be snapped. With SKIP 1, the first occurrence of the specified command is snapped; SKIP 2 snaps the second occurrence of the specified command. Be careful, if you are logging system file commands, you need to include these in the SKIP.

Dbid, Fnr

Optional three-digit database number. Leading zeroes are optional. Allows selective logging of commands issued against the specified file number.

Low-Rsp High-Rsp

Optional three-digit response codes. Leading zeroes are optional. Allows logging of commands that receive a response code in the specified inclusive range.

A Sample TEST & TEST DBLOG Session

The following program is used to demonstrate the DBLOG facility. A trace of the ADABAS calls is initiated by typing 'TEST DBLOG ?'. The execution output of the example program follows.

```
0010 DEFINE DATA LOCAL
0020 1 CAR-OWNERSHIP (A22)
0030 1 EMP VIEW OF EMPLOYEES
0040   2 PERSONNEL-ID
0050   2 FIRST-NAME
0060   2 NAME
0070   2 SEX
0080   2 CITY
0090   2 DEPT
0100     /* END OF VIEW EMPLOYEES-VIEW
0110 1 CAR VIEW OF VEHICLES
0120   2 PERSONNEL-ID /* CNNNNNNN
0130   2 MAKE
0140   2 MODEL
0150     /* END OF VIEW VEHICLES-VIEW
0160 END-DEFINE
0170 *
0180 RD. READ EMP BY CITY
0190   DISPLAY FIRST-NAME NAME CITY (AL=20) SEX DEPT
0200   LOOK. FIND CAR WITH PERSONNEL-ID = PERSONNEL-ID (RD.)
0210      WRITE 'Car owned:' MAKE 'Model:' MODEL
0220   END-FIND
0230   IF *NUMBER (LOOK.) = 0
0240      CAR-OWNERSHIP := 'No automobile on record.'
0250   END-IF
0260 END-READ
0270 END
```

```
MORE
Page       1                                                97-04-01   12:34:56

     FIRST-NAME                NAME                  CITY              S DEPARTMENT
                                                                       E    CODE
                                                                       X
-------------------- -------------------- -------------------- - ----------

WILLIE                SENKO                AIKEN                   TECH10
Car owned: FORD                 Model: GRANADA
ANNIE                 GODEFROY             AIX EN OTHE.            COMP02
No automobile on record.
FRANCOIS              CANALE               AJACCIO                 TECH03
Car owned: RENAULT              Model: R5
MICHAEL               PLOUG                ALBERTSLUND             ADMA02
No automobile on record.
JOE                   HAMMOND              ALBUQUERQUE             SALE40
No automobile on record.
ROB                   ROLLING              ALBUQUERQUE             MGMT30
No automobile on record.
VENUS                 FREEMAN              ALBUQUERQUE             MGMT30
Car owned: CHRYSLER             Model: PLYMOUTH
Car owned: GENERAL MOTORS       Model: CHEVROLET
JEFFERSON             LINCOLN              ALBUQUERQUE             TECH10
```

Figure 6.2. Program execution output.

A subsequent entry of the basic command DBLOG will result in a summary display of the *recent* ADABAS calls issued.

```
16:18:36              ***** NATURAL Test Utilities *****            97-05-15
User APDDBS                    - DBLOG Trace -                 Library APDDBS
M No   Cmd DB  FNR Rsp     ISN        ISQ      CID  CID(Hex) OP   Pgm      Line
_    1 L3  240 143       76276                 NLBP D5D3C2D7  V ATEST      2260 Top
_    2 RC  240 143                             NLBP D5D3C2D7 SF ATEST      2260
_    3 RC  240                                      00000000 F  ATEST      2530
_    4 L3  240   4         627                 ?-?? 01800101 HV FINDCAR    0180
_    5 S1  240   5         609           1 ? ?? 02000101    FINDCAR    0200
_    6 RC  240   5                              ? ?? 02000101 SI FINDCAR    0200
_    7 L3  240   4          14                 ?-?? 01800101 HV FINDCAR    0180
_    8 S1  240   5                              ? ?? 02000101    FINDCAR    0200
_    9 L3  240   4         114                 ?-?? 01800101 HV FINDCAR    0180
_   10 S1  240   5          39           1 ? ?? 02000101    FINDCAR    0200
_   11 RC  240   5                              ? ?? 02000101 SI FINDCAR    0200
_   12 L3  240   4         467                 ?-?? 01800101 HV FINDCAR    0180
_   13 S1  240   5                              ? ?? 02000101    FINDCAR    0200
_   14 L3  240   4         510                 ?-?? 01800101 HV FINDCAR    0180
_   15 S1  240   5                              ? ?? 02000101    FINDCAR    0200
_   16 L3  240   4         516                 ?-?? 01800101 HV FINDCAR    0180
_   17 S1  240   5                              ? ?? 02000101    FINDCAR    0200

Command ===>
```

Figure 6.3. DBLOG ADABAS command trace output

The number of commands you get depends on the NATURAL DSIZE parameter. If you want to see activity against the NATURAL system files, 'LOG=ON' must be specified as a parameter at LOGON.

The columns of data are:

M — Mark with one of the following parameter values:
C to see the Control Block for the specific command
F to see the Format Buffer (maximum 80 bytes)
R to see the Record Buffer (maximum 80 bytes)
S to see the Search Buffer (maximum. 80 bytes)
V to see the Value Buffer (maximum 80 bytes)
I to see the ISN Buffer (maximum 80 bytes)
If a particular buffer has not been logged in first place, a message will be issued.

No — represents the commands generated by the application program in sequence

Cmd — is the ADABAS Command generated by NATURAL

DB — is the database ID against which the command is issued

FNR — is the file number against which the command is issued

Rsp — is the ADABAS response code

ISN — is the internal sequence number of the record successfully returned

ISQ — is the quantity of ISNs resolved to meet the selection criteria of a FIND

CID — is the Command Id displayed in both hex and alphanumeric. The first two bytes is the line number of the NATURAL program which contains the ADABAS access verb 01, 02 are the command options 1 and 2

CID(Hex) — the Command ID in hexadecimal

Pgm — the *PROGRAM value which is issuing the ADABAS calls
OP values stored in command options 1/2

Line — source line where the command is coded

Entering the letter 'F' under the 'M' column displays the format buffer for the S1 (FIND) command (only if you requested this on the initial trace screen)

```
Seq No ..      7   Format Buffer
0000 * C1C36BF0 F0F86BC1 6BC1C46B F0F2F06B * AC,008,A,AD,020, * 0000
0010 * C16BC1C5 6BF0F2F0 6BC16BC1 C76BF0F0 * A,AE,020,A,AG,00 * 0010

0020 * F26BE44B 00000000 00000000 00000000 * 2,U.             * 0020
0030 * 00000000 00000000 00000000 00000000 *                  * 0030
0040 * 00000000 00000000 00000000 00000000 *                  * 0040
Global FID .... 0000000000000000
```
Figure 6.4. DBLOG Output, format buffer display

The program was changed (with deliberate problems) slightly to look like the following:

```
0010 DEFINE DATA LOCAL
0020 1 CAR-OWNERSHIP (A22)
0030 1 EMP VIEW OF EMPLOYEES
0040   2 PERSONNEL-ID
0050   2 NAME
0060   2 CITY
```

```
0070     /* END OF VIEW EMPLOYEES-VIEW
0080 1 CAR VIEW OF VEHICLES
0090   2 PERSONNEL-ID /* CNNNNNNN
0100     /* END OF VIEW VEHICLES-VIEW
0110 END-DEFINE
0120 *
0130 RD. READ EMP BY CITY
0140   LOOK. FIND NUMBER CAR WITH PERSONNEL-ID = PERSONNEL-ID (RD.)
0150   IF *NUMBER (LOOK.) = 0
0160      CAR-OWNERSHIP := 'No automobile on file.'
0170   END-IF
0180   DISPLAY NAME CITY (AL=20) CAR-OWNERSHIP
0190 END-READ
0200 END
```

Typing the command TEST DBLOG in response to the MORE or NEXT prompt will invoke the DBLOG facility. The execution of the example program produces the following output. My suspicion was that more people than 'SENKO' should have an automobile on file.

```
MORE
Page      1                                                    97-04-01  12:34:56


        NAME                 CITY               CAR-OWNERSHIP
-------------------- -------------------- ----------------------

SENKO                AIKEN
GODEFROY             AIX EN OTHE.         No automobile on file.
CANALE               AJACCIO              No automobile on file.
PLOUG                ALBERTSLUND          No automobile on file.
HAMMOND              ALBUQUERQUE          No automobile on file.
ROLLING              ALBUQUERQUE          No automobile on file.
FREEMAN              ALBUQUERQUE          No automobile on file.
LINCOLN              ALBUQUERQUE          No automobile on file.
GOLDBERG             ALFRETON             No automobile on file.
FERNANDEZ            ALICANTE             No automobile on file.
MARTINEZ             ALICANTE             No automobile on file.
PEREZ                ALICANTE             No automobile on file.
PUERTOLAS            ALICANTE             No automobile on file.
GOMEZ                ALICANTE             No automobile on file.
LLAGUNO              ALICANTE             No automobile on file.
RODRIGUEZ            ALICANTE             No automobile on file.
MUNOZ                ALICANTE             No automobile on file.
FERNANDEZ            ALICANTE             No automobile on file.
```

Figure 6.5. Program executionoutput

Here is the trace result of typing 'TEST DBLOG' after initiating the request (notice no '?' this time).

```
12:34:56                 ***** NATURAL Test Utilities *****               97-04-01
User APDDBS                       - DBLOG Trace -                    Library APDDBS
M No    Cmd DB FNR Rsp      ISN         ISQ      CID   CID(Hex) OP   Pgm       Line
_     1 L3 240 143        76276                 NLBP D5D3C2D7  V ATEST     2260 Top
_     2 RC 240 143                              NLBP D5D3C2D7 SF ATEST     2260
_     3 RC 240                                       00000000 F  ATEST     2530
_     4 S1 240 145         7601            1         00000000              0000
_     5 S1 240 145         8582            1         00000000              0000
_     6 L3 240   4          627                 ?-?? 01900101 HV GOMEZ     0190
_     7 S1 240   5          609            1 ???? 02200101    GOMEZ     0220
_     8 RC 240   5                              ???? 02200101 SI GOMEZ     0220
_     9 L3 240   4           14                 ?-?? 01900101 HV GOMEZ     0190
_    10 S1 240   5                              ???? 02200101    GOMEZ     0220
_    11 L3 240   4          114                 ?-?? 01900101 HV GOMEZ     0190
_    12 S1 240   5           39            1 ???? 02200101    GOMEZ     0220
_    13 RC 240   5                              ???? 02200101 SI GOMEZ     0220
_    14 L3 240   4          467                 ?-?? 01900101 HV GOMEZ     0190
_    15 S1 240   5                              ???? 02200101    GOMEZ     0220
_    16 L3 240   4          510                 ?-?? 01900101 HV GOMEZ     0190
_    17 S1 240   5                              ???? 02200101    GOMEZ     0220

Command ===>
```

Figure 6.6. DBLOG ADABAS command trace output

DBLOG Snapshot

A snapshot is invoked by typing '`TEST DBLOG ?`'. The first block which is displayed by DBLOG snapshot is the Control Block. The other buffers can be viewed by typing the appropriate code (FB, RB, SB, VB, IB) on the screen or pressing the appropriate PF key. The snapshot of the example program produces this control block listing

```
12:34:56                 ***** NATURAL Test Utilities *****               97-04-01
User APDDBS                        - DBLOG Menu -                    Library APDDBS


                    Code  Function
                    ----  ------------------------------------
                     B    Begin Logging of ADABAS Commands
                     E    End and Display Log Records
                     S    Snapshot of Specific ADABAS Commands
                     .    Exit
                    ----  ------------------------------------
            Code .. s

       Command .. s1      Skip .......  _____     Program ....  ________
       DBID ..... _____   FNR ........  _____     Line from .. 0000
       Low Resp . _____   High Resp .. 9999_      Line to .... 0000

       Optional Buffers for Code B
         FB .. _ RB .. _ SB .. _ VB .. _ IB .. _

Command ===>
Enter-PF1---PF2---PF3---PF4---PF5---PF6---PF7---PF8---PF9---PF10--PF11--PF12---
      Help  Print Exit  Begin End   Snap                                  Canc
```

Figure 6.7. DBLOG Menu, request to snap

The snap of the format buffer explains why NATURAL 2.2 is better with FIND NUMBER than 2.1. It contains '.' only.

```
_       Seq No ..      7   Format Buffer
0000 * 4B000000 00000000 00000000 00000000 *
0010 * 00000000 00000000 00000000 00000000 *

0020 * 00000000 00000000 00000000 00000000 *
0030 * 00000000 00000000 00000000 00000000 *
0040 * 00000000 00000000 00000000 00000000 *
```

Figure 6.8. DBLOG Snapshot, display of format buffer.

To understand the DBLOG snapshot screen, you should know the ADABAS control block fields and their lengths/formats. Here is the layout of the control block.

ADABAS Control Block Definition			
Field	Position Within Control Block	Length in Bytes	Format
Reserved	1 - 2	2	
Command Code	3 - 4	2	Alphanumeric
Command ID	5 - 8	4	Binary/Alphanumeric
File Number	9 - 10	2	Binary
Response Code	11- 12	2	Binary
ISN	13 - 16	4	Binary
ISN Lower Limit	17 - 20	4	Binary
ISN Quantity	21 - 24	4	Binary
Format Buffer Length	25 - 26	2	Binary
Record Buffer Length	27 - 28	2	Binary
Search Buffer Length	29 - 30	2	Binary
Value Buffer Length	31 - 32	2	Binary
ISN Buffer Length	33 - 34	2	Binary
Command Option 1	35	1	Alphanumeric
Command Option 2	36	1	Alphanumeric
Additions 1	37 - 44	8	Binary/Alphanumeric
Additions 2	45 - 48	4	Binary/Alphanumeric
Additions 3	49 - 56	8	Alphanumeric
Additions 4	57 - 64	8	Alphanumeric
Not Used	65 - 72	8	
Command Time	73 - 76	4	Binary
User Area	77 - 80	4	

Figure 6.9. Control Block Layout

None of this information pointed to why I saw so many with 'No automobile..'. I decide to trace the program execution itself. This time I invoke the Test facility with 'TEST' and initiate the trace with option 'T'.

```
19:50:55               *** NATURAL TEST UTILITIES ***
Test Mode OFF             - Debug Main Menu -                97-04-01
                                                       Object FINDCAR

                    Code  Function
                    ----  -----------------------------
                     T    Set Test Mode ON
                     E    Debug Environment Maintenance
                     S    Spy Maintenance
                     B    Breakpoint Maintenance
                     W    Watchpoint Maintenance
                     C    Call Statistic Maintenance
                     L    List Object Source
                     V    Variable Facility
                     ?    Help
                     .    Exit
                    ----  -----------------------------
            Code .. T     Object Name .. FINDCAR_



Command ===>
```

Figure 6.10. Test facility - turn on Test mode for program FINDCAR

After entering the 'L'ist onject source option, I set a breakpoint immediate for execution. A breakpoint refences an entry in the execution code you want to stop executing and to initiate the Test facility to do further investigation.

```
12:34:56               *** NATURAL TEST UTILITIES ***
Test Mode ON             - List Object Source -              97-04-01
                                                       Object FINDCAR
 C  Line Source                                           Top of Data
 -- ---- ---------------------------------------------   Message
 __ 0010 DEFINE DATA LOCAL                               ------------
 __ 0020 1 CAR-OWNERSHIP (A22)                          :
 __ 0030 1 EMP VIEW OF EMPLOYEES                        :
 __ 0040   2 PERSONNEL-ID                               :
 __ 0050   2 FIRST-NAME                                 :
 __ 0060   2 NAME                                       :
 __ 0070   2 CITY                                       :
 __ 0080   2 JOB-TITLE                                  :
 __ 0090     /* END OF VIEW EMPLOYEES-VIEW              :
 __ 0100 1 CAR VIEW OF VEHICLES                         :
 __ 0110   2 PERSONNEL-ID /* CNNNNNNN                   :
 __ 0120     /* END OF VIEW VEHICLES-VIEW               :
 __ 0130 END-DEFINE                                     :
 __ 0140 *                                              :
 SE 0150 RD. READ EMP BY CITY                           :
                                                        : FINDCAR0150
```

Figure 6.11. Test facility list option

Upon executing the porgram, after reaching the breakpoint, this window appears, in which I enter 'L':

```
   >>> Debugging Facility <<<

Break by Breakpoint FINDCAR0150
at line   150 of object FINDCAR

     G  Go
     L  List Break
     M  Debug Main Menu
     R  Run (Test Mode Off)
     V  Variable Facility

Enter Code ..: L
```

The 'G' option would allow the program to continue executing beyond this breakpoint.

I can now step through the program one execution at a time with PF key 2. Doing this, I realized my message was never re-initialized (how trivial). I could also have invoked the variable facility (Fig. VI.12). Within this screen, you can either '**MO**'dify or '**DI**'isplay variables, then continue execution to see the affects of your request.

```
12:34:56                 *** NATURAL TEST UTILITIES ***                 97-04-01
Test Mode ON          -  Display Variables (alphanumeric)  -    Object FINDCAR
                                                                            All
 C  L Variable                           F Len  Contents              Message
 -- - ---------------------------------- - ---- -------------------- ---------
 __ 1 CAR-OWNERSHIP                      A   22 No automobile on fil
 __ 1 EMP
 __ 2 EMP.PERSONNEL-ID                   A    8 50016600
 __ 2 EMP.FIRST-NAME                     A   20 ANNIE
 __ 2 EMP.NAME                           A   20 GODEFROY
 __ 2 EMP.CITY                           A   20 AIX EN OTHE.
 __ 2 EMP.JOB-TITLE                      A   25 COMPTABLE
 __ 1 CAR
 __ 2 CAR.PERSONNEL-ID                   A    8
 __ 0 *ISN                               P 10.0 14
 __ 0 *COUNTER                           P 10.0 2
 __ 0 *NUMBER                            P 10.0 0
 __
 __
 __

Command ===>
```
Figure 6.12. Test facility variable option

A good debugging technique is to wrap an invocation to DBLOG around a portion of code in which you want to check the ADABAS accesses. In the following example, I wanted to see, as I suspected, that I would not issue calls to hold records unnecessarily if the flag #ACTUAL-RUN-FLAG is not set to 'A'.

```
0510 FETCH RETURN 'TEST DBLOG'
0520 IF #ACTUAL-RUN-FLAG = 'A' THEN DO
0530    RD-5.
0540       READ TB5-P BY TABLE-KEY
0550         = #KEY THRU #END-KEY
0560          ADD 1 TO #REPLACE-COUNT
0570          DELETE (RD-5.)
0580          WRITE 'DELETED PRODUCTION KEY:'
0590             TABLE-KEY (RD-5.) /
0600                 'DELETED PRODUCTION TABLE RESULT-1:'
0610             TABLE-RESULT-1 (RD-5.) /
0620                 'DELETED PRODUCTION TABLE RESULT-2:'
0630             TABLE-RESULT-2 (RD-5.) /
0640         IF *LINE-COUNT > 20 THEN NEWPAGE
0650       LOOP (RD-5.)
0660     DOEND
0670 ELSE DO
0680       RD-5B.
0690       READ TB5-P BY TABLE-KEY
0700         = #KEY THRU #END-KEY
0710          ADD 1 TO #REPLACE-COUNT
0720          WRITE 'CURRENT PRODUCTION KEY:' TABLE-KEY /
0730                'CURRENT TABLE RESULT 1:' TABLE-RESULT-1 /
0740                'CURRENT TABLE RESULT 2:' TABLE-RESULT-2 /
0750          SKIP 1
0760          IF *LINE-COUNT > 20 THEN NEWPAGE
0770       LOOP (RD-5B.)
0780     DOEND
0790 FETCH RETURN 'TEST DBLOG'
```

NOTE

Command IDs are data which appear in the Control Block and serve an important function during ADABAS command execution. NATURAL's READ, FIND, HISTOGRAM, GET, GET SAME, STORE and UPDATE commands require a Format Buffer. This is translated into internal format buffers and stored in a pool. The Command IDs are associated with each translation and are stored on a list called the Table of Sequential Commands. Upon executing any of these commands, ADABAS checks the table to see if the Command ID is present. If so, then no translation into internal format buffers is done. Instead, it uses the existing internal format buffers for that Command ID. The Command ID is maintained in either the Table of Sequential Commands, the Internal Format Buffer Pool or the Table of ISN Lists. NATURAL releases them at the right time.

The execution of the program produced the online output shown on the next page.

```
PRODUCTION TABLE BEING REPLACED…TEST RUN

ADALOG started

CURRENT PRODUCTION KEY: TAK43700
CURRENT TABLE RESULT 1: NO ACTIVITY THIS SEM
CURRENT TABLE RESULT 2:

CURRENT PRODUCTION KEY: TAK43710
CURRENT TABLE RESULT 1: REG FORM PRE-PRINTED
CURRENT TABLE RESULT 2:

CURRENT PRODUCTION KEY: TAK43720
CURRENT TABLE RESULT 1: DROP/ADD-NO REG FORM
CURRENT TABLE RESULT 2:

CURRENT PRODUCTION KEY: TAK43740
CURRENT TABLE RESULT 1: REG FORM PROCESSED
CURRENT TABLE RESULT 2:

CURRENT PRODUCTION KEY: TAK43750
CURRENT TABLE RESULT 1: REG FORM WITH D/A
CURRENT TABLE RESULT 2:

   37 RC  215   5   0   21743      0 ?-?? 068000201        COPY2270
   36 L3  215   5   0   21743      0 ?-?? 068000201 H   V  COPY2270
   35 L3  215   5   0   21742      0 ?-?? 068000201 H   V  COPY2270
   34 L3  215   5   0   21741      0 ?-?? 068000201 H   V  COPY2270
   33 L3  215   5   0   21740      0 ?-?? 068000201 H   V  COPY2270
   32 L3  215   5   0   21739      0 ?-?? 068000201 H   V  COPY2270
   31 L3  215   5   0   21738      0 ?-?? 068000201 H   V  COPY2270
   30 OP  215   5   0     203      0 ?-?? 068000201     V  COPY2270
```

VII. Some Common ADABAS/NATURAL Errors

Subcodes to the ADABAS respnse codes are now available in the rightmost two bytes of the Additions-2 field in the ADABAS control block

Error 3001 (ADABAS response code 1).

1. Subcode 2: Security by value disallowed the record retrieved via an Sx command.
2. Subcode 3: LS database parameter may need increasing. An S2 command ran out of room.

Error 3002 The data in the ET buffer exceeded 2K. (ADABAS response code 2)

Error 3007 An Sx command was interrupted due to maximum time permitted has been exceeded. (ADABAS response code 7)

Error 3008 The current user's command was interrupted to prevent a WORK overflow because of a pending backout operation. (ADABAS response code 8)

Your application is probably a victim, not the culprit. Your database administrator may have to resize the LP parameter, or a runaway application may be issuing too many file modifications without a BT or an ET command.

Error 3009 Transaction backout (ADABAS response code 9)

This error occurs when ADABAS has issued a BT against your session. This happens in timeout situations; however, this is not the only reason for getting 3009. What has happened is that ADABAS identifies you as not at ET status with an open transaction. In its attempts to respond to your file modification(s), you do not close the transaction in sufficient time. Other reasons for response code 9 are:

(1) UQE (User Queue Element) is still around, and another user with another external id but with the same UQE will get a response code 9

(2) if the work data set is full ADABAS attempts to back out the oldest user to release enough blocks to continue

(3) if a security check (200,201,202) occurs while a user is not at ET status a response code 9 can occur on the next call.

Recently, a unusual situation in regards to timeouts had occurred in one of BU's on-line systems. COM-PLETE's SD files were used as a storage medium for an application. The data base was read for all of the query/update needs. Once the data requirements were fulfilled, the user perused the SD files without doing any more ADABAS calls. ADABAS' non-activity timer was exceeded even though the application looked fine. The remedy was to reissue a dummy ADABAS call periodically to keep ADABAS aware of the function's existence (UGH!).

3009 Subcodes

Subcode 1	Hold queue filled
Subcode 2	TT exceeded
Subcode 3	Non-activity timer exceeded (TNAE, TNAX, or TNAA)
Subcode 4	User removed by STOPU / STOPI

Subcode 15	Pending work overflow (LP needs increasing)
Subcode 62	OP issued w/o user id
Subcode 63	OP issued for user not in ET status (repeat OP)
Subcode 64	OP issued for ETID which already exists
Subcode 66	OPENRQ=YES and user issued first command ne 'OP'

Error 3017 Invalid file number - (ADABAS response code 17)

This error implies that the file is not there or you are locked out. This happens a lot when an application is expecting to read a file and the file is currently being loaded or it has been deleted before the load procedure.

ADABAS response code 17 subcodes

Subcode 1	Access system file 1,2 and no OP
Subcode 2	Access to system file 1,2 and user is not authorized
Subcode 4	Invalid file number specified
Subcode 5	Not loaded, or locked by another UQE with UTI status
Subcode 6	Delete w/o valid file number
Subcode 7	'LF' on system file 1 , 2
Subcode 8	File accessed not in OP file list w/ 'R' option
Subcode 9	File completely locked
Subcode 10	Program not authorized to access specified file. File locked by another user with EXU status.

Error 3021 Invalid Command ID - (ADABAS response code 21)

This error is returned particularly if the Command ID is lost. This occurs if an RC is issued for a particular Command ID, or an RC with Command ID of binary zeroes. This wipes out all Command IDs and any entry on the Table of Sequential Commands and the Table of Search Results. This scenario has reared its ugly head in calling COBOL from NATURAL and returning to NATURAL. This occurs during a timeout scenario (TNAA) for an access only user and ADABAS has lost reference to your Command Id since it has deleted associated control blocks. Another reason is that a direct call was issued for an Lx sequence without the appropriate Additions-1 field.

Error 3041 Format buffer error - (ADABAS response code 41)

A request was made by NATURAL through the Format Buffer for a field on the file for your program, but it IS NOT ON THE FILE ! This can occur if a field was dropped or its ADABAS name changed and the new information is not reflected on the DDM. Another reason for getting response code 41 involves problems with MU/PE field references in the Format Buffer.

Error 3042 Format Buffer too long - (ADABAS response code 42)

This error reflects that the Format Buffer construction was too long. ADABAS could not find the space in the format pool to translate your Format Buffer to its Internal Format Buffer. This may not be directly a programmatic error but it is fixable by the systems/DBA staff. They may have to raise the LFP parameter (length of format buffer pool) for your installation. This can also happen if your NATURAL program invokes a non-NATURAL routine which issues ADABAS calls (version 4) with blank CID values. This can cause fragmentation of the format pool unnecessarily.

Error 3044 For UPDATE - (ADABAS response code 44)

This error points out that the Format Buffer as constructed is not proper for executing an UPDATE command. This happens when the buffer contains 2 field references to the same ADABAS field. This happens with synonyms and references to MU / PE fields BEFORE an OBTAIN statement if it is a version NATURAL 1.2 program. You must remove the multiple references in the Format Buffer. This can be done by not using aliases or placing the OBTAIN immediately after the access verb (better yet, convert it to NATURAL 2 if possible). You can check the buffer contents using DBLOG.

Error 3047 (ADABAS response code 47). The application has exceeded ¼ of the ADABAS parameter NH – 1. Too many records are being held without a BT or an ET being issued.

Error 3048 For UPDATE - (ADABAS response code 48)

Your program wants to open a file which has been opened by another user with conflicting usage. This often happens if a file is accessed by a ADABAS utility.

3048 Subcodes

Subcode 2	The FNR conflicts w/ same file for another user
Subcode 3	The 'OP' issued by a utility cannot be performed; an online ADASAV is in progress.
Subcode 4	User id in OP command is already in use
Subcode 6	The utility requires exclusive use and the User Queue is not empty.
Subcode 8	The logical user id already exists for another non-ET user or for another ET user who has been active for the last 60 seconds.
Subcode 9	The user requested 'UPD' status for a file which is locked by another user who is 'EXU' or 'UTI'
Subcode 10	The user requested 'EXU' or 'UTI' status for a file which is locked by another user who is 'UPD'

Error 3049 Exceed maximum length of compressed record (ADABAS response code 49)

The length of the compressed record that you are trying to add or update is longer than the maximum record length for the file. You can specify the max compressed record length on a file by file basis. The default maximum is determined by the block size of your DATA component. ADABAS V5 has a default for data storage block size on a device, 5064 on a 3390 device type for example. You cannot put a record larger than that (and possibly some % shy of it) or you generate this error.

Make ADABAS block sizes bigger to hold larger compressed records,

Make changes to the application/file design to store smaller records.

Change the data type of numeric fields to binary to make smaller records.

Maybe it can be normalized, or it contains multiple record types and can be split into more files.

Eventually, you will be faced with an issue of redesigning the file.

Error 3052 Illegal conversion (ADABAS response code 52)

An error occurred while ADABAS was processing the Record Buffer. It implies that a value that should be packed/unpacked was not in the proper format. I have seen this on UPDATE commands; it is possible that there was a disparity between the ADABAS FDT and the DDM generated for NATURAL's use.

Error 3055 Illegal conversion (ADABAS response code 55)

This ADABAS error occurs whenever data conversion is necessary between what is actually in physical storage and what is requested by the Format Buffer. In the attempt to construct the Record Buffer, ADABAS might have to convert, e.g., there is binary data on the file and you request it as packed data and it cannot be packed. This discrepancy can occur between what is on the DDM and what has been defined to ADABAS. Also, a numeric value cannot fit into a field described in the Format Buffer. Also, a field defined with the SQL NC option contained a null, and the format buffer counter indicated.

Error 3060 Format or Search Buffer error - (ADABAS response code 60). There are several subcodes that describe this failure.

Error 3061 Search Buffer error - (ADABAS response code 61)

A similar problem to 3041. In this case a field was used as a key and ADABAS returns an error to NATURAL implying the field is NOT a descriptor. Again, if the DDM says a field is a descriptor then it may be out of synch with the physical file. Compare the DDM view and the physical file layout.

This error message is also generated if an attempt is made to read a file logically (READ ... LOGICAL which issues L3/L6 commands) using a (super)descriptor constructed from a PE group element. This is currently illegal. To overcome this, you can issue HISTOGRAM using the key in question, retrieve the key value, and issue a FIND using this value. Your application issues extra ADABAS calls, but this is weighed against the requirement to create results requiring a sort order on this key

Error 3070 Table Overflow (ADABAS response code 70)

This occurs whenever the TABLE of Sequential Commands overflows. This happens when not enough RC's are being issued or the appropriate ADABAS parameter needs increasing.

Errors 70 through 74 all deal with overflows of one of ADABAS' queues.

Error 3075 No more file space (ADABAS response code 75). Your program may be adding more records than the space defined for the file can contain. The file probably needs reallocation of either the associator or data storage space.

Error 3079 Illegal hyperexit reference. (ADABAS response code 79). The database may have been brought up without knowledge of the hyperexit.

Error 3113 Invalid ISN - (ADABAS response code 113)

This error denotes that an illegal ISN was presented to ADABAS for the retrieval of a record - either it doesn't exist or the physical data storage sees duplicate ISNs in different data storage blocks for the file. The second possibility implies physical damage to the file and will require DBA attention.

The first can occur under a couple of situations. For one, a FIND on a large number of descriptors can take a good amount of CPU time to be constructed. For example, suppose your program constructs a list of 150,000 ISNs (Ugh, I have seen this !!). If during the construction and processing another program comes along and deletes your 100,000th ISN before you process it you will get a NAT3113! (Of course, if this involved processing on a single descriptor and the program performed read logical instead of FIND this would NOT happen !!).

Also, you should check to see if the program tried to add records to a file protected by security-by-value.

Error 3145 Hold conflict - (ADABAS response code 145)

This occurs under two situations. First, a record is being held you wish to also put on hold (done automatically by NATURAL for and UPDATE/DELETE issued against a record accessed by READ/FIND). Second, a filled ADABAS hold queue also returns this error code. You can use NATURAL's retry capability or alter the WH parameter in either NATURAL parameter or in the definition of the application in NATURAL SECURITY. See your loving DBA.

Error 3148 ADABAS not available - (ADABAS response code 148)

The error message needs no explanation. However, the cause can be mysterious. Be careful of moving programs modules between data bases. There is a possibility that the object program might reference another data base due to information attached at compile time.

Error 3198 Unique key exists - (ADABAS response code 198)

This one isn't in the NATURAL Error Message Manual and is issued when you attempt to store a record with a unique key value already assigned to another record. This generally is a program logic problem. Any application with a defined error transaction handler (e.g., NATURAL's parameter ETA should be aware of error 3198).

Error 3200, 3201, 3202, 3203 - (ADABAS response codes 200, 201, 202, 203)

All are related to ADABAS security. Either you did not provide the file password (which means it is blank), the incorrect one for the correct access level, or just an incorrect password (not defined for this database at all). This can happen when new files are added to a series of programs or a different access level is required for a file(s) than previously allowed.

Error 3254 (ADABAS response code 254)

You either have encountered an attached buffer overflow or you have exceeded the CT parameter.

Error 3255 (ADABAS response code 255)

This ADABAS response code indicates that there were no attached buffers available for your command. This may imply that the NABS ADABAS parameter needs increasing by your staff DBA.

Case Study #1: FIND NUMBER vs. HISTOGRAM

```
0010 DEFINE DATA LOCAL
0020 1 #KEY          (A1)
0030 1 FILE-1 VIEW OF BLDG-FILE
0040   2 FIELD-1
0050 1 #SEARCH-FIELD (A1)
0060 END-DEFINE
0070 *
0080 H1. HISTOGRAM FILE-1 FOR FIELD-1 STARTING FROM #KEY
0090  #SEARCH-FIELD = FIELD-1
0100  WRITE 'COUNT OF KEY VALUES = ' *NUMBER
0110  ESCAPE BOTTOM
0120 END-HISTOGRAM
0130 *
0140 WRITE 'LOOP ITERATION COUNT =' *COUNTER (H1.)
0150 IF #SEARCH-FIELD = #KEY THEN
0160    WRITE 'MATCH ON KEY VALUE'
0170 ELSE
0180    WRITE 'NO MATCH ON KEY VALUE'
0190 END-IF
0200 END
```

```
COUNT OF KEY VALUES =      22692
LOOP ITERATION COUNT =         1
MATCH ON KEY VALUE
```

Command logging request:

COMM CODE	ASSO I/O	DATA I/O	WORK I/O	TOTAL I/O
L9	42	0	0	42
RC	0	0	0	0

```
0010 DEFINE DATA LOCAL
0020 1 #KEY          (A1)
0030 1 FILE-1 VIEW OF BLDG-FILE
0040   2 FIELD-1
0050 1 #SEARCH-FIELD (A1)
0060 END-DEFINE
0070 *
0080 FIND NUMBER FILE-1 WITH FIELD-1 = #KEY
0090 WRITE 'COUNT OF KEY VALUES = ' *NUMBER
0100 *
0110 IF *NUMBER > 0 THEN
0120    WRITE 'MATCH ON KEY VALUE'
0130 ELSE
0140    WRITE 'NO MATCH ON KEY VALUE'
0150 END-IF
0160 END
```

```
COUNT OF KEY VALUES =      22692
MATCH ON KEY VALUE
```

Command logging request:

COMM CODE	ASSO I/O	DATA I/O	WORK I/O	TOTAL I/O
S1	42	0	0	42

```
0010 DEFINE DATA LOCAL
0020 1 #KEY           (A1)
0030 1 FILE-1 VIEW OF BLDG-FILE
0040   2 FIELD-1
0050 1 #SEARCH-FIELD (A1)
0060 END-DEFINE
0070 *
0080 #KEY = 'X'     /* INITIAL VALUE FOR KEY VALUE
0090 *
0100 HISTOGRAM FILE-1 FOR FIELD-1 STARTING FROM #KEY
0110  #SEARCH-FIELD = FIELD-1
0120  WRITE 'COUNT OF KEY VALUES = ' *NUMBER
0130  ESCAPE BOTTOM
0140 END-HISTOGRAM
0150 *
0160 WRITE 'LOOP ITERATION COUNT =' *COUNTER (0100)
0170 IF #SEARCH-FIELD = #KEY THEN
0180    WRITE 'MATCH ON KEY VALUE'
0190 ELSE
0200    WRITE 'NO MATCH ON KEY VALUE'
0210 END-IF
0220 END
```

```
LOOP ITERATION COUNT =          0
NO MATCH ON KEY VALUE
```

Command logging request:

COMM CODE	ASSO I/O	DATA I/O	WORK I/O	TOTAL I/O
L9	5	0	0	5

NATURAL does not issue an RC if EOF is detected (ADABAS response code 3).

```
0010 DEFINE DATA LOCAL
0020 1 #KEY           (A1)
0030 1 FILE-1 VIEW OF BLDG-FILE
0040   2 FIELD-1
0050 1 #SEARCH-FIELD (A1)
0060 END-DEFINE
0070 *
0080 #KEY = 'X'     /* INITIAL VALUE FOR SEARCH KEY
0090 *
0100 FIND NUMBER FILE-1 WITH FIELD-1 = #KEY
0110 WRITE 'COUNT OF KEY VALUES = ' *NUMBER
0120 *
0130 IF *NUMBER > 0 THEN
0140    WRITE 'MATCH ON KEY VALUE'
0150 ELSE
0160    WRITE 'NO MATCH ON KEY VALUE'
0170 END-IF
0180 END
```

```
COUNT OF KEY VALUES =            0
NO MATCH ON KEY VALUE
```

Command logging request:

COMM CODE	ASSO I/O	DATA I/O	WORK I/O	TOTAL I/O
S1	5	0	0	5

So far, there are no surprises.I decided to then repeat my tests using the FILE-2 file on production and WHAM! The results were at first puzzling. Here are the results of command logging for the corresponding tests.

```
#KEY = 'A'
HISTOGRAM FILE-2 FOR FIELD-2 STARTING FROM #KEY
```

COMM CODE	ASSO I/O	DATA I/O	WORK I/O	TOTAL I/O
L9	31	1 (??)	0	32
RC	0	0	0	0

```
#KEY = 'A'
FIND NUMBER FILE-2 WITH FIELD-2 = #KEY
```

COMM CODE	ASSO I/O	DATA I/O	WORK I/O	TOTAL I/O
S1	30 (??)	1 (??)	0	31

```
#KEY = 'Q'
HISTOGRAM FILE-2 FOR  FIELD-2 STARTING FROM #KEY
```

COMM CODE	ASSO I/O	DATA I/O	WORK I/O	TOTAL I/O
L9	485	1 (??)	0	486
RC	0	0	0	0

```
#KEY = 'Q'
FIND NUMBER FILE-2 WITH  FIELD-2 = #KEY
```

COMM CODE	ASSO I/O	DATA I/O	WORK I/O	TOTAL I/O
S1	5	0	0	5

"Why?" you ask. The answer lies in knowing the breakdown of the of the descriptor FIELD-2 on this particular file.

```
KEY VALUE =   COUNT OF KEY VALUES =        19
KEY VALUE = A COUNT OF KEY VALUES =     12668
KEY VALUE = D COUNT OF KEY VALUES =     25467
KEY VALUE = L COUNT OF KEY VALUES =     32073
KEY VALUE = N COUNT OF KEY VALUES =     17934
KEY VALUE = P COUNT OF KEY VALUES =       364
KEY VALUE = R COUNT OF KEY VALUES =    295241
KEY VALUE = S COUNT OF KEY VALUES =      1937
```

The problem with a large amount of ASSO I/O required to answer the question for the HISTOGRAM is due to what the command wants to do. Since values are important and a starting point in the key list was provided, ADABAS reads all the ASSO blocks for a value to get the next value. This method accounts for fragmentation in the RABN structure. Below is a chart summarizing various key list situations vs. the ADABAS command.

Situation with I/O requirements

	1	2	3
COMMAND	I/O for existing descriptor	I/O for Non-existing descriptor at EOF	Non-existing Descriptor before entry with large ISN QTY
HISTOGRAM (L9)	SAME	SAME	HIGHER
FIND NUMBER (S1)	SAME	SAME	LOWER

Since I believed that 99.8 % of lookups fell into categories 1 and 2, I used to lean toward coding the HISTOGRAM versus the FIND NUMBER. However, I try to live up to my namesake. Since FIND NUMBER no longer does extra I/O to the ADABAS WORK data set and NATURAL's loop processing generates RC commands (as long as EOF - ADABAS response code 3 - is not encountered), I use FIND NUMBER to answer a question "Is this value on file?". It is also good for counts of ranges and counts depending on complex criteria. However, I use the HISTOGRAM whenever a key value contains a code and / or information for decoding. The bottom line is **"KNOW THY DATA"**.

Another question: Would you ever update the same record you just stored?

I have seen scenarios where this is possible. First, you may be writing a control field into a record to guarantee uniqueness. Some people use a number generated by the application, or the record's ISN.

Summary of key points:

1. There is no one set answer to "Do I use READ or FIND?". **KNOW THY DATA** is the key. The READ can use more CPU since it requires multiple passes of the inverted lists out of the Associator.
2. Beware of potential misuse of comparison operators, particularly GE or LE. Coupling these with FIND inappropriately can lead to heavy CPU and excessive I/O.
3. Beware of READ LOGICAL which may have no end point. This can lead to iadvertantly processing the entire file when not required.
4. Beware of FIND (1). The construction of the FIND request might lead to excessive inverted list processing to retrieve one record.
5. Be judicious of FIND...SORTED BY but do not believe it must be banned.
6. Monitor your application for potential excessive ADABAS communication. A careful review of the design of the transaction will help here.
7. Monitor the environment for excessive ADABAS calls to the NATURAL system files. You may want to periodically review your steplib architecture and the buffer pool performance.
8. Use Prefetch whenever possible (see the discussion in chapter 7).
9. Avoid GET SAME processing if possible.
10. Live in the ADABAS nucleus as little as possible. Watch for excessive I/Os.

Chapter 7

NATURAL in Batch

I. Batch NATURAL JCL in MVS/XA Environment

Here is an example of a Batch NATURAL JCL stream which I use at Boston University to execute NATURAL in batch.

```
//NATURAL PROC SYSOUT='*',  SYSOUT CLASS
//         PRT01='*',                       DEFAULT OUTPUT
//         PRT02='*',                       DEFAULT OUTPUT
//         PRT03='*',                       DEFAULT OUTPUT
//         PRT04='*',                       DEFAULT OUTPUT
            ...
            ...
//         PRT32='*',                       DEFAULT OUTPUT
//         WKF01=NULLFILE,      DEFAULT WORKFILE DSN
//         WKF02=NULLFILE,      DEFAULT WORKFILE DSN
//         WKF03=NULLFILE,      DEFAULT WORKFILE DSN
//         WKF04=NULLFILE,      DEFAULT WORKFILE DSN
            ...
            ...
//         WKF32=NULLFILE,      DEFAULT WORKFILE DSN
//         SRTSPCE=5,                       SORT WORK SPACE
//         ENVIRO=DEVL,             DATABASE ENVIRONMENT
//         DUMP='*'                         SYSUDUMP DATA SET
//*********************************************************************
//*  NATURAL  - MULTI-USER PROCEDURE TO EXECUTE
//*  SECURED NATURAL V2.2 BATCH JOBS, ENVIRONMENT SPECIFIC
//*  03/02/93 - JTW - NEW
//*********************************************************************
```

```
//NATURAL EXEC PGM=NAT&ENVIRO,TIME=120,REGION=6000K
//*
//DDCARD    DD DISP=SHR,DSN=ADABAS.&ENVIRO..SUPT.CARDLIB(MULTI)
//*
//*  USER LIBRARY FOR CALL/CALL FILE MODULES GOES HERE
//*
//SORTLIB   DD DISP=SHR,DSN=SYS1.SORTLIB
//*******************************************************
//*    DD STATEMENTS FOR USE WITH THE OS SORT PRODUCT    *
//*******************************************************
//DDSORTIN     DD DISP=(NEW,PASS),DSN=&&SORT,UNIT=SYSDA,
//              DCB=RECFM=FB,SPACE=(CYL,(&SRTSPCE,&SRTSPCE))
//DDSORTUT   DD DISP=(OLD,DELETE),DSN=*.DDSORTIN,VOL=REF=*.DDSORTIN
//SORTWK01   DD UNIT=SYSDA,SPACE=(CYL,&SRTSPCE)
//SORTWK02   DD UNIT=SYSDA,SPACE=(CYL,&SRTSPCE)
//SORTWK03   DD UNIT=SYSDA,SPACE=(CYL,&SRTSPCE)
//SORTWK04   DD UNIT=SYSDA,SPACE=(CYL,&SRTSPCE)
//SORTMSG     DD SYSOUT=&SYSOUT
//SORTOUT     DD DUMMY,DCB=BLKSIZE=80
//*******************************************************
//*     PRINTERS                                         *
//*******************************************************
//CMPRINT    DD SYSOUT=&PRT01,DCB=BLKSIZE=133
//CMPRT01    DD SYSOUT=&PRT02,DCB=BLKSIZE=133
//CMPRT02    DD SYSOUT=&PRT03,DCB=BLKSIZE=133
//CMPRT03    DD SYSOUT=&PRT04,DCB=BLKSIZE=133
//CMPRT04    DD SYSOUT=&PRT05,DCB=BLKSIZE=133
              ...
              ...
//CMPRT31     DD SYSOUT=&PRT32,DCB=BLKSIZE=133
//*******************************************************
//* USER WORK FILES                                      *
//*******************************************************
//*
//CMWKF01    DD DSN=&WKF01,DISP=OLD
//CMWKF02    DD DSN=&WKF02,DISP=OLD
//CMWKF03    DD DSN=&WKF03,DISP=OLD
//CMWKF03    DD DSN=&WKF04,DISP=OLD
//CMWKF04    DD DSN=&WKF05,DISP=OLD
              ...
              ...
//CMWKF31    DD DSN=&WKF32,DISP=OLD
//*
//*CMOBJIN    DD DUMMY,DCB=BLKSIZE=80
//CMSYNIN     DD DDNAME=SYSIN
//SYSOUT        DD SYSOUT=&SYSOUT
//SYSUDUMP  DD SYSOUT=&DUMP
//ABNLIGNR   DD SYSOUT=&DUMP
```

The notion that a batch program must be created in batch is fallacious. In fact, I personally find that creating NATURAL programs other than online is not a desirable way to work. Especially with the addition of session managing using NET-PASS (a Software AG product), I can create, compile, even test a batch program online and transfer to the job submission facility to check its batch operation and output(s).

If you allow any FNAT file without NATURAL security in your shop, you must compile a version of NATURAL with the parameter DYNPARM=OFF. This will disallow anyone from trying to invoke NATURAL with any potentially damaging parameter override. A second approach is to link edit NATURAL into a library which has access secured under TOP SECRET, RACF or some VTAM level security package. This is important if you want to keep your environment secure. Batch NATURAL parameters may be different for batch NATURAL than for your on-line environment.

Here are some specific Batch NATURAL parameters which are useful for batch. Below is a brief description of some of them.

```
SL=72               NUMBER OF CHARS./SOURCE STATEMENT
LS=132              LINE SIZE
PRTBLK=(6118,133)   REPORT BLOCK/RECORD SIZE(S)
MSIZE=180           SIZE OF DYNAMIC STORAGE AREA
MT=0                MAX CPU TIME PER NATURAL PGM, DEFAULT NO LIMIT
PRINTER=10          MAX NO OF PRINTERS FOR COMPILES
WORK=10             MAX NO OF WORKFILES FOR COMPILE
PS=60               REPORT PAGE SIZE
SORTMAX=160         MAX K FOR CORE  FOR SORT
INTENS=1            NO OVERSRIKE FOR PRINTING
DYNPARM=OFF         NO DYNAMIC PARMS ALLOWED
MAINPR=31           REDESIGNATION FOR CMPRINT FILE
CC=ON               ACTION IF ERROR IN BATCH OCCURS
```

The MT parameter has an interesting twist. A high value may allow NATURAL programs to loop without failure, whereas too low a setting might necessitate an override. I have seen new production failures because the override was not added or because the proper override was not provided. Also, I have seen programmers write it into the program forgetting that it takes effect upon COMPLETION of the program. Some shops set MT=0, turning off the NATURAL timer and allowing MVS timers to control job step and job run times. This is more of a procedural issue than a NATURAL issue.

II. Print Files

When executing NATURAL programs in batch, writing to print files or reading/writing work files require references in JCL in an IBM environment. To compile a NATURAL program in batch (ADHOC or CREATE) a reference to print files (CMPRTnn) must be specified in the JCL stream or a NAT0304 will result.

CMPRINT is the primary output report file. Any DISPLAY, WRITE or PRINT without a print file number results in output to this file. DCB information is optional. I believe in supplying specific DCB information in regards to any print file 1 thru 31 and defaulting to the system message class for CMPRINT. Allowing reports to write to CMPRINT is appropriate for adhoc programming, but not for a production batch system.

If any DCB information is supplied, here are some basic rules which are applied:

RECORD FORMAT	RECORD LEN	BLOCK SIZE
RECFM=UA	133	133
RECFM=VA RECFM=VBA RECFM=VBSA	137	JCL BLKSIZE overrides default of BLKSIZE (default of 133) parameter in NATPARMS
RECFM=FA RECFM=FBA RECFM=FBSA		The value of JCL BLKSIZE is used. If not BLKSIZE in specified, defaults to BLKSIZE NATPARMS in NATPARMS. If RECFM=FA, block size equals record size.

RULE 1: Never send specific production reports to CMPRINT. Always designate a specific print number. This supports standardization or makes electronic distribution easier.

The MAINPR parameter can be set to redirect CMPRINT output to another printer designated in JCL. If parameters are passed during startup, this is one way to accomplish this without changing programs or JCL. Any program which issues I/O without a printer number does not have to be changed to include a printer number as long as the JCL includes the printer reference.

RULE 2: Always make LRECL for a print file at least the same as the LS setting for the NATURAL report plus 1. It should include the extra byte for carriage control which is not part of the LS specification.

III. Work Files

NATURAL work files; are read from or written to by specifying the a READ WORK FILE or a WRITE WORK FILE statement in your program. DCB information should always be provided; so there is no guess work when determining print requirements. If no DCB is supplied then the default is as follows:

DCB=(RECFM=VB,LRECL=as per program ,BLKSIZE=WORKBLK NATURAL parameter) The default WORKBLK parameter is 4628 bytes. The LRECL for any data set created with this default is 4624 bytes. Setting the WORKBLK to 0 allows your disk management system to optimally block files. BLKSIZE in JCL overrides any default.

There are other rules which are applied if partial DCB information is specified. Indeed, NATURAL 2.2.8 produces unexpected output for forms which exceed 132 characters and the DCB is not specified. Fewer problems will result if you supply complete DCB information.

The READ/WRITE WORK FILE n statements provide I/O to work files. When multiple WRITE statements to the same work file appear in a program, if the total buffer length needed to hold the fields are of different lengths, a compile time error of NAT0364 is generated. NATURAL considers this a requirement that you want to write a variable length record. This is sometimes an accidental error because the incorrect file number was written in the WRITE WORK statement.

READ WORK SELECT is the default for NATURAL.

The difference between is that data type checking is performed. This can be deceiving since many programmers write 'READ WORK FILE n' not expecting data checking to occur. If a data error is detected at run time, you will get the error NAT1505. You should immediately check your numeric data fields for non-numeric data.

To debug this situation, print out a key field or a count since NATURAL stops at the first record it encounters with a data problem. If you do NOT want data checking, you must write 'READ WORK RECORD'.

As one of our fellow SAG users remarked on SAG-L, you can encounter a second problem even if you use the READ WORK 1 #DATA RECORD option. They noted how this option allows the input of records without a NATURAL abend due to a data check error, but still might encounter a NAT954 attempting to perform COMPUTEs, etc., with fields that are assumed are in packed format and are not. He tried:

```
'IF #FIELD = MASK(NNNNN)'
'IF #FIELD = MASK(ZZZZZ)'
'IF #FIELD = MASK(HHHHH)'
'IF #FIELD-A IS (N3)'
'IF #FIELD-A IS (N5)'
'IF #FIELD-A IS (P5)' etc.....
DEFINE DATA LOCAL
01 #FIELD(P5)
01 REDEFINE #FIELD
02 #FIELD-A(A3)
```

which didn't work. He assumed that there ought to be a more straight-forward process and not such a lengthy, convoluted mess and asked what was overlooked and how could he solve this problem once and for all in NATURAL. Many responses followed, but two stood out as definitive.

Barry Kropt provided an excellent discussion on packed fields, which you can find in Chapter 3. To summarize, he introduced the 'Z' mask (Zoned mask) which checks for a valid signed digit. He stated it could also be used to validate non-packed signed numeric fields.

The 'Z' mask definition character is a check for valid signed digit, which, if any sign appears, is always in the rightmost character. The 'Z' (Zoned) mask definition can also be used to validate non-packed signed numeric fields. Barry summarizes:

"To pack a number:

Number is (as we would see it on a NATURAL display) -121 In a 3 character zoned decimal format: 12J

As a 3 character hex format: F1 F2 D1, then switch the nibbles in the rightmost byte, throw out high order bits, low order bits saved

Number is (as we would see it on a NATURAL display) -121.

As a 2 byte Packed Number - hex format:12 1D `Try this program:`

```
DEFINE DATA LOCAL
1 #P   (P3)   INIT <-121>
END-DEFINE
WRITE #P #P (EM=HHHH)
END
then test #P = MASK (NNN), and MASK (NNZ)"
```

Rick Crowder of Software AG and myself summarized the method as follows:

"1. Always use the Packed field name in the MASK statement.
2. Do NOT use the name of an Alpha field which is either a redefinition of the packed field, or the field the Packed field redefines. (NATURAL realizes this is a Packed field by it's definition, and DOES check things differently)
3. In the MASK, specify the total number of digits the Packed field can contain, making the last one a "Z" in the MASK."

For example,

```
1 #C (A2)
1 REDEFINE #C
2 #D (P3)
END-DEFINE
#C := H'100F'
IF #D = MASK (NNZ) THEN
```

A violation of the above three points will result in an incorrect result. Also, the use of the 'Z' mask may not be adequately documented, so this is another issue for the NATURAL trainers out there.

A non-NATURAL solution offered was:

Run the sequential dataset thru the ADABAS Compression utility ADACMP. With the appropriate control cards, you'll get 2 output datasets:

- good records with valid packed values (now you can send them to NATURAL with no worries).
- bad records with non-packed values (which you have isolated without abending NATURAL). It will execute efficiently, a utility written in assembler.

This is an interesting idea. This would allow you to process records with valid data with no problem and to ship off records with invalid data to a sequential dataset for further analysis.

WRITE WORK FILE may now pass a variable number of occurrences of an array, both beginning and ending range values may be variable:

```
WRITE WORK FILE 1 VARIABLE ARRAY(1:V,1:V)
```

IBM Partitioned Data Sets presented an issue which at first I did not understand. Simply, you cannot issue more than one one WRITE WORK to two members of the same PDS. In my environment, you get a system error 213-30. NATURAL is attempting to issue two OPENs for write to a partitioned data set without a close. This situation should be different in the future when NATURAL provides OPEN/CLOSE logic. This doesn't, however, preclude you from doing multiple READs to different members or READ/WRITE to two different members.

A question that often comes up is how to WRITE/READ the same work file and avoid a NAT1511. I'm sure the 1511 is related to the session parameter WFOPFA. Let's assume WFOPFA = OFF and you start a NATURAL session to execute the following sequence of events.

Main program:

```
DEFINE DATA LOCAL
1 #CTR (N4)
1 #WORK (A60)
END-DEFINE
* CLOSE WORK 1
CALLNAT 'TESTWRK2'
READ WORK 1 #WORK
WRITE 'A' #WORK
END-WORK
END
```

Subprogram TESTWRK2:

```
DEFINE DATA LOCAL
1 #CTR (P4)
1 #WORK (A60)
END-DEFINE
WRITE 'Entering subpgm writing to WORK
file 1'
FOR #CTR 1 TO 10
      MOVE #CTR TO #WORK
      WRITE WORK 1 VARIABLE #WORK
END-FOR
END
```

The WFOPA parameter controls when an open to the work file is issued. The above execution will get NAT1511 since the file was opened for READ at execution but was written to in the subprogram. A CLOSE WORK FILE statement need to be executed to avoid this, or set the WFOPA parameter to

ON. Issuing Opens for READ/WRITE re-positions the record cursor to the top of the file.

No doubt you can write NATURAL programs to write out records which are to feed the ADABAS compression utility. There are two things you must remember if you wish to do this.

First, check with your resident DBA/DA about how information about files is stored within Predict, in respect to MU/PE structures. If the ADAWAN definition includes counts for MU/PE fields, then you write out the fields in group order. If there are no counts, then you write out the fields and include a one byte binary count field as part of the data record.

As a performance issue, is it always more efficient to code READ WORK group-name as opposed to READ WORK data-element-list. Using a data-element-list forces NATURAL to perform individual data element moves between IBM's I/O buffers and NATURAL's data buffers. Alternatively, you may code READ WORK view-name to achieve the same savings in overhead. This eliminates the need to code individual data elements to store values which are in turn moved into view data elements.

Arrays may be specified with variable indexes when read from a work file.

```
READ WORK FILE 1 #INFIELD(1:#VAR)GIVING LENGTH #LEN
```

The GIVING LENGTH option identifies avariable which contains the actual length of the record read from a workfile. That variable must be defined as I4.

Under normal circumstances, work files are closed when the program EOJs. CLOSE WORK FILE explicitly closes the work file specified from within a NATURAL program:

```
CLOSE WORK [FILE] (work-file-number)
```

An interesting question was asked on the SAG-L listserv. One user had the requirement to write an output file in ASCII from the mainframe VSE/ESA environment. The wanted to know if NATURAL provided something similar to COBOL, which supports 'ASCII SEQUENCE' reserved words for the SPECIAL-NAMES and FD sections for the EBCDIC to ASCII conversion. Two responses followed, one directly involving NATURAL, the other IBM.

The NATURAL suggestion involves using the EXAMINE TRANSLATE statement. It has an option whereby you can define an ASCII to EBCDIC translation table. The IBM technique, described by Russell Nash, from Canberra, Australia. involves creating ASCII files using a parameter on the DCB card (Data Control Block) that defines the output file in the batch. The parameter is OPTCD=Q.

Some sample JCL for an unlabelled, ASCII tape at 6250 BPI look like this:

```
//CMWKF01 DD DSN=<filename>,UNIT=TAPE,
//           VOL=SER=<tapeid>,LABEL=NL,
//           DCB=(RECFM=FB,BLKSIZE=1024,DEN=4,OPTCD=Q)
```

IV. Global Settings

There are two ways in which the global settings can be modified for a NATURAL session. First, the program can contain a SET GLOBALS statement. Second, the job stream can have a GLOBALS command specified to NATURAL. With the first method, the setting takes effect only after the program is concluded. In the second method, it takes effect immediately for the entire session. Here is an example of this confusion I once saw in a batch run.

The program contained:

```
0010 SET GLOBALS MT=900
```

The program died at the default 4 minute mark in test. It was believed that the program would take 15 minutes.

Many programmers set LS / PS settings in Batch programs. For any program which runs as one of many in a batch session, it is important to note that the LS / PS must be set for the largest program in the stream. These parameters must be set at the beginning of the session or I/O buffer errors may arise.

Remember to do all arithmetic in packed format. However, define display numeric fields as unpacked.

V. Performance Issues

The ET/BT commands are resource intensive commands. Major system resources are required and therefore should not be overused. All too often ETs and/or BTs are issued to release held records, or to attempt to lower ET time. Avoid using these commands for this purpose.

It is undesireable to hold records which are not intended for update. If the application can afford to not keep a record on hold (I have seen such applications in other shops which require this), then it is better to reread the record than to fill up the hold queue. One can also turn the RI NATPARM on to release any rejected record which fails an ACCEPT (or qualifies for a REJECT) or a WHERE clause. However, by turning on this parameter you will increase the number of ADABAS calls.

There is a better approach to handling the transaction timer in ADABAS. Programmatically, if you are maintaining an ET counter, you might find that the transaction time limit expires before an ET is issued. The usual practice is to lower the ET counter. I have seen ET counters set to one. That is, in a batch program, an ET is issued every UPDATE or STORE or DELETE. This is not necessary and is very inefficient for large volumes of records. Instead, using an approach I learned from Cornell University, it is better to programmatically keep track of the time passed since the last ET. If the time is about to expire, then issue an ET (as long as you are not disrupting a logical transaction) and reset your program's timer. See *TIME-OUT.

Remember to do all arithmetic with packed operands. Define your display numeric fields as unpacked.

Follow the ADABAS guidelines laid out in Chapter 6. Get an understanding of the record distribution in order to design the proper selection technique.

If after an access a record is to be re-read, you can minimize retrieval time by having stored the value *ISN in a variable or an array and use GET processing which does not need descriptor information.

This next issue is as much an issue of standards. Jobs with no error checking or no defined ETA can cause excess ADABAS consumption if you must restart from the top. For report purposes there may be no other way. However, posting "checkpoints" with END TRANSACTION data to ADABAS or to any other file repository can provide restartability. NATURAL will not post an ET if no record is on hold, but END TRANSACTION data is issued regardless.

VI. PREFETCH

PREFETCH is meant to reduce the amount of overhead associated with long iterations of L1 through L6 calls (those generated by NATURAL READ statements, direct calls, or Software AG's sunsetted product ADAMINT macros READSET/SEQREAD) in the batch environment. It is a request to the database to read ahead. This reduces the number of ADABAS calls and hence reduces the number of router calls (SVCs). The PREFETCH feature is invoked only if MODE=MULTI is specified. ADARUN loads the module ADAPRF along with ADALNK. The path of communication starts with the application program with control passing to ADAPRF which invokes ADALNK.

PREFETCH works by altering the Lx call made to the database. On the first Lx call, ADAPRF builds a "PREFETCH pool" for the application in the program's own region space. In this PREFETCH pool two buffers are maintained. The first holds the records that will be passed back to the application program, and the second holds the records coming from ADABAS. On the second Lx call made with the same Command ID and file number, the command will be converted to PREFETCH format. This means that ADABAS will read as many records as needed to fill the first buffer. Once this buffer has been filled, the application program will get the first record in the buffer. Subsequent Lx calls will get the remaining records in the buffer. Meanwhile, ADABAS will fill the second buffer area with records that will eventually be requested by the application program. Finally, once the first buffer has been depleted and ADABAS has filled the second buffer with records now needed by the program, PREFETCH will swap the two buffer areas and the process continues. Before ADABAS 5.2, there was a little known fact that the buffer must be sized so that at least eight records must fit into the constructed buffer (the ISN buffer); if not, PREFETCH logic was ignored. This limitation is not true in 5.2 and beyond. As few as two records can fit into the PREFETCH buffer.

As an example, suppose an application were to read 1000 records by an L3 sequence (READ LOGICAL / SEQREAD). When PREFETCH is in-voked, the first L3 call will cause a PREFETCH buffer pool to be allocated for this application. On the next L3 call, ADABAS would fill one of the two buffer areas with as many decompressed records as it determines. The two buffers will now be swapped. ADABAS errs on the safe side. It uses an algorithm to slowly fill the buffers in case you retrieve too many before issuing an RC command. The application will begin reading the decompressed records while ADABAS fills the second buffer area with the next chunk of decompressed records. Here is a summary of a job which without PREFETCH would issue 4,603 L3 commands (Figure 2.1).

The first set is the result of:

```
PREFSBL=PREFTBL=32000
The second set is the result of:
PREFSBL=PREFTBL=64000
```

Note that each iteration will appear as a single L3 entry on the command log. In this example, there will be 50 L3 entries for this job; however, the I/O required to obtain each batch of records would be reflected in each entry.

JOB ENTRY TIME	MIN IBL	NO.CMDS	IBL	MEAN IBL	MAX IBL
6,188,237	0	1	0	0.00	0
6,188,237	1,982	2	1982	1,982.00	1982
6,188,237	7,930	1	7930	7,930.00	7930
6,188,237	15,860	31	15860	15,860.00	15860
111,174	0	1	0	0.00	0
111,174	3,982	2	3982	3,982.00	3982
111,174	15,930	1	15930	15,930.00	15930
111,174	31,860	15	31860	31,860.00	31860

Figure 7.1

PREFETCH produces significant improvements in a number of types of batch jobs. The categories of jobs that benefit most are:

- Report jobs that issue only L1 , L2, or L3 calls against a single ADABAS file (such as jobs that produce OS file extracts of ADABAS files). In NATURAL, these calls are equivalently READ by ISN, READ PHYSICAL, and READ LOGICAL.
- Update jobs that issue only L4, L5, or L6 calls against a single file to obtain the records to be updated. In NATURAL, these calls are equivalently READ by ISN, READ PHYSICAL, and READ LOGICAL with the HOLD option to eliminate simultaneous updates.

Jobs that intersperse other ADABAS calls within the Lx sequence tend to produce sporadic results. Software AG's current version of PREFETCH minimizes any negative effects for PREFETCH. When a call other than Lx is issued, the buffering process is stopped until the other command is processed. If the buffering process is continually halted so that other commands can be processed, PREFETCH benefits will be greatly minimized.

Invoking PREFETCH is in most cases a trivial matter. In any batch application that is not "hard-linked" to ADALNK, a JCL override to the standard ADABAS load library and the specification of one to three ADARUN cards are all that is necessary. For NATURAL, this means adding ADARUN PREFETCH cards to DDCARD. The following ADARUN cards will invoke PREFETCH with the default values:

```
ADARUN PREFETCH=YES
ADARUN PREFSBL=3000
ADARUN PREFTBL=15000
```

The PREFSBL card refers to the PREFETCH buffer pool. This area contains PREFETCH control information as well as the two buffer areas that hold decomp-ressed records for the application. The value of PREFSBL will determine how many records are passed back to the application with every Lx call, and can be calculated by the following formula:

```
PREFLSB = 120 + FBL + ((16 + DRL) * NR * 2)
```

where:

- FBL = Format Buffer Length of the Lx call
- DRL = Decompressed Record Length
- NR = Number of Records to return with every Lx call

There are some problems with not sizing this buffer correctly. First, its default size is 3,000 bytes. Its maximum size cannot exceed 64K. More precisely, you get an error NAT0146 if

```
ACBL + FBL + RBL + SBL + VBL + 1/2(PREFLSB) > 64K
```

or an error NAT0152 if you exceed the LU parameter for the session. All of this will show up in the batch run as a USER 0007. Other causes of this response are:

1. PREFETCH library not ahead of other
2. In DOS when SIZE= parameter for EXEC not big enough
3. Generically, an unknown parameter (check the spelling)

The value 120 in the first part of the formula is made up of 28 bytes for the ADABAS parameter address list, 80 bytes for the ADABAS control block, and 12 bytes for the PREFETCH addresses. The 16 in the second part of the formula takes into account the header attached to each decompressed record. The sum of these two numbers is multiplied by the number of records (NR) to return with every Lx call. Lastly this amount is multiplied by two, because there are two buffer areas, one for records going back to the application and one for ADABAS to fill with decompressed records.

The value of PREFTBL is the value of PREFSBL times the number of concurrent Lx sequences in the application. In most of our cases, how-ever, PREFSBL and PREFTBL are the same since only one Lx sequence is active at a time. Most of the time, you can get maximum results by simply using the 64K maximum as the default for PREFSBL (i.e., 65,536).

There are other parameters to better control the PREFETCH process. Consider the following example:

```
           FIL                          MEAN
JOBNAME2   NUM  CMD    NO.CMDS          IBL
AALBATXY    61  L1       4,998          4.00
AALBATXY    62  L3     270,671          0.00
AALBATXY   249  L3          45          0.00
```

Before looking at any code, it is possible the program in this job is doing coordinated lookups on file 62 for records found on file 61 (not quite true, but for the example's sake let's presume this is true). Suppose you want to PREFETCH the records retrieved on the second file and not the first. This may be accomplished with ADARUN cards which look like the following:

```
ADARUN PREFETCH=YES
ADARUN PREFSBL=64000
ADARUN PREFTBL=64000
ADARUN PREFXFIL=61
ADARUN PREFXCMD=L1
```

These options allow you to exclude a specific command sequence. This parameter are dependent to invoke PREFETCH correctly. The two together will allow for the exclusion of a particular command sequence for a specific file. If you provide only the XFIL option, you get a response code U0035. If you provide only the XCMD option, your PREFETCH performs as if you did not provide exclusionary information.

There is a special technique to exclude a specific file on a database other than the primary database. First, convert both the DBID and the FNR to hexadecimal , e.g. DBID=215, FNR=6 translates to DBID=D9 and FNR=06. Concatenate them together giving D906 and convert to decimal, i.e. 55558. Use this new number as your FNR:

```
ADARUN PREFXFIL=55558
```

In my testing, any job that fit into one of the two "best case" categories experienced an improvement of 10% to 65% in run time. CPU time was reduced only slightly.

There are occasions when not to use PREFETCH. If the program issues nested READs, where the inner READ retrieves a few records, a buffer filled with records for the inner READ is wasted. Remember that the number of buffers = PREFTBL / PREFSBL. If you size PREFTBL = PREFSBL, then you will cause only one buffer to be created by PREFETCH and this is associated with the first active sequential command encountered. You can use a single buffer for two files if they are accessed in a network manner. This means:

```
READ
END-READ...

READ
END-READ
```

If non-sequential commands are interspersed among sequential READ commands, there can be a degradation in performance. Also, watch out for:

- the application program releasing CIDs or ISNs in the midst of sequential processing.
- the application program changing the format buffer in the midst of sequential processing.
- the application program flipping the hold and no hold status of sequential processing.

Case Study

Two questions I would expect someone who wants to use PREFETCH to ask is "Will my batch job benefit from using PREFETCH?" and "How do I determine the optimum ADARUN PREFETCH parameters?". The first question can be handled by doing two things.

First, if it isn't obvious from the code, you should restrict your job step to a few iterations and perform a trace using Insight or, if possible, on-line using the TEST trace facility. This trace is important to determine if a reduction of accesses can be accomplished with PREFETCH.

The second is more complicated.

You can take the simple attitude and decide to run the job w/o PREFETCH and again with PREFSBL maxed out. But sometimes the job is a little more complicated with several files issuing L1, L2 or L3 sequences. A current production job - AAD3190 - was analyzed and I determined it was not straight forward.

The first set of results below is a summary of the job with no PREFETCH; the second set applied PREFETCH on file 233, and the last set included file 140. Runtime for the jobs also reduced with PREFETCH. Although, at first, it seemed that PREFETCHing file 173 should yield results, the number of calls to this file actually increased. However, the appli-cation was not passed extra records which would result in a logical inconsistency. I have yet to ascertain why this happened and I continue to research this phenomenon. The cards which defined the "best" PREFETCH scenario are listed below.

JOB ENTRY DATE	JOB ENTRY TIME	FIL NUM	CMD	SUM CMD COUNT	SUM ASSO IO	SUM DATA IO
94,268	4,623,804	140	L3	330,846	9,595	74,890
94,268	4,623,804	166	L1	2,980	6	98
94,268	4,623,804	173	L3	228,776	6,990	18,013
94,268	4,623,804	233	L3	722,174	59,998	64,249
94,268	4,623,804	239	L3	24,250	578	10,274
94,268	6,253,440	140	L3	330,846	9,436	74,859
94,268	6,253,440	166	L1	2,969	15	107
94,268	6,253,440	173	L3	228,776	7,074	18,083
94,268	6,253,440	233	L3	361,675	61,732	68,331
94,268	6,253,440	239	L3	24,250	587	10,272
94,269	872,943	140	L3	1,502	9,920	75,004
94,269	872,943	166	L1	2,969	14	107
94,269	872,943	173	L3	228,776	7,185	18,029
94,269	872,943	233	L3	410,081	61,542	66,958
94,269	872,943	239	L3	24,250	550	10,273

```
ADARUN PREFETCH=YES
ADARUN PREFSBL=65536
ADARUN PREFTBL=65536
ADARUN PREFXFIL=173
ADARUN PREFXCMD=L3
ADARUN PREFXFIL=239
ADARUN PREFXCMD=L3
ADARUN PREFXFIL=249
ADARUN PREFXCMD=L3
```

A fellow colleague asked the following question on the SAG-L listserv discussion group:

> "We have always stayed clear of using PREFETCH on files where updating of the records is being performed. However we have a large file where we read sequentially and update, during the read loop, records every so often when a particular criteria is met, for the sake of

> discussion every 100th record is updated. According to my ADABAS internal notes the PREFETCH buffer is released after an ET. Does anyone know if the PREFSBL or the PREFTBL is released ?."

Several responses followed which I will summarize here.

In this situation, I would consider weighing PREFETCH and re-reading the records for update vs the extra ADABAS calls to re-read the records. This would allow the use of PREFETCH without the problem expressed in the next point.

First, as has been rightfully pointed out by a few, the ability to exclude will allow you to 'tune' your PREFETCH request much better than what we used to be able to do. You can also exclude by command sequence coupled with a specific file. The PREFETCH buffers as defined by PREFSBL and PREFTBL are released upon an RC command for that sequence, not ET commands. Your total PREFETCH buffer is part of the ISN buffer and is attached to the ADABAS buffer and handled by the PREFETCH piece of ADARUN.

You can adjust your PREFETCH buffers not to hold so many records and accommodate the ADABAS session without adjusting parameters affecting everyone. Remember, ADABAS will only put on hold the records you READ ahead. This doesn't say anything about how your logical transaction might be affecting the scenario as well.

Be aware that you might exhaust either the ADABAS nucleus NISNHQ parameter or the NH parameter. Make sure your application doesn't require adjusting before asking the DBA to do so. The solution to an application problem with ADABAS does not always imply it is one of the nucleus session parameters that is too low. Work together to solve the problem.

If you use PREFETCH for batch update jobs, a problem may result. You may put hundreds of records on hold, based upon how much data can fit in the PREFETCH buffers. You may have to adjust the size of hold queue or lower the size of PREFETCH buffers.

When you use PREFETCH in a read plus update loop you have to adjust the NISNHQ parameter (an ADABAS nucleus parameter). This parameter specifies the maximum number of records that can be placed in hold status at the same time by a user.

Read with PREFETCH. When an update is required, write the update data with the record ISN to a work file. Have a second program that does a GET, overlay the update data and ET however often your site requires. This change will cut 70% off a 10 minute CPU job.

One more thing to watch out for is nested READ's.

Where the file in the main loop has a one-to-many relationship to another file, and you take the key value from the main record and initiate a READ ... BY... FROM... THRU... or any logical equivalent like READ... IF field GT end-value, ESCAPE BOTTOM, PREFETCH can ruin you, because it pre-reads hundreds of records beyond the THRU value. Depending on your requirements and file structures, you maybe can use FIND.., THRU just as well, if the sort order doesn't matter or exclude the linked file from PREFETCH. But basically READ ... THRU... doesn't mix well with PREFETCH.

PREFETCH does not release buffers on an ET. Instead, PREFETCH passes to ADABAS a list of the records to be kept on hold (or placed back on hold) after the ET is done. This is so that records already read by PREFETCH, but not processed in the program yet, are not released from hold.

PREFETCH has logic to re-read any records that are already in the PREFETCH buffer, but have not yet been processed by the pre-fetch read/find logic, and which were updated within the program. Thus, when you do updates, there is overhead in doing this processing. If you are doing a lot of updates, you still do lots better with PREFETCH. If only a small percentage of records end up being updated, it is much better to write out the ISNs and later re-read and update just those ISNs. While that will markedly improve the PREFETCH/update process, the PREFETCH/update process is almost always faster than the plain read/update.

VII. Coordination with OS File Handling for Restartability

This is a serious issue if you plan to write batch jobs that require OS file writes coordinated with writes to ADABAS. This is especially true for jobs that UPDATE an ADABAS file. The program's ET/BT logic determines the nature of the transaction logic. The programmer should not write to an OS file unless ADABAS writes. There are a few ways you can guarantee the efficiency of the coordination.

If the OS file is for restart purposes, the batch application can be written utilizing the ADABAS checkpoint file. If you are writing a COBOL application, C3 type user records can be considered. If you are writing a NATURAL application, you can write directly to the checkpoint file (see Restart Considerations in this chapter). Beware, however, that this is NOT possible if no record is held. This is because currently NATURAL does not issue any ET command if no record is held. It assumes that there is no real transaction logic in progress so why bother! This can negatively affect an ET data command.

If the information is to be passed to other job steps, then you must be careful to write when ADABAS writes. Therefore, you should buffer your records in core and write them out as often as ADABAS. If your job dies in NATURAL, (and this can be checked by testing for problems internally and using TERMINATE with a condition code), you can use condition code checking in JCL to determine if the job stream should continue. If you do not do this, then any reconciliation between a batch job's OS file and the actual work performed by ADABAS is left to the designer of the job or in most cases, the support staff.

VIII. How to do Online Behavior in Batch

Some of you may have wanted to run an on-line program in batch. This is not impossible, and possibly necessary. Program logic can be written to circumvent certain program logic in the event the program is running in batch. The system variable *DEVICE; contains the value 'VIDEO' or the value 'PC' when the session is online and the value of 'BATCH' when the session is batch.

Sometimes, maps are passed data in batch. How to do this is simple. Each input statement needs its own line in SYSIN to supply data to the map. That means if you need to 'pass through' two maps, then you need to supply two in-stream data lines. Data can also be supplied through CMOBJIN DD, but I find it just as easy to put everything in one place - SYSIN.

Also, notice the inclusion of the line 'GLOBALS ID=, IM=D'. First, it makes it possible for the character ',' be the delimiter character for data passed through the stack to an INPUT statement. Second, the IM parameter makes NATURAL understand that input is to be accepted positionally with the stated delimiter. This is necessary for supplying data in batch to INPUT statements.

In batch, '%Q' means do not echo the screen, '%' means as the last non-blank character of a record treat the next input as continuation, and '%*' means do not echo data for the next input value.

IX. RUN vs. EXECUTE

This is very simple. You should as much as possible EXECUTE programs in batch. Some people have invested in dynamic parms and global variables and want these type programs to get reconciled ar runtime. Dynamic code should also be avoided due to the overhead required for the run-time compile requirements. Other scenarios are possible than to resort to dynamic code. At Boston University, the Production environment uses NATURAL Security and command mode is generally not allowed in batch, curtailing the use of the RUN command.

X. READ PHYSICAL/SORT vs. READ LOGICAL

This approach should be considered for large files where reading logically processes a large percentage of the file. Another good application of this approach is for large reporting jobs that produce multiple reports. I have seen controlling programs which were written to traverse ADABAS file(s) multiple times to create multiple reports. One should consider creating a download job which can then run off the extracted data. This can greatly reduce the overhead to ADABAS and place the burden of report creation on writing NATURAL to access an OS physical sequential file.

Another issue is reading a file by descriptor because of sorted orders needed for reporting. If the reading traverses a large percentage of the file, it would be better to read it physically and to do a non-ADABAS sort, e.g. Syncsort, or use NATURAL's SORT statement.

Another consideration is table lookups. Many times batch reports need to lookup coordinated information in regards to processing master records. If the process is driven by processing 250,000 master records, and each record requires 4 table lookups, the application program would require 1,000,000 table lookups. It would be better to download the tables into local storage (if the core is available) once, and do the necessary lookups in memory.

XI. Mass Purge

Using NATURAL and ADABAS' ADALOD utility and other Techniques.

How many of us have seen purge programs written against moderate to large files and watch them run for hours. This fact necessitates looking at better methods for mass purge than programmatic deletes. Some of the material has to do more with ADABAS and its utilities than with NATURAL.

NATURAL Deletes

For a small volume of data, a NATURAL program will suffice to delete the records. However, since programmatic deletes require ADABAS calls and I/O, the ADABAS utility ADALOD can reduce the system requirements to get the same result.

ADALOD

This utility can be used to either add compressed records to or delete records from an ADABAS file. The delete process requires that a list of 4 byte binary ISNs be supplied. The list can be created by a NATURAL program which reads the file, makes the decision on what record is to be deleted, and records the ISN on a work file. The utility uses the work file produced by the NATURAL program as input.

I did a study to compare this approach to programmatic deletes. The tests compared running NATURAL programs to select records for deletion. Programs ran generating records of 1 ISN, 50 ISNs, and 2,000 ISNs. The ADABAS ADALOD utility used these various files of ISN/record ratios. Since my original study, I have learned a few facts which can improve total job performance. First, the utility checks to see if the list of ISNs is sorted. If one is detected out of sequence, it sorts the list in the LIP area. This can be CPU intensive if the list is long. Therefore, it is important to provide a sorted list to the utility to prevent this performance degradation. Secondly, this approach is more useful if you have only between 20% - 50% of the file to delete.

Shadow Deletes

Another possible method is to keep fields on the file to help signify shadow deletes - flags that are switched to imply the record is logically not present, although it is physically on the file. The definition of the field(s) includes null compression. In this way, at the appropriate time of the year, the ADABAS file modification utility is run to build a descriptor or superdescriptor that points to the records to be kept.

These records can be unloaded by specifying a SORTSEQ=field, allowing you to unload only those represented with a non-null value or component.

User Exit

Another technique is to use ADABAS user exits. User exit 6 allows you to intercept records which are to be processed by the ADABAS compression utility (ADACMP). The user exit can be used to pass whatever records you wish to the compression utility. This means that a file could be unloaded, decompressed, and compressed using user exit 6 to eliminate records and reload with the accepted records. Work with your staff DBAs to determine the best strategy to use. Similarly, user exit 9 is useful to process records in the unload phase. Either way, you can filter records into two datasets - those to be kept and those to be archived/deleted.

There are two more facts which must be understood about ADALOD to use it effectively and efficiently. First, it reorders the associator component of a file each time it runs. Since on any run of a purge job the number of ISN files might differ, one would not want to run a step if there is no file for input. The number of ISNs one can delete on a run is controlled by the LIP value. To calculate the size of LIP multiply the number of stand alone ISN's by 4 and add it to the result of the number of ranges multiplied by 8.

Step 1. A NATURAL program selects the records for deletion. Write the ISNs to a work file, one per record. This step is linked to PREFETCH, if possible.

Step 2. A Syncsort step to order the ISN list. With the parameter card:

```
SORT FIELDS=(1,4,BI,A)
```

Step 3. NATURAL program reblocks the ISN list into records with up to thousands of ISNs per record. Create ISN file with no more than 250,000 ISNs per file, or for whatever limit your systems group has allowed for above the line region the ADALOD utility can use.

Step 4. The ADALOD MASS UPDATE is used to delete large numbers of ADABAS records. Remember to size the ASSO/TEMP files sufficiently for the reorder and the LIP and LWP requirements to manipulate the ISNs and the file.

NATURAL Alternative

Another method for mass deletion is useful when you want to delete a large percentage of the file (70% or higher). Consider writing a NATURAL program to read the file physically or logically, if suited, with PREFETCH. You can write the records you wish to preserve to a work file. Next, use the compression utility to reform the unloaded records for the final step, which is to use the ADABAS load utility. If the unloaded records are to be ordered, you can include a SORT step before compressing and loading.

Mass Purge Summary

In any case, if mass purge requirements are being studied, work cooperatively with the resident DBA to determine the best technique to satisfy the requirement. No matter what technique is used, the file should be backed up before the process begins.

For a substantial number of records to be deleted, NATURAL and ADALOD provide a better means of getting the job done than NATURAL alone. Another fact that became clearer is that the PREFETCH option works well for programs requiring Lx processing. Also, JCL modification using 'OPTCD=C,BUFNO=nn' in DCB information can vastly improve performance.

This approach improves performance for the following reasons:

- Fewer ADABAS calls
- Lower system overhead (e.g. reduced SVC demands)
- Less ADABAS internal resources
- Reduction in I/O requirements.

XII. Terminate with Condition Codes

NATURAL allows a batch program to issue a condition code for condition code checking in the job stream. The TERMINATE statement is modified to allow:

TERMINATE *nnn*, where *nnn* is either a packed or unpacked value in a user variable or an integer constant. The value must be in the range of 32 to 255. Here is a very simple program and the OS job stream entry.

```
0010 #X (N2) = 10
0020 #Y (N2) = 10
0030 IF #X = #Y THEN TERMINATE 200
0040 END
```

The program was run in batch and here is the output in the batch JCL report.

```
IEF142I DRMJTW21 NATTEST STEP1 - STEP WAS EXECUTED - COND CODE 0200
```

XIII. Batch ETA

There isn't anything special about a Batch ETA versus any other error transaction program, but there are some extra considerations one can consider. This is activated either because of the NATURAL parameter 'ETA=' is set or the batch user/application library NATURAL Security definition might include the definition of an error transaction program. This provides a standard mechanism for terminating all your batch jobs without having to change any specific NATURAL programs. It provides the opportunity to TERMINATE with condition codes. Also, you may want to perform special behaviors as a result of invoking an ETA program in batch, such as writing WTO messages for automation, forcing an abend where NATURAL normally might end a session or write an error log to analyze later to get you closer to a fault tolerant zero batch environment.

Addendum - Discussion of ET logic on SAG-L

An interesting discussion ensued on SAG-L in regards to ET logic. Several people contributed some good rationales and philosophy. Who would have expected that such an innocent question would raise such a response.

User #1:

> "Since I've had this question bounced off of me at least three times within 3 weeks. What is ET logic ? and what is the optimum for this feature of NATURAL? Is there some formula for the optimum # of ET's for a particular file or is there a buffer limit for the number of records and the number of fields/and size of fields for this operation? I know as a general rule we use 50 records but inquiring minds want to know and I would really like to hear the 2 cents from all platforms on this subject since I'm sure they are different."

First, here is a response from Steve Robinson, consultant.

> "ET logic is mainly an ADABAS "thing" not a NATURAL thing. It is ADABAS's implementation of what is called rollback recovery; the seminal work for which was done at the University of Texas at Austin by a professor and two graduate students.
>
> The purpose was very simple. Up to that point recovery of a database followed the old philosophy from tape days of old master + transactions ==> new master. Basically a DBA would take a snapshot of the ADABAS database before the system came up in the morning; then keep the transactions during the day on the log tape. If anything happened to the database, it was back to the backup followed by a lengthy restore to duplicate the log tape activity.
>
> If a database went down at 4 in the afternoon, after being up for 8 or 9 hours; recovery could be very lengthy.
>
> ET logic allows every job to establish its own rollback "intervals". The idea is that if the database goes down at 4; one would like to restore the database to , say, 3:55 and work forward from there. This is what rollback is all about.
>
> In on-line mode the purpose of et logic is to allow you to synchronize database activity with business logic. Suppose a clerk enters in an order on-line. Their "unit of work" is the order. Your program , suppose, must add a record to the order file, update the customer record, then update inventory records. If the system were to "go down" in the middle of all these updates the database would not "balance". The purpose of et logic, in on-line mode, is to ensure that either all the updates that comprise one

business activity (the clerk entering the order), get to the database, or none of them get there.

In batch mode the counter to 50 serves a slightly different role. Here, the counter to fifty (or 100, or whatever) places an upperbounds on the number of updates that would have to be redone (for that application) in the event of a system crash."

An additional point from another user:

"I think that there is a difference between batch and on-line transactions. Doing on-line work you want to work with small units and commit the data A.S.A.P. Doing batch work you want to get much more work done between ETs."

A transaction is a transaction, and its design greatly affects the answer to this question. A transaction requires as many resources as the designer ascribes to it. The database requirements however may look different, depending on the database of choice.

The counter approach is not generally a good one in my belief. This sets up a batch job for potential failure without any changes to the batch job. I would suspect the Work part I wrap around issue (which is hard to do in ADABAS 5) affects a victim more than the actual culprit. This means the mix of jobs can have a negative affect on a particular batch job at any point in time. You may be not helping the environment with your ET work, but it also may not be your fault. The job gets a response code 9 subcode 15 and it has a history of never dying for the two years it has been running. As an on-line program you would never lump together a group of logical transactions into one ET block, but you do not mind doing it in batch. Why?

I suspect old habits. BTs are expensive commands (less than ETs but maybe not as bad as some S2 or S8 I have seen). So we grew up as ADABAS junkies proclaiming "Hey programmers, reduce the workload - both CPU and I/O - by issuing fewer of them". But is it any longer fair, or even required? How many updates are performed before an ET/BT is issued? Since you are as good as your last ET/BT, someone decides that you can you lose an arbitrary number of transactions and still recover. But this has nothing to do with ET logic. The resources for one job which ETs every 50 may be equivalent to another which ETs every 5, or another which ETs every 100.

In addition, doesn't the question of more or less ETs become moot, at least in MVS/ESA land with larger pool areas and LFIOFP and ADABAS' scheduling of ETs/BTs in blocks as it is? Why force the program in today's technology to worry about it? I do not know about ADABAS-D, ADABAS under UNIX, but ideally, what is wrong with defining your logical transaction, execute your updates/deletes/stores and issue ET/BT? Let the database worry about the resources for one transaction per user, not the other way around. And yet, I still write programs that count and have never understood why I keep doing it.

Some might suggest that the counter approach is done for performance reasons.

If you wish to use the counter approach, then I highly suggest you couple it with checking the ADABAS timer check as well. In this case, you probably want some standard routine which can compare against your current TT timer for that database. For programming practice, the ET limit should probably come from a table or card image so you can control it from run to run if need be. I want to reiterate my concern about counters: Work part I is a shared resource, and keeping so many records without ET/BT and having many users not at ET status can at a point in time cause undue response code 9's. If you single stream your batch, ok, but if not, this can be an issue.

Skip Hansen, consultant, responded:

> "The only thing I would expand comments on is on-line vs. Batch.
>
> I think the crucial difference is in recovery. In on-line a user finishes their "transaction", then that is the "logical transaction" and that (at the minimum) is when the ET must be done. Because the user expects that last piece of work to be there tomorrow even if the systems crashes immediately afterwards.
>
> In batch, you have to have a recovery mechanism. Either restartable via ET data, rerunable, or restore and rerun. Thus the ET now longer is really used to coordinate "logical transactions", but is used to take a checkpoint. This checkpoint has performance implications that others have discussed. Therefore you do not do it every record but every N records. What the optimum for N is a good question. To make things easy on us poor ol programmers, we usually just set some arbitrary limit that will not cause the Hold queue to get filled or the WORK to wrap. Usually 50 is a good number. For some very large files 20 might be more reasonable. Also, the holding of records causing other users to wait (or resp145) may be an issue, but generally batch transaction whip through pretty fast."

Another regular on SAG-L responded to this discussion as follows:

> "There are other considerations which you may or may not have forgotten as "good practice" has been ingrained. The purpose of a database has changed since those days of yore such that multiple accesses of database files/records are quite common place. If one reduces the batch ET count to a reasonable amount then the likelihood of another user getting a 145 is reduced.

An example is that of a batch suite updating the credit card master file whilst customers are waiting at the garage to pay for their petrol. Should we fail the credit card authorization because his record is held by another user issuing ET's every 5000 updates ?"

A different twist was added to the discussion:

"When doing batch updates to a file with many records of which you are only selecting a few for updating, that if your ET checkpoint logic (i.e.. 50 records) is too large, or if the system is REALLY busy at the time, your transaction could very possibly time out.

I wonder if one could come up with a formula for calculating an optimal number of transactions to perform before an ET/BT in order to minimize the potential for problems. The formula could take into account things like record length, approximate percentage of the total records that are expected to be updated, time of day (for peak/off-peak times). "

A final point worth mentioning:

"From a theoretical point of view, the ET must be issued in a the point where your logical unit of work is done. If, however, the logical unit of work is to update 16 Meg records, this gets impossible.

Our guidelines on MVS, VM, VSE and UNIX states 500 as the maximum records to be held by any user-program (Our HQ-sizes are from 2000 - 8000 entries depending on database usage).

Older experiences told us not to issue too many ET's due to performance issues. This - again - caused programs that violated the 'logical unit of work' - rule by clustering multiple logical units into one physical unit to avoid issuing too many ET's. Since then, ADABAS has been reworked and has implemented features, where ADABAS will scan the CQ for other ET's and then execute all of these together.

My point : **Consider data integrity issues before performance issues."**

Response code 9 in batch and the designing of a batch transaction.

Another discussion thread on response code 9 in batch began on SAG-L. Before reading this, you should read the above discussion on batch transactions as well as the info in chapter 6 about ADABAS response codes.

A user asked:

> "I have a batch job that runs 75 minutes, reading vsam files and loading a tape. At the very end, there is a single record read and update to an ADABAS file that contains only 1 record. The program successfully processes all the way through the tape and closes it. Next it does:
>
> ```
> R1. READ (1) AUTO-DATAWH-SYSTEM-1-REC-VIEW          <--- blows here
> ```
>
> It abends on the read with the error message: 'NAT3009 -- The last transaction has been backed out of the database', pointing to the read statement. It also returns 'NAT9987 error occurred during execution/compilation'.
>
> I have set the parameter MT=0 (to prevent CPU timeouts) in the JCL. I have tried adding an End Transaction every 5 minutes but that didn't work.
>
> I have checked my security, looking for restrictions.
>
> I have tried putting the offending read code in a subprogram and that didn't work.
>
> I have tried putting the offending code in a separate program and running it after the first program (on the next JCL line) and still get the same error message.
>
> If I run the job and limit the tape file processing to 5 or 10 minutes, there is no problem.
>
> What's wrong?"

Here is a partial list of interesting responses:

> "Your problem is caused by the long time it takes until you get to the code at the end of your program. As far as ADABAS knows, you start working (Reading the NATURAL System file for you program), then you are doing nothing (As far as ADABAS is concerned) for a long time. Hence the 3009. The reason inserting dummy ET's in the program didn't help is that NATURAL is not issuing the ET's to ADABAS when no records are held for you (You see - he is a smart guy).
>
> What you should do is either ask your DBA to increase TNAE parameter (He/She might as well increase TNAA and TT), or make sure your ET's will be sent to ADABAS by reading and holding a record before the ET. This can be done by something like :

```
...
RD. READ (1) .....

IF 1 = 0 THEN
UPDATE (RD.)  /* dummy update so that the record will be held
END-IF
END TRANSACTION
```

Of course, this should be inserted inside your main loop, but not too frequently or the overhead will kill you."

From Darrell Davenport:

"As already mentioned the timeout is because you leave ADABAS for so long and NATURAL is not sending the ET (or BT) command to ADABAS. You have several options in addition to putting a record on hold just so you can do an ET/BT.

If you are not getting a timeout from keeping a record on hold the whole time, you should be able to keep ADABAS's attention with a simple GET command every few minutes.

-or-

By calling CMADA you can do direct calls to ADABAS. With an OP command (and the appropriate values in the ISN-Q ISN-Lower-limit fields) you may be able to set the ADABAS timer(s) yourself for just this job (no one else's timer changes).

-or-

You can go ahead and get the 3009 and recover with an intellegent error handler (either use ON ERROR or specify it with MOVE 'myerrta' to *ERROR-TA); which does the UPDATE/STORE for you.

Another response was added into the fray:

"The problem is that you access ADABAS at the start of the program in order to load the program into the buffer pool. Then when ADABAS gets no more commands from you for 75 minutes, the 1st command it does get is a rsp cd 9. One solution is to put the ADABAS update in a separate program. Read the record once at the end of the current program. This will usually trigger the rsp cd 9. But whether it does or not, fetch the program to do the ADABAS update (passing along any needed data in a special GDA just for these tow programs or via the stack; i prefer the GDA). Use On Error to trap the response code 9."

My good analyst suggested:

"I would like to suggest that you might want to take advantage of this situation by adding information to this or another record.

Use a 'table file' record to store progress information. For example you could update the same record over and over with date/time, record counter, totals...etc. Dedicate a record to this job or run and store this progression information until the job is done. Then you can Do an update on the AUTO-DATAWH-SYSTEM-1-REC-VIEW record.

This offers the added advantage of being able to interactively check on the progress of the job (also potentially has advantages for restarting)."

The programmer was charged with writing a program to READ/PROCESS Vsam data, then update an ADABAS control record. This job fell into a category of jobs which involve communicating to databases in large time spans. The problem generally is the TNAE setting. But even if an organization resets TNAE to 6000 to match the other parms, which by the way may force DBAs to re-consider LU, WORK PART II,III and any other areas which are affected by allowing users to "stick around" longer, what happens if the job executes in the future and it exceeds 6000 seconds?

You could do dummy calls, but this forces the construction of programs to do other than the original two tasks. And this whole approach, in my opinion, begs the question. *What should a programmer do in the future if tasked again to write program solutions which require multiple database access?*

The first question I would ask is are the files placed correctly? Notice he got this problem because the NATURAL System File lives on the same database as his control record file. This problem would not exist if this were not true. Let's assume he has no choice. Then why not write a two step job, one for each task, condition coding the second step to the first if necessary.

Voila, no response code 9.

Another scenario which can cause this to happen is a little more subtle and harder to address. What if access to the NATURAL system file itself is the problem - a CALLNAT that occurs light years from the original startup? One suggestion is to check periodically to see if you are about to exceed the timer for where the NATURAL FUSER file lives. If so, then execute a dummy CALLNAT, or consider increasing the TNAE/TNAA for this database (which I only like if NATURAL lives on its own database), or, if necessary, lift the CALLNAT code into the upper block.

Most of all, I would try to stay away from dummy updates and record holds. This doesn't seem productive; and I think it begs to real question(s).

Another user suggested issuing ET data. I like this idea in that it supports restartability. One philosophical question arises - why engineer solutions to get around what is essentially an environmental issue? The real problem is how to design a logical unit of work which fits into the environment or how to modify the environment to fit the logical unit of work. Constructive thought is always required to answer this.

Notes

Chapter 8

Structured Programming in NATURAL

I. Application of Structured Tenets to NATURAL

In software design the programmer/analyst is concerned with matching the business requirements to a functional design, designing data structures, and developing the overall system architecture. In the later stages of the software development process, specifications are converted into executable code in a specified language. Armed with the output from design tools such as Structured English, data flow diagrams and structure charts the programmer sets about the task of implementing the design, in the case of SAG USERS, in NATURAL.

This chapter discusses how one can increase productivity and write much simpler, easier to maintain NATURAL code with the application of structured programming ideas.

Structured programming deals with the formulation and the application of control structures to achieve program structure. Top-down design, structured design, and Jackson's approach are three methodologies which form subclasses of structured programming. Structured programming associates decomposition methods to a problem to make it more manageable. The result is a hierarchically organized solution. The definition I prefer is that of Jacob Wirth:

> *"Structured programming' is the formulation of programs as hierarchical, nested structures of statements and objects of computation."*

Why use structured techniques ? The objectives include

- to lower costs for software development
- to provide a framework to aid in attaining solutions to complex problems
- to increase programmer productivity.
- to minimize maintenance efforts

We need to develop systems with a longer life. We need to develop systems in a shorter time frame which more accurately reflect user requirements, are easier to maintain, and perform efficiently. Structured programming can help to accomplish these goals.

I would like to examine the task of constructing programs. It is imperative to answer the question "How is the design of NATURAL programs accomplished ?". First, Section 2 discusses program control structures. These structures are the fundamental components for writing NATURAL code. The use of these structures enforces the development of structured programs. They can also be graphically represented so that one can visualize the structure of a program. Second, functional decomposition forces the construction of NATURAL subprograms/ subroutines.

I have always attempted to do functional decomposition to derive the hierarchy of modules for a function. However, I have always struggled with answering certain questions.

How should I define the limits of a module? How should I determine what the module should do? What are the limits and scope of the module? How big should it be? What means of inter-module communication should be established? What constitutes a good module from a bad one? The answers to these questions are not obvious.

I propose that the use of Structured Analysis and Design and of Structured Programming techniques and guidelines resolves these issues.

Building Blocks of Structured Programming

Foundations by Bohm, Jacopini

Three basic control structures were described by Bohm and Jacopini in May 1966[1]. They demonstrated that only three basic control structures were necessary for expressing any program flowchart.

[1]C. Bohm and G. Jacopini, "Flow Diagrams, Turing Machines, and Languages with Only Two Formation Rules", Communications of the ACM,9,5,May, 1966, 366-71.

The bottom line became clear. These constructs form a sufficient set of blocks which when translated to a computer language can make it possible to write any program. Whether or not this set is optimum is arguable. Later, programming pragmatists added the need for an escape mechanism to this list. NATURAL 2 indeed does support these four constructs.

However, NATURAL currently has other constructs or attributes which are anti-structure. For example, the absence of block structure, artificial iteration constructs not under the programmers control and implied GOTOs add to the difficulties of building structured programs. However, in spite of this, structured programming is achievable with NATURAL.

It is only fair to point out that the absence of structures listed above does NOT mean one cannot write structured NATURAL programs. It does mean that one has to be attentive to implementing the analysis of systems in NATURAL.

Structured Programming Control Structures

Based on the work of Jensen[2], I put forth seven fundamental programming control structures necessary to form well structured NATURAL code. They fall into four classes which reflect the work of Bohm and Jacopini. The classes are identified as sequence, selection, iteration, and exit. The four classes are subdivided into several control structures.

The classes and their corresponding structures are:

1. Sequence class - concatenation
2. Selection class
   1 IF-ELSE
   2 IF-ORIF-ELSE
   3 CASE
3. Iteration class
   4 WHILE
   5 DO-UNTIL
4. EXIT class
   6 QUIT
   7 CONTINUE

Jensen explains that these seven control structures, which I have renamed, are reducible to three lowest level graphic representations that describe program structure. One uses these representations to describe the logical flow of a program through a "program logic flowchart".

[2]Randall W. Jensen, "Structured Programming", Software Engineering, Prentice-Hall, Inc. pp. 221 - 328

Implementation of Jensen's 7 Control Structures in NATURAL

First, the sequence structure is a concatenation of well-formed processes. A well-formed, but improper process is acceptable ONLY if the abnormal exit leads to the exit of the program block or segment.

Second, the selection class is implemented with NATURAL's IF-THEN-ELSE statement. The nested structure is supportable but not ideal beyond three levels. The IF-ORIF-ELSE structure is also supported by writing IF-ELSE IF-ELSE IF-ELSE in NATURAL. Or better yet, the DECIDE FOR structure accomplishes this requirement as is much easier to read and to maintain. The CASE structure exists in the form of the DECIDE ON statement.

For the iteration class, the WHILE structure is supported using NATURAL's REPEAT-UNTIL / WHILE statement. NATURAL's FOR statement is a WHILE structure using the integer data type.

Summary of NATURAL's Capability to Support Control Structure

Fundamental Structure	NATURAL Implementation

IF-ELSE	IF .. THEN
	...
	ELSE
	...
	END-IF

IF-ELSE IF-ELSE	IF .. THEN
	...
	ELSE IF
	...
	ELSE IF
	...
	ELSE
	...
	END-IF
	END-IF
	END-IF

	DECIDE FOR
CASE	DECIDE ON
WHILE	REPEAT .. UNTIL FOR <var> = init-val thru final-val STEP step-val (only good for counting)
DO-UNTIL	WHILE equivalent
QUIT	ESCAPE, ESCAPE ROUTINE, ESCAPE TOP
CYCLE	Not implemented for loops in general. Applicable to data base access loops, READ WORK, FILES, or SORT coupled ACCEPT/REJECT statement.

Lastly, the EXIT class can be implemented in part in NATURAL. The unconditional QUIT can be written in NATURAL using the ESCAPE statement (surprise). It is up to the programmer to see that the structure is implemented judiciously. The CYCLE structure does not have a direct corresponding statement in NATURAL for ALL of its loop constructs. For NATURAL's implied loops - READ, FIND, HISTOGRAM - the ACCEPT/REJECT statement provides this capability. This is also applicable to SORT and READ WORK FILE.

Developing "Good" Modules - Definition of "Good Module"

I use the word module to mean a functional unit, whether it is a function or a procedure subprogram, invoked by a PERFORM or CALLNAT respectively. A good module is one which has certain characteristics. They include:

1. This sounds obvious, but it should do ONLY what it is intended to do. No side effects, no plans for the future!

2. The scope of effect should lie entirely within the scope of control. This is very important. The scope of control encompasses a module and all of its subordinate modules. This is evident from a hierarchy chart or a structure chart. The scope of effect encompasses any module which is affected by the decisions of a module. It is proper to maintain the scope of effect within the scope of control.

3. It should be PREDICTable, i.e. two identical inputs should generate identical output.

4. The size should not be very large. The executable code should not be larger than 150-200 lines. I have heard programmers say that 2 screens (from the Program Editor) is large enough (excluding definitions and comments). Be careful of fragmenting too much. It can cause the buffer pool to thrash.

5. Error handling should be done within the modules themselves. The passing of error flags increases coupling. This is not however a firm rule. I observe it as much as possible. This rule applies if there is no central error handler. The problem with NATURAL central error handlers is the inability to handle errors and return in place.

6. Using the "black box" is ideal. This implies one input line and one output line. A black box is a segment that can be removed and treated as an independent unit.

7. Modules reflect transformation functions depicted on structure charts.

8. Modules should have a high degree of cohesion. One application of this rule is for a module not to affect multiple objects, e.g.,. GET STUDENT-RECORD and HOUSING-RECORD is a no-no.

9. Modules should have a low degree of coupling.

Coupling

Coupling is defined as a "measure" of the interconnections that are established between modules. The less coupling there is, the better the module is. If module A invokes module B, to know how module A works should not require an understanding of how module B works. If so their dependence is increased. A question is: how does one measure the coupling?

One good way is to analyze the interconnections between the two modules. The fewer there are, the better! The total number of bytes and the number of data items are indicators as to what degree modules are coupled. Normal connections, i.e.,. through passed parameters, is ideal. References within one module into another are called "pathologically connected". *This is NOT desirable* *. Modules which access central data stores (e.g., ADABAS files) are inherently coupled. They are referred to as "pathologically connected indirectly".

Last, data passed between modules as control information should be used as sparingly as possible. Control coupling occurs when module B expects to alter the path of execution of module A. If the sending module is not intending to control the target module there is no problem.

Cohesiveness

This is a term of great subjectiveness. It means "how well do the components" work together. A high degree of cohesion is desired. Structured design theorists describe different levels of cohesion. They are, in order from least to highest level of cohesiveness:

- coincidental association
- logical association
- temporal association
- procedural association
- communicational association
- sequential association
- functional association

I will concentrate on the higher end of the scale since that is most desirable. Sequential and functional association are described next.

- The program cross reference is a form of documentation and is discussed in Section VI - Documentation. Much of the documentation can be stored in PREDICT along with cross-reference data.

Sequential Association

This occurs when the data output from one module serves as data input into the next. The two modules are successive processes connected on a single path. Each process serves as a transformer of the data passed to it.

Functional Association

This association is the highest order of cohesiveness. It implies that all of the internal elements are necessary for it work and no unnecessary of extraneous references exist. One mark of a good module is that it can be described with one English sentence. The sentence requires one action verb and one object. Often, the action is clearer if the object is clarified with an adjective. Here is a list of useful verbs.

Verbs

Retrieve	Build	Compute	Scan
Add	Edit	Calculate	Verify
Modify	Display	Change	Print
Store	Set	Restore	Lookup
Load	Compare		

The actions operate on objects such as record, table, file, field, switch, flag, text, or key. I try to avoid intransitive verbs as determine, analyze, and any verb that is a NATURAL keyword (e.g., FIND) unoless it is the actual operation being done.

II. NATURAL Program Structure

Overview - Need for a Standard

Many shops may have already adopted standards for NATURAL program structure, even if unofficially. Figure 8.1 is an outline of the components of a NATURAL program.

GDA Usage

The GDA is better suited for data required for sharing across fetched modules. It can also hold common tables. The GDA is also useful for function key definitions, current record(s) information, any special security information, menu stack, and any other data necessary for the current transaction to restart of to be shared. It is best to plan the data requirements carefully and try fitting the requirements into the buffers allocated for that purpose.

PROGRAM DOCUMENTATION
VARIABLE DECLARATIONS
Global Data Area Definition
Parameter Data Area Definition (Subprogram, Subroutine, Helproutine)
Local Data Area(s) Definition
SESSION PARAMETERS
PF Key Settings
Report Formats
MAIN PROGRAM BLOCK
NATURAL Machine Code
Transfer of Control
On Error Routine, If no ETA
Stop
Write Title Section
Internal SUBROUTINES
PROGRAM CROSS-REFERENCE*
Documentation of Program flow
END

Figure 8.1. NATURAL Program Structure

Using an external Local Data Area to define program constants is a good idea. One, checking for flag values or making comparisons and assignments of data constants is handled consistently. How often have I assigned 'YES' for a flag in one place and 'Y' in another. The debugging nightmare was not worth it. The following definition established above for CONSTANTS alleviates this problem.

```
1 CONSTANT-CHAR
2 #BLANK                      A   1
2 #ASTERISK                   A   1 INIT<'*'>
2 #QUESTION-MARK              A   1 INIT<'?'>
2 #DASH                       A   1 INIT<'-'>
2 #COLON                      A   1 INIT<':'>
2 #SEMI-COLON                 A   1 INIT<';'>
2 #DOLLAR                     A   1 INIT<'$'>
2 #PERCENT                    A   1 INIT<'%'>
2 #EXCLAMATION                A   1 INIT<'!'>
2 #PLUS                       A   1 INIT<'+'>
2 #MINUS                      A   1 INIT<'-'>
2 #HYPHEN                     A   1 INIT<'-'>
2 #AMPERSAND                  A   1 INIT<'&'>
2 #EQUAL                      A   1 INIT<'='>
2 #CENT                       A   1 INIT<'¢'>
2 #QUOTE                      A   1 INIT<''''>
2 #SLASH                      A   1 INIT<'\'>
2 #COMMA                      A   1 INIT<','>
2 #PERIOD                     A   1 INIT<'.'>
2 #DECIMAL                    A   1 INIT<'.'>
```

```
    2 #YES                        L       INIT<TRUE>
    2 #NO                         L       INIT<FALSE>
```

LDA/PDA Usage

Often, data is stored in variables for a program's use. Local data requirements can be defined in one or more LDAs. I advocate adopting defining variables in a DEFINE DATA block, and cease defining variables anywhere in a program in the very old NATURAL 1.2 tradition.

I have written much on LDAs and PDAs in Chapter 4. There is an interesting characteristic that has a very good side effect. Some of you have noticed that an externally defined PDA can be treated as an LDA. This provides the capability of consistency and accuracy in defining data for transfer to lower levels. This reduces, if not eliminates, occurrences of NAT0936.
As a suggestion of my DA colleague, data can be defined in an external PDA with up to three groups - #PDA-subroutine·name-IN, #PDA-subroutine·name-IN-OUT, and #PDA-subroutine·name-OUT. These definitions can be stored in an external PDA and treated as LOCAL or '.I' into the parent block. Additionally, you can define the appropriate AD= attribute. The attribute at the group level applies to all the elements in a group.

Although it is not strictly a program structure issue, naming conventions needs discussion in relation to establishing the identity of variables. The table in Fig. 1 suggests modifications to any standard. Any adopted prefix / suffix notation coupled with a general naming convention should express a variable's purpose for existence, its format as well as its contents. Also, any abbreviations should follow a corporate standard, possibly supported through the dictionary (PREDICT can be extended to do this). This makes impact analysis easier.

Session Parameters

Session parameters which affect output and session control are described fully in the NATURAL Reference Manual. The parameters can be affected programmatically with SET GLOBALS, FORMAT DISPLAY, INPUT, and various WRITE statements. Warning - the SET GLOBALS takes effect after the program terminates. This has important ramifications in batch.

VARIABLE CATEGORY	PURPOSE	PREFIX
GLOBAL	Data required for navigation in applications, especially if support business functions are invoked.	#NAV- or + for AIVs
GLOBAL	Data retention for applications. To be available across programs	#APP- or + for AIVs
GLOBAL	Data retention for programs. To be available upon reentering programs, for example saving states.	#PGM- or + for AIVs
Map Variables	Non ADABAS Map Fields	#SCR-
Control Variables	Identify Control Variables	#CV-
Main User Variable (simple arrays)	Immediate values, computational fields, I/O etc.	#
Redefined Variable (in GDA)	Maintain group structure for global data.	##,###
Local Variables for In Line Subroutine	Simulate block structure Mandatory for construction of a "good" module.	# and sub.prefix

Figure 8.2. NATURAL Parameter Hierarchy.

The need for globally accessible data in an application

This issue never seems to die in SAGland. As a result, I have collected several alternate techniques to make data globally available. I would not rely heavily on the NATURAL stack for this reason. There are better alternatives.

First, if you have the Entire System Server (AKA NATURAL Process), there is a view called COMMON-DATA. So, your applications could create and share data values across all modules, regardless of application, database, or even platform. All they need is the ability to issue ESS calls (disguised ADABAS calls). Or you could use another ESS view (EVENTING) to send messages (data) to other modules, applications, or platforms. This works like a SUBSCRIBE/POST queue.

Secondly, we should not forget the obvious data store of all - ADABAS. If an application needs to keep lots of data about a user session and make it globally available, it might be easy to maintain the data in an ADABAS record and retrieve it when needed.

Thirdly, *COMM holds a small amount of data which is available globally for a single session.

Fourth comes a suggestion from Darrell Davenport. A system programmer or a good hacker can add some user defined space at the end of the IOCB DSECT and write a shared NATURAL subprogram that uses CMMPP to save/read the data. It would function like a freeform GDA. It would increase the thread requirements by the size you define.

Mainline Code

The mainline of a program outlines the logical flow of the total program. Statements range from single operations (e.g., MOVE, COMPRESS) to calls to external routines and performs of subroutines.

The main should reflect the logical structure depicted at the top level of a program hierarchy chart. Most important, the code should reflect the proper control structures. Use of the proper control structures along with a structured methodology, such as structured design, provides the basis for implementing good structured code. Section 6 discusses the implementing the logical control structures. In Section 1 I discussed guidelines for evaluating the "goodness" of modules.

Program Transfer

The successful completion of a transaction or function may require procedures to be executed over one or more programs. By program transfer I mean changing modules at level 1 between program objects. The transfer mechanism is FETCH, FETCH RETURN, or STACK, which is discussed in Chapter IV.

The FETCH mechanism should be done only at level 1. FETCH RETURN is the most expensive method of doing program transfer; therefore I suggest designing with external subprograms (invoked via CALLNAT) instead. Data stored in the GDA is available across level 1 FETCHes. The one situation I might acknowledge using FETCH RETURN is a menu system to direct the selection of FETCHed objects and return to the main menu. This requirement can also be satisfied by defining the main menu program as the startup program to NATURAL Security, continuing to FETCH the selectable option, and executing STOP/END to return to the main menu.

Error Handling/Program Termination

Programs do occasionally end abnormally. Errors other than ADABAS/NATURAL should be handled within the module of importance. This lowers the need to couple modules with control flags to prevent ADABAS/NATURAL errors which can cause abends or prevent the logical flow of the function from progressing, you can use NATURAL/SECURITY's application definition or an external module assigned to *ERROR-TA to handle those errors. NATURAL also allows you to define an ETA - error transaction handler - for an entire session.

Normal termination can happen with the execution of NATURAL's STOP or TERMINATE statements. The STOP will cause all commands in the stack to be executed from the top down. This should happen at only one point in the NATURAL program. The TERMINATE statement allows batch programs to issue user codes which can be examined in JCL.

III. Modularity in NATURAL Programs

To achieve structured programs is an enviable goal, but the path is sometimes covered with thorns. Structured programs imply modularity. This modularity can be accomplished using NATURAL sub-module definitions. Using the fundamental structures discussed earlier along with NATURAL sub-modules, one can create structured programs. However, one must consider several deficiencies in NATURAL when using sub-modules to promote structure.

Classification of Sub-modules - Subprograms and Subroutines

Sub-module Type	Abbr	Description
Functional	FS	Computational in nature. Usually requires at least one argument to generate a computed output. An example is to compute standard deviation. This type is highly cohesive and low in degree of coupling.
Procedural Logic	PLS	Sub-task not related to a specific computation but to a sub-function. Sometimes requires at least one input parameter. An example is a physical data base access module. The implementation of this type presents problems in constructing a module with low coupling and high cohesion.

Procedural Logic modules can be broken into several subgroups by functionality. I have identified some as follows:

Input/Output	IO	This procedural type primarily functions as an I/O module.
Logical Access	LAS	This procedural type establishes the logical relationships for data base lookup.
Physical Access	PAS	This procedural type does the actual data base lookup.
Common Modules	ACM	Subprogram modules which are created external to a program but are common to an application. An example is to build SD files for an application.
Subprogram	SSR	Subprogram modules which are created external to a program but are common to our entire system. An example is the subprogram which presents help documentation in response to help request for any system.

These designations are not mutually exclusive. For example, a random number generator is a functional subprogram which might reside in a system subprogram library. Such a routine should be designated as FS/SSR. Although these designations are of my own invention, their purpose is to provide meaningful documentation and an overview of a system or function through the documentation.

Shared routines are of utmost importance to software development. NATURAL provides the capability of calls to external NATURAL and noon-NATURAL modules. This further supports the approach of developing shared code. Also, well-constructed black boxes allow the programmer not to have to reinvent the wheel and thus to be more productive.

Let me clarify the definition of the terms *shared* vs. *common*.

Shared routines are just that - routines which can used by multiple systems. At invocation by program A and program B they are using the same version. A change in the logic should not effect any routine which invokes it. Only a change in data requirements would require a recompile by units which invoke it. Subprograms are typical shared routines.

Common code is source that is standardized and the same in any unit which uses it. INCLUDE COPYCODE is well suited to regulate common code. However, a change in the COPYCODE member would not reflect operationally in the unit that invokes it without recompiling that unit. The possibly for non-synchronization exists.

NATURAL Sub-modules

Subprograms and Subroutines

NATURAL subprograms should be thought of as having the relationship to the "outer" block which invokes them as is established in Pascal or C, for example. NATURAL subprograms are foremost mini-programs. For subprograms which are common to an application or to the system, their support documentation should be developed the same as program documentation. A sample subprogram header is listed in Section VI - Documentation.

Subprograms are very useful tools for the application designer. They are maintained independently; they are loaded only when invoked. If used judiciously, this can help out the NATURAL program buffer pool. Data can be passed via a parameter list - NATURAL passes by address, not by name. Parameters can be defined as modifiable or not, although be careful at further lower levels. GDA data is available if the address(es) is passed through the PDA. Admittedly, I use subprograms whenever it makes sense to build sub-modules.

Subroutines should have a structure like that of a program. Documentation should be included first. This would include stating the purpose, a description of any algorithms, "passed" arguments if any, etc. Next, all the local data definitions are declared. After the definitions is the code for the subprogram, and lastly the RETURN.

NATURAL subroutines can be external as well as subprograms. These modules, however, do not have access to the parent LDA when external, only the current GDA and its own local data. Personally, I find very few situations for which an external subroutine is useful. I am NOT saying that external subroutines are not useful. They can reduce source code size, and data is more localized; I just have not found much use for them. I have written SORT subroutines, but if I were to externalize them, I would convert them to subprograms. Maintenance should be reduced if they are well designed and fully debugged, but I would not want to be a maintenance programmer who had to support a system with a large number of external subroutines.

A local data area is definable to a subroutine only if it is external. Any data required by an inline routine must be defined in the DEFINE DATA block of the parent block. Adopting a prefix/suffix notation to identify variables specific to inline subroutine is a good idea. For example, if the name of an inline subroutine is SORT-VARIABLE-LIST, variables specific to the routine would start with the identifier # (to denote a user definition) and SVL, which is based on the subroutine name. This helps to minimize side effects. Data in the GDA is available to a subroutine. If needed, data can be transmitted to an external subroutine through the STACK/INPUT mechanism.

NATURAL maintains the relationship between the logical tag for a subroutine (DEFINE SUBROUTINE name) and its physical tag (SAVE/STOW name). This relationship is maintained on the NATURAL FUSER file (where NATURAL stores its objects) and in the object code for the referencing object. In the event the external subroutine physical tag is changed, any object referencing the external routine will have its object code changed as well.

External Maps

Mastering all of the capabilities of external maps is a course unto itself. This is why I have included an entire chapter devoted to map discussions. I tend to use them all of the time. However, if you have a complex map which requires two or three dimensional arrays, the map editor may not help you to do what you want. In this case, an internal map may be the only way to solve your needs.

IV. Other NATURAL Design Structures

COPYCODE

COPYCODE is good for repeated standard inline routines. A change in any COPYCODE requires a recompile of the code units which utilize it. This is good for making changes to a standard routine - you only make the change in one place.

COPYCODE allows you to reduce source code size. This is especially good for source which is functionally common to many programs in an application. Some good examples of COPYCODE is standard PF key settings, standard navigation logic, standard ON ERROR processing if an ETA is not used.

COPYCODE is a good alternative to external subroutines. The subroutine can be coded into block for INCLUDE, thus reducing the source and still achieving the result of using standard common code. The negative is that any changes will require a recompile.

However, it should not be viewed in the same way as a subprogram or a subroutine. It does represent common code which is standardized; it is not, however, shared code. For any logical change to take effect the module which invokes it must be recompiled.

HELPTEXT

HELPTEXT is definable in the map editor. It is easy to use. However, it is limited. It may be useful for short text display. It cannot access the GDA and is limited in passed parameters, but you also do not have worry about window control definition and can be shared amongst different maps. However, I believe as much in a centralized help manager as I do a centralized error handler. Remember to set the correct line size and page size in the map settings screens for your HELPTEXT.

HELPROUTINE

HELPROUTINEs offer full functionality and do have access to the active GDA and whatever is passed through the two passed parameters. They are defined in the program editor with SET TYPE 'H'. HELPROUTINEs do not increment the level. Help can be developed for different scenarios. There is application help, functional help, and field help. Field help falls into categories of legitimate field values, available values and field definition.

Whatever form of help you provide, you should return to the screen from which the help was requested.

I have tended to stay away from HELPROUTINEs and HELPTEXT but to use SYSHELP or a complete help system developed for the application that coordinates all help resources and manages them. This central help facility operates primarily with subprograms or links to standard help.

V. More Design Considerations

Processing Rules/Validation Rules

Processing rules can help maintain data integrity. At this time, there are some reasons why I do not use processing rules. One, they are tied to maps. If I want the process rule to include fields NOT officially part of the map, this is not possible without "hiding" the field on the map. They are invoked even if the attribute tag is set to output only.

If I use processing rules as edit/validation and attach them to fields in PREDICT, they are invoked only in map processing. If a batch program does not use a map (very likely), the rules cannot be used.

If I wanted a processing rule to validate, I would have the process rule invoke a validation subroutine/subprogram and do any REINPUT if required. This way, the routine is available for both online and batch processing, you do not rely on NATURAL's traceback across levels from REINPUT, and you use your organization's standard routines.

PF Keys

There is controversy over whether or not PF keys should be employed, mainly because of different terminal keyboards which an organization might have. The combination of key strokes needed to imitate a PF key stroke can therefore cause confusion. However, this does not mean that a shop standard should be adopted which says no PF keys. It may be more flexible to adopt an environment which is tailored to a specific user. It may be more programming and a change in style but making applications which are user sensitive just might be worth it.

If PF keys are desirable but you do not wish to take up room on map displays, consider defining a pop-up window that displays function key values. Also, if you are intent on using PF keys, remember to provide alternatives so that an application does not misfire because some users may not have function key support.

Modifiable User Interface.

It is better to design for a broad range of users than for a specific group. This can make all users more effective and generally more pleased with the systems you develop for them. A user profile could include such issues as direct commands understood by the user, user modified error messages, PF key profile, the level of the user (novice, moderate, expert), etc. The level would signify to what extent a user needs to be prompted to do their work. Enough information could be maintained to tailor the environment for their use.

Error Handling and Error Analysis

Error recovery presents a very interesting problem for the systems designer. It is easy enough to capture an error condition, report on it and restart the user to the current transaction or the entry point to the application. I would like to pose a question to the reader. How often do we take action to minimize or eliminate the situations that cause errors? Do we actively work towards a zero fault tolerant environment? Do we even know over time what errors are plaguing the users in their use of the system?

One approach we can take is to first decide that all your systems use a central error transaction handler. One of the purposes is obvious: you can develop a common response to errors which provides a standard interface to users. But also you can design the error handler to record data about error occurrences. Later, reports on the nature of the errors can be analyzed to develop a strategy to diminish the number of error occurrences.

Errors can occur for many reasons. Typing mistakes, misunderstanding of input data, illegal requests, database errors, NATURAL errors, system resource unavailability, mechanical problems, user unfamiliarity are just a few. What the designer needs to know is how to categorize the problem events and commit to a strategy to correct them.

User reactions to errors can also be minimized if the messages they see are their own. No matter what approach you decide to take, remember the success of any system relies on the good word of the users for whom it is intended. When they have the feeling that in spite of problems that the systems developed keep them in mind, then the few extra CPU cycles and I/Os are worth it.

Library of Shared Subprograms/ Subroutines

The importance of a library of shared code can not be underestimated. Common routines aid in developing well-structured programs more easily and quickly. Lowering development time allows for more to be accomplished in the same amount of time with the same people resources. Also, in developing programs programmer/analysts can concentrate on the kernel of the function while the subsidiary or support functions are invoked from central libraries.

Here are a couple of examples of subprograms which are functional. The first example, computation of standard deviation, uses MOVE INDEXED. I realize that this is not only un-optimized, but the statement is considered forbidden. But I raise the question: How does one construct general routines in NATURAL which function for any size array without the capability to pass dynamically the dimensions of the array? This routine is tied to the format of the elements (in this case N8.3), but by passing the start address and the length of the array the routine can function for many programs.

```
0010 *  SUBPROGRAM: STNDDEV
0020 *    FUNCTION: COMPUTE STANDARD DEVIATION
0030 * ---------------------------------------------------------
0040 * ALGORITHM: STAND. DEV. = SQUARE ROOT (VARIANCE)
0050 *            VARIANCE    = SUM OF SQUARE OF DIFFERENCES BETWEEN
0060 *                          EACH DATUM AND AVERAGE
0070 * ---------------------------------------------------------
0080 * EJECT
0090 DEFINE DATA PARAMETER
0100 1 #ARR-ELT       (N8.3)   /* INPUT
0110 1 #ARRAY-LENGTH (N5)      /* INPUT
0120 1 #AVERAGE       (N10.3)  /* INPUT
0130 1 #STND-DEV      (N10.3)  /* OUTPUT
0140 1 #VARIANCE      (N10.3)  /* OUTPUT
0150             LOCAL
0160 1 #NUM           (N8.3)
0170 1 #TOTAL         (N10.3)
0180 1 #I             (P5)
0190 END-DEFINE
0200 * SKIP 1
```

```
0210 FOR #I = 1 TO #ARRAY-LENGTH
0220     MOVE INDEXED #ARR-ELT <#I> TO #NUM
0230     #TOTAL =
0240       #TOTAL + (#AVERAGE - #NUM) ** 2
0250 LOOP (0210)
0260 * SKIP 1
0270 #VARIANCE = #TOTAL / #ARRAY-LENGTH
0280 #STND-DEV = SQRT (#VARIANCE)
0290 ESCAPE ROUTINE
0300 END
```

```
0010 DEFINE DATA
0020   PARAMETER
0030 1 #RANDOM-NUMBER (N9)
0040 1 #SEED          (N10.6)
0050 *
0060   LOCAL
0070 1 #HOLD-TIME      (N7)
0080 1 #HOLD-NUMBER    (N9)
0090 1 #TEMP           (N10.6)
0100 1 #RAN-FRAC       (N10.6)
0110 1 REDEFINE #RAN-FRAC
0120   2 #INTEGER      (N10)
0130   2 #DECIMAL      (N0.6)
0140 1 #START-RANGE    (N9)      CONST <1>
0150 1 #END-RANGE      (N9)      CONST <899999999>
0160 END-DEFINE
0170 *
0180 IF #SEED = 0 THEN DO
0190    MOVE *TIMN TO #HOLD-TIME
0200    #SEED = #HOLD-TIME / 1000      /* N3.4 FOR SEED
0210    DOEND  /* (0180)
0220 #SEED = 9821.0 * #SEED + 0.211327
0230 #RAN-FRAC = #SEED
0240 #SEED = #DECIMAL
0250 #TEMP = (( #END-RANGE - #START-RANGE + 1) * #DECIMAL ) + #START-RANGE
0260 #RANDOM-NUMBER = INT (#TEMP)
0270 ESCAPE ROUTINE
0280 END
```

VI. Documentation

Program Cross-Reference

The use of a subroutine documentation cross-reference listed at the beginning/end of a program can be valuable. And let us not forget the best CREF of all, that which is captured automatically within NATURAL via the XREF=ON parameter. The elements for any cross-reference should include:

- An indented listing to display the logical block structure of the module.
- Line number references to display the physical location of the program components.
- Any stacked commands which establish transfer points out of the program. These are marked conditional if decision structures are needed.
- Module type identification.

However, we shouldn't document for its own sake. It should prove valuable, reliable and maintainable. It might be useful to investigate PREDICT's capability to incorporate standard documentation.

Here is an example of a program subroutine documentation cross reference.

```
3600 ************************************************************
3610 *         PROGRAM INDENTED LOGIC STRUCTURE CHART     *
3620 ************************************************************
3630 * GLOBAL DEFINITIONS                            (0320) - (0490)
3640 * MAIN DECLARATIONS & DEFAULTS                  (0680) - (0780)
3650 * MAIN ROUTINE                                  (0870) - (0980)
3660 *   $$STACK ##TO-PROGRAM                        (0980)
3670 *   DETERMINE-PRINT-SUBFUNCTION(PLS)            (1010) - (2340)
3680 *     $$STACK ##TO-PROGRAM(C)                   (1290)
3690 *     DETERMINE-PGM-OR-DDM(PLS/LAS)             (1440) - (2340)
3700 *       FIND-PGM(PLS/PAS)                       (3110) - (3350)
3710 *       FIND-DDM(PLS/PAS)                       (3370) - (3590)
3720 *     RESOLVE-PARMS(FS)                         (2720) - (3090)
3730 *   DISPLAY-SCREEN (PLS/IO)                     (2360) - (2700)
3740 *     RESOLVE-PARMS(FS)                         (2720) - (3090)
```

Bibliography

Berztiss, A.T., *Data Structures, Theory and Practice*, New York & London, Academic Press, 1971

Davis, William S., *Systems Analysis and Design, A Structured Approach*, Reading, MA, Addison-Wesley Publishing Company, 1983

Dijkstra, Edsger W., *A Discipline of Programming*,Englewood Cliffs, NJ, Prentice-Hall, Inc., 1976

Jackson, M.A., *Principles of Program Design*, New York & London, Academic Press, 1975

Jensen, Randall W., Tonies, Charles C., *Software Engineering*, Englewoods Cliffs, NJ, Prentice-Hall, Inc., 1979

Yourdon, Edward, *Techniques of Program Structure and Design*, Englewood Cliffs, NJ, Prentice-Hall, Inc., 1975

Yourdon, Edward, Constantine, Larry L., *Structured Design, Fundamentals of a Discipline of Computer Program and Systems Design*, Englewood Cliffs, NJ, Prentice-Hall, Inc., 1979

Wirth, Niklaus, *Algorithms + Data Structures = Programs*, Englewood Cliffs, NJ, Prentice-Hall, Inc., 1976

Notes

Appendix A

Random Meanderings and My Thoughts on NATURAL.

1. Eliminate the use of MOVE INDEXED.

2. Use array processing and eliminate GET SAME processing.

3. If there is doubt whether to use subroutines or subprograms, use subprograms.

4. Standardize your maps.

5. Eliminate NATURAL 1.2 global styled variables.

6. Structured Mode is the way to go; but I still on occasion use Report Mode with DEFINE DATA structures. That doesn't make me a bad person.

7. Why is deciding my application library search path an issue for NATURAL security? An application developer should be able to decide their search path dynamically.

8. Use shared code as much as possible, whether it be standard subroutines or subprograms.

9. Study maps/windows well; talk to colleagues about using them. They provide a far more varied user interface than was possible in 1.2.

10. Think carefully about implementing verification rules. Many of us find them difficult to control and manage. They are also not useable in batch processing.

11. Build your LDA's externally. The source can always be included by using the .I option. If your source document gets too long you can back out the number of source lines by removing the LDA entries.

12. Do not get carried away with all of NATURAL's capabilities of defining objects. Plan your applications well, do structured walkthroughs. The decisions you implement will be supported in the future. Remember the poor maintenance programmer!

13. I do love SET CONTROL 'N'.

14. I love DEFINE WINDOW even more.

15. Define models for standard programming activities by your staff. A model by definition allows simple replacements and *voila*, you have a working program.

16. Do not use subroutines recursively; if you must, you better know what you're doing.

17. Do not build hierarchical program units with too many levels.

18. If your NATURAL object begins to get close to 1,000 lines, you better start taking notice.

19. If a subprogram has a large list of variables passed, is it possible that the subprogram is doing too much work?

20. I hope you never have to debug NAT1132, NAT0886, NAT0888, NAT0936, or NAT9969.

21. My kingdom for user-defined data types. The helps me to define objects and relationships closer to real life. Imagine a loop that reads like this:

```
FOR DAY = MONDAY TO FRIDAY
      WHERE DAY CAN HAVE VALUES MONDAY, TUESDAY,...,SUNDAY
             OR
IF DAY = WEEKEND
      WHERE WEEKEND HAS ONLY VALUES SATURDAY AND SUNDAY
```

22. PF Key recognition is a good candidate for an external subroutine or INCLUDE COPYCODE.

23. Turn XREF on for all applications.

24. Master the TEST facility. The time is well worth spent.

25. Do not allow for REINPUTs to be separated too far by levels and routines from the map for which it is intended. It is an implied GOTO and that means anything can happen that wasn't intended.

26. Whether or not the work area should retain its contents after a LOGON is controllable under NATURAL Security. Establish the option accordingly for your shop or specific users.

27. Establish an editor profile for yourself. Autosave is a handy feature.

28. Do yourself a big favor and learn ADALOG and get familiar with PF9 in the DUMP utility.

29. If you decide to use PF keys, make all the keys program sensitive and avoid unPREDICTable behaviors with undefined PF key strokes. Also, consider them as part of state at any level. Learn to save and restore states as you traverse levels in your application.

30. Numeric referbacks are a pain in the rear. Labels are much more useful.
    (Note from the editor: qualification to the level 1 structure or viewname is even more useful increasing the readability and removing any ambiguities within the code).

31. Save yourself some headaches in batch in IBM land. Always specify DCB for a new data set.

32. A WRITE TITLE generates a NAT0316 if it is placed in a subroutine.

33. You may have to periodically force individual objects out of the NATURAL buffer pool. Speaking of the buffer pool, you might consider not running one global buffer pool for all TPs, especially mixing production on-line environments with application testing environments. There still seems to be a synchronization problem sporadically when it is filled. Keep your production environment as pristine as possible.

34. If you define your data requirements first, either in a program or in an external data area, you can then bring them into the map editor. You will find you will have fewer occurrences of NAT0247 or NAT0936.

35. Restart data does not have to be identified by user-id or jobname. You can define an ETID through the NTSYS parameter or assign a random value through NATURAL Security.

36. Advice for those who like to know more about NATURAL. Study the BB area (Benutzer Block).

37. DEFINE PRINTER is great for directing output to printers in both on-line and batch modes.

38. Do we really need INPUT...NO ERASE any longer?

39. Upon unloading and reloading the FNAT system file, remember to specify USERISN in the load step. You must also load to the same file number from which it is unloaded.

40. If the mapname of an INPUT statement ends with an '&', the language code currently in use for the session is substituted as the last character of the mapname before the map is invoked. This way you can build language sensitive applications without special program logic.

41. Whenever invoking pop-up windows, remember to issue the statement SET CONTROL 'WB' to prevent subsequent screen outputs from using the window definitions accidentally.

42. For upper/lower case translation, the session must be in mixed mode setting. Then on any map, any field can be altered to upper case with the (AD=T) parameter. If the session is in upper case mode, all fields are translated *regardless* of the AD parameter.

43. Do not use views with fields unreferenced in the program. Check L X verify application option.

44. This point may seem obvious, beware of naming the COPYCODE members, or any NATURAL object, with a reserved word. It is identified at compile time as NAT0200 or NAT0280, which masks the real problem.

45. My kingdom for SET PFn-m CLR

46. Why can't technical editors be context sensitive. If I type RAED can't it recognize the best guess is its probably READ and change it? If I type READ , when I hit <ENTER> can't it check the mode, see structured, interpose END-READ and open the screen for 10 more lines?

47. Do I love .G in the editor (thanks to Construct).

48. Having seen Construct in action, models are extremely helpful. Isn't it possible to use this system and take a tip from Microsoft and develop NATURAL wizards?

49. On occasion, even NATURAL has suffered from the phantom program change that has disappeared. Or am I sometimes living on another plane of reality?

50. Don't you love the error message 'Check program and correct error' when you have been working for 30 minutes trying to correct the error? Better yet, how about 'contact your NATURAL administrator' or 'contact your database administrator' when you ***are*** the NATURAL Administrator Database administrator (or both)?

Notes

Appendix B

Coding for Performance.

An interesting discussion appeared on SAG-L on this subject. It arose due to a discussion of whether to use MOVE, MOVE BY NAME, MOVE EDITED to solve a particular problem. Several people brought forth very good insights to the subject of why we make the choices we do and why we should make certain choices. I want to share them here, since I did not discuss this issue anywhere else in this book.

One user noted:

> "To every simple problem, there are some really complex solutions. The writers of those solutions never consider them complex, just "ingenious", but the longer I've been in this business, the more convinced I am that simplicity and obviousness of intent are important values in coding, because they significantly reduce the chance of error, especially of hidden errors. For example, the intent of MOVE EDITED is, to me, a lot clearer than moving substrings and somewhat clearer than MOVE BY NAME.
>
> My time is worth more than the CPU's time, in almost all cases. It's worth knowing that packed numbers compute faster than unpacked, that FOR loops are slower than REPEAT loops (especially if the index isn't packed), that EXAMINE is relatively slow compared to other ways of checking field contents. But in the typical business application, raw CPU usage for internal moves and computations is a small factor

in performance, relative to database access and other I/O functions. Even if I make it four times worse by using selected statements, the effect on overall performance will be slight.

The relative risks and costs of bad database access paths are so much higher, so that's where designers should concentrate on understanding efficiency issues. For internal statements the rule should almost always be that if it works, and it's clear why it works, then it's good code. Yes, I have known NATURAL programs which are CPU-bound oinkers, and **for those programs** it makes sense to do some statement-level tuning; but **only** where the circumstances dictate. Our first task is to get a finished, maintainable product working, rather than agonize over the fine details."

A second user chimed:

"If you know that a particular statement/technique is more efficient than another then you should use it. It may not be quite as readable but then what are programmers paid for and what are comments for. How often do you hear the statement 'I will go back and fix it later' and how often do you hear "I never have time to go back and fix it". Do it as efficiently as you can the first time because you probably won't get the time later. Most programmers know the syntax, good programmers know how to use the language effectively and efficiently. True that I/O is the killer but every little bit helps. Remember that someone is still paying for the CPU time even if not directly then in the cost of buying a new machine."

As a programmer, some points to remember are:

(1) Simplicity as a value to coding
(2) Clarity between choices
(3) Some options are clearer than others
(4) Do not trade off efficiency, if known, since that is part of what programming is about.

And an opinion from a consultant I highly respect over the years, Skip Hansen.

"That is part of the problem with arguing about maintainability - code style can be a matter of preference and mostly - familiarity.

There seems to be an assumption among some (not necessarily the sender of the above) that creating standards that promote coding structures that are more efficient is in direct opposition to maintainable and/or easy to write code (more often than not it is the latter without the former). I strongly disagree. Steve Robinson's comments regarding performance are very appropriate. Programmers can learn any of the code fragments submitted without much trouble - so everything else being equal I would suggest using the more efficient solution.

I would also suggest that comments be liberally sprinkled throughout the code and the data structures are what make a program most maintainable."

Notes

Index

B

C

D

E

O

P-Q

R

T

U

V

W

X-Y-Z

Comments and Suggestions

Any suggestions you have regarding the general content or layout of this publication would be greatly appreicated. Please forward your comments and suggestions to:

Editor
WH&O International
P O Box 812785
Wellesley, MA 02181-0025

email who@tiac.net
Telephone (617) 239-0822
Fax (617) 239-0827

Publication: NATURAL Tips & Techniques________________

Topic: ______________________________

Page: _____ ☐ Incorrect ☐ Not complete ☐ Not clear

Comments: ______________________________
____
____
____
____
____
____
____
____
____
____
____

Date: ______________________________

Name: ______________________________

Company: ______________________________

Address: ______________________________
____
____
____
____
____

Telephone: ______________________________

Email: ______________________________

Additional Comments

Comments and Suggestions

Any suggestions you have regarding the general content or layout of this publication would be greatly appreicated. Please forward your comments and suggestions to:

Editor
WH&O International
P O Box 812785
Wellesley, MA 02181-0025

email who@tiac.net
Telephone (617) 239-0822
Fax (617) 239-0827

Publication: NATURAL Tips & Techniques______________________

Topic: ______________________

Page: ____ ☐ Incorrect ☐ Not complete ☐ Not clear

Comments: ______________________

Date: ______________________

Name: ______________________

Company: ______________________

Address: ______________________

Telephone: ______________________

Email: ______________________

Additional Comments